The Universe Natural History Series

THE INVERTEBRATE PANORAMA

The Universe Natural History Series

The Invertebrate Panorama

Editor and contributor:

J. E. SMITH

Other contributions:

J. D. CARTHY
GARTH CHAPMAN
R. B. CLARK
DAVID NICHOLS

UNIVERSE BOOKS
New York

Published in the United States of America in 1971
by UNIVERSE BOOKS
381 Park Avenue South, New York City, 10016

Library of Congress Catalog Card Number: 79-103104
ISBN 0-87663-154-5

Printed in Great Britain

Contents

List of Plates

Acknowledgements

The authors and publishers would like to acknowledge the following sources for the drawings used in this book: 9.1 R.B. Clark, *Dynamics in Metazoan Evolution*, Clarendon Press, Oxford, 1964; 9.2 G. Chapman, *J. Exp. Biol.*, 27, Cambridge University Press, 1950; 9.3 G. Chapman, *The Body Fluids and their Functions*, Arnold, 1967; 9.4 E.A. Robson, *J. Exp. Biol.*, 38, 1961; 9.5 D.M. Ross and L. Sutton, *Proc. Roy. Soc. B.*, 155, Cambridge University Press, London, 1961; 9.6 R.B. Clark and J.B. Cowey, *J. Exp. Biol.*, 35, 1958; 9.7 J. Gray and H. Lissman, *J. Exp. Biol.*, 15, 1938; 9.9 G. Chapman and W.B. Barker, *Zoology for Intermediate Students*, Longmans, Green & Co. Ltd., London, 1964; 9.10 H. R. Wallace, *Ann. App. Biol.*, 46, Cambridge University Press, London, 1958; 9.11 and 9.12 S.M. Manton, *J. Linn. Soc. Zool.*, 41, Academic Press, 1950; 9.13 S.M. Manton, *J. Linn. Soc. Zool.*, 42, 1952; 9.14, 10.7, 10.11, 10.12, 10.15 T.H. Waterman (ed.), *Physiology of Crustacea*, Academic Press, New York, 1961; 9.15 J.E. Smith, *Symp. Soc. Exp. Biol.*, 4, Cambridge University Press, London, 1950; 10.1 *Traite de Zoologie*, Vol. V, Masson, 1959; 10.2 J.D. Carthy, *An Introduction to the Behaviour of Invertebrates*, Allen and Unwin, 1958; 10.3 *Traite de Zoologie*, Vol. IV, 1961; 10.4 L.J. Milne and M. Milne, *Handbook of Physiology*, Section 1, 'Neurophysiology', Vol. I, American Physiological Society, Washington; 10.5, 10.13, 11.19, 12.18 M.J. Wells, *Brain and Behaviour in Cephalopods*, Heinemann, 1962; 10.6, 12.19, 12.20 J.Z. Young, *A Model of the Brain*, Oxford University Press, 1964; 10.8 J.D. Carthy, *Animal Navigation*, Allen and Unwin, 1956; 10.9 *Traite de Zoologie*, Vol. VI, 1949; 10.10 J.D. Carthy, *op. cit.*, from Seefert; 10.14, 11.15, 11.18, 11.20, 14.2, 14.11, 14.15, 14.16, 14.17, 14.19 Borrodaile, Eastham, Potts and Saunders, *The Invertebrata*, Cambridge University Press, 1958; 10.16, 11.1, 11.2, 11.3, 11.5, 11.6, 11.11, 11.12, 11.13, 11.16, 11.17 Bullock and Horridge, *Structure and Function in the Nervous Systems of the Invertebrates*, Vol. I, W.H. Freeman, 1965; 10.17 Dethier, *The Physiology of Insect Senses*, Methuen, 1963; 10.18 C. Hoffman, *Z. vergl. Physiol.*, 54, 1967; 11.4 W.S. Hoar, *General and Comparative Physiology*, Prentice Hall, New York, 1966; 11.7, 11.8 J.E. Smith, *Phil. Trans. roy. Soc. B.*, 1957; 11.9 G.A. Horridge, *Proc. roy. Soc. B.*, 1963; 11.10 J.A. Ramsey, *Physiological Approach to the Lower Animals*, Cambridge University Press, 1962; 11.14 E.J.W. Barrington, *Invertebrate Structure and Function*, Nelson, 1967; 12.1, 12.2 J.D. Carthy, *An Introduction to the Behaviour of Invertebrates*, Allen and Unwin, 1958; 12.3 J.D. Carthy from Evans; 12.4 J.D. Carthy, *The Behaviour of Arthropods*, Oliver and Boyd, 1965, from Bainbridge; 12.5 J.D. Carthy from Pardi and Grassi; 12.6 J.D. Carthy from Williamson; 12.7 J.D.

ACKNOWLEDGEMENTS

Carthy from Baldus; 12.8 J.D. Carthy from Drees; 12.9 J.D. Carthy from Weber; 12.10 J.D. Carthy from Millott; 12.11 J. Crane, *Zoologica* 42, 1957; 12.12, 12.13 J. Crane, *Zoologica* 33, 1948; 12.14 L. Gidholm, *Zoologiska Bidrag fran Uppsala*, 37, 1965; 12.15 J. Harker, *The Physiology of Diurnal Rhythms*, Cambridge University Press, 1964; 12.16, 12.17 R.B. Clark, 'The Learning Abilities of Nereid Polychaetes', suppl. I 'Learning and Associated Phenomena in Invertebrates' in *Animal Behaviour*, 1965; 14.1 after MacBride, 1914; 14.3, 14.4, 14.14, 14.18 after Barrington; 14.5 after Fraser, taken from Thorson, 1946; 14.7 after MacBride; 14.8 after Parker and Haswell, 1936; 14.9 after Fraser; 14.10 after Gurney, 1942; 14.12, 14.13, after Gurney; 14.20 after Parker and Haswell, 1947; 15.1 after Sidgwick and Jackson, 1952; 15.2 after Phillips Dales, 1966; 15.3 after Davenport, Bamougis and Hickok, 1960; 15.4 after Henry, 1966; 15.5 after Rothschild and Clay, 1952; 15.6 after Dawes, 1947; 15.7 after Baer, 1951; 15.8 after Smith, Woods *et. al.*, 1923; 15.9 after Lees, 1948; 15.10 after Smyth, 1962; 15.11 after Baer, 1951.
Acknowledgement is also due to the following for the plates used in this book: 1a, Dorothy Rutherford and Dr G. Horridge; 1b, Dr G. Horridge; 3, 8, Dr A. Hughes; 4, Dr J.D. Carthy; 5, Bullock and Horridge; 6, 9, 12, 13, 14, 15, Douglas Wilson; 7, Professor F. Papi; 11, Heather Angel; 17, 18, 19, 20, Dr Laverack; 22, D.M. Ross, The Zoological Society of London; 23, Glaessner and Wade (1966); 27a, by courtesy of the State Department of History and Archives, Iowa; 28a, by courtesy of the American Museum of Natural History.

Chapter 1

Introduction

IT HAS LONG been the custom to classify animals, at the first sorting, within one or other of two general categories. In the one category are included animals with backbones made up of a series of articulating vertebrae (the mammals, birds, reptiles, amphibians and fishes); to the other are assigned all the many different types of creatures that lack a backbone of this kind, of which snails, octopuses, starfishes, jelly-fishes, worms, insects, crabs and spiders are but a few examples.

Although the vertebrate/invertebrate dichotomy is, at the best, a rough and ready division that has little meaning in a modern classification of animals, zoologists have good reasons for wishing to preserve it. For, to put the matter shortly, the dichotomy, having provided in the history of the the investigation of animals the only practicable framework within which to work out their classification and systems of relationship, is now recognised – its historical usefulness exhausted – as a convenient way of dividing zoology into two complementary fields of study each concerned with its own special aspects of animal form and function and, in large measure, productive of its own viewpoints, generalisations and principles.

This volume of the *Natural History* takes as its subjects the invertebrate animals though it omits from its review two large and important groups, the insects (which are treated separately in an earlier volume of the series, *The Life of Insects*) and the minute single-celled Protista. And we will begin our survey of the invertebrates by considering some implications of the past and present uses of the dichotomy in so far as they bear on the nature of the animals with which we are to deal, the tempo and progress of their discovery, and the manner in which it is proposed to present them.

The dichotomy was first given scientific expression in the pioneer classification of animals formulated in the fourth century BC by the Greek philosopher Aristotle, pupil of Plato and tutor to the young Alexander the Great. Aristotle's descriptions of animals are recorded, probably in a much edited form, in the series of books known as the *Historia animalium*, *De partibus animalium* and *De generatione animalium* (an account of animals and of their anatomy, reproduction and development). The *Historia* includes references to more than five hundred kinds of animals, almost all of them native to the countries of the eastern Mediterranean or found within the waters of the Aegean Sea. Many of the descriptions are either too terse or too ambiguous for the positive identification of the animals to which Aristotle is referring, and there are in the *Historia* a few travellers' tales of near-legendary beasts. There are, on the other hand, many detailed and accurate anatomical descriptions – as of the jaw apparatus of sea-urchins to this day known as Aristotle's lantern – and observations on the reproduction and development of animals – as of the octopus – that are of unadulterated originality and genius. For the fact of the matter is, to put Aristotle's achievements in their true perspective, that no one had previously looked at animals as he did purely out of interest to discover what they are, how they live and what they do; and no one was to do this again with the same degree of insight and understanding for almost another two thousand years.

It is probable that many of Aristotle's observations on animals and most of his thinking about them were concentrated within the three years or so that he spent at Mytilene on the island of Lesbos in the company of his friend the botanist Theophrastus. When Plato died in 348/7 BC Aristotle, finding himself out of sympathy with the developments in the Academy at Athens, journeyed to Assos in Asia Minor to join two friends and former members of the Academy who lived under the protection of the ruler of the region Hermeias, himself a Platonist. While in Assos, Aristotle married the niece and adopted daughter of Hermeias, but her early death and the murder of Hermeias by the Persians caused him to find refuge in Lesbos. Aristotle's three years in Mytilene almost certainly explains his particular interest in marine animals and there is little doubt that, in pondering the principles of animal classification, he received much help from his discussions with Theophrastus who was later to become his literary executor.

When Aristotle described an animal he gave to it an identifying name (the *eidos*) and in classifying it he assigned it to a more general category (the *genos*), in much the same way as we would say that a frog is an amphibian and a horse a mammal. He recognised in all eleven general categories of animals and he divided these into two major groupings. In the one grouping he included the six categories of man, the cete (whales, dolphins and porpoises), the viviparous quadrupeds (mammals other than man and the cete), the oviparous quadrupeds (reptiles and amphibians) the birds and the fishes; and the most remarkable of Aristotle's many achievements was the discovery that the animals of all six of these categories have a fundamentally similar type of construction. For on dissecting them or in other ways observing them he found that all have a backbone made up of a series of bony rings, a skull, a heart, a system of blood vessels, and indeed many other features and organs so alike in their form and placement within the body as to leave no doubt of the essential similarity of the general design. Aristotle entitled these animals of a like construction the Enaima (animals with red blood) and, in so doing, established them for all time as a well-founded classificatory grouping.

All other types of creatures were grouped by Aristotle within the five categories of the cephalopoda (octopuses, squids and cuttlefishes), the malacostraca (the larger crustaceans), the insecta (insects, spiders, scorpions, etc.; worms with ringed bodies), the testacea (shell-bearing molluscs; starfishes, sea-urchins; sea-squirts, etc.) and the acalephae (corals, anemones, jelly-fishes; sponges). To these animals he gave the general title of the Anaima (animals lacking red blood). This was a mistake, for some anaiman animals have blood and others not, and of those that have blood there are a few in which it is red. There is in fact no semblance of common similarity within the immense diversity of the anaiman styles of construction to justify their linkage within a single category of relationship. So what Aristotle's classification really means is this. In sorting animals by their similarities and differences of structure we may, as a first separation, divide them into two major groups. One grouping (the mammals to the fishes) consists of animals which can be classified together on the basis of a common and identifiable style of construction. All other creatures are, by contrast, organisms of many different and puzzling designs. They are, therefore, merely for the convenience of later sorting, best kept in a nondescript container. This is the historical meaning of the enaiman and anaiman categories, later and more appropriately named the vertebrates and the invertebrates.

The difficulties met with in the further census and sorting of the anaiman animals and the extent to which they were subsequently neglected is evident from the circumstances of their relabelling more than two thousand years later. In 1793, during the later phases of the Revolution, the National Convention of the First Republic of France included in its programme of educational reform the establishment of a number of professorships in the natural sciences, of which two were in zoology. One of the two chairs was instituted for the study of the enaiman animals and was entrusted to the anatomist Geoffroy Saint-Hilaire. To the other was appointed a one-time serving soldier, journalist and author of a Flora of

France, Jean Baptiste Lamarck, who, by the terms of his appointment, was required to curate the collections of anaiman animals in the Museum of Natural History in Paris and to give a course of lectures on the natural history of 'insects, worms and other little-known animals, including microscopic organisms'.

In the preface to his *Philosophie Zoologique*, published in 1809, Lamarck recalls how, in his first series of lectures given in the spring of 1794, he had told his class of his intention to rename the Enaima and the Anaima as the vertebrate and the invertebrate animals (the animals that are not vertebrates) in order to replace Aristotle's misleading title of the Anaima by one made purposely nondescript in recognition of the heterogeneity of the invertebrates. The synopses of Lamarck's lectures show, moreover, how year by year he had reviewed the growing content of the invertebrates and had progressively refined their classification. He began by using the classification which the eminent Swedish systematist Linnaeus had evolved some fifty years earlier in the successive editions of his *Systema Naturae* and which, in its broader delineations of the invertebrate groupings, is little superior to Aristotle's system; and he produced finally a classification which, though it contains combinations and separations which would not now be acceptable, includes many precisely defined groups with names of Lamarck's invention (for example, echinoderms, annelids, arachnids, crustaceans, cirripedes) which are familiar categories within a present-day classification.

Lamarck's contributions to systematic zoology have been mentioned in some detail for they represent the first attempt to survey the invertebrates as a whole and to arrange them within an extended series of groupings. They mark, moreover, a vantage point in history from which to view the progress of invertebrate zoology and the changing tempo of discovery. For, to put the matter shortly, the early years of the nineteenth century stand at an historical divide. Behind them, and following the publication of the *Historia animalium*, stretch the two thousand years and more during which the growth of understanding of the nature and variety of the invertebrate animals had made but slow and spasmodic progress; ahead lay the expanding revelation of the nineteenth and twentieth centuries upon which, to all intents and purposes, our knowledge of the invertebrates is founded. Whereas Aristotle had described about three hundred anaiman animals, and Linnaeus had listed in the tenth edition of his *Systema Naturae* (1758) some three thousand of their species, the number of invertebrates that have been described and given a name is, at the present reckoning, more than a million. When set within the framework of their present-day classification, this vast and heterogeneous assemblage, comprising at least 95 per cent of all the known kinds of animals, can be grouped within some twenty to thirty major categories (or phyla), each one of which is as distinctive in its style of architecture as is the one group of the vertebrates and, in most instances, as variable in its modulations of design.

The invertebrates, as we have seen, are not a classificatory grouping. They were named as the area of the animal kingdom which until comparatively recent times was little understood and virtually unexplored; and they have come to contain practically the whole of the animal kingdom, and to encompass almost the entire range of animal design and grade of organisation from unicells to organisms of highly elaborate construction.

Here then is the directive to our survey and the reason for entitling it *The Invertebrate Panorama*. For if, in its present usage, the grouping of the invertebrates means anything at all it is that it provides an opportunity for viewing the great variety of animal architecture, the varied character of the functional machinery of animals, the multiplicity of their environmental adaptations and the many facets of their systems of relationship in the broad perspectives of a general survey. In surveying the invertebrates, we shall therefore avoid the conventional textbook approach of a phylum by phylum review. We shall, instead, repeatedly view them in their entirety, each time looking at different aspects of their form and presentation. And we shall report our impressions under the successive chapter headings as listed in the table of contents.

It is hardly necessary to add that our intention to view the invertebrates in the broad perspectives of a panoramic survey will preclude a detailed inspection of any part of it. Each chapter deals in a few pages with aspects of the life of the invertebrates about which many books have been written. A few selected references to the main sources of information and some suggestions for further reading are included in a short bibliography at the end of the book.

Chapter 2

A history of the study of invertebrates

'FACTS,' SAYS Bertrand Russell, 'can only be defined ostensively. Everything there is in the world I call a fact.' And it is the purpose of science to discover the facts and to formulate a systematic body of knowledge about them.

The facts of zoology are animals and the past, present and future history of zoology has, is and will be concerned both with the discovery of the different kinds of animals there are in the world and with the understanding of them. This means not only finding out as much as possible about their individual kinds but recognizing things about animals that are true generally and not only applicable to particular cases.

Zoology at the present day spans many aspects of the study of animals and is tending to become increasingly specialized and to be concerned with particular aspects of the lives of invertebrates rather than with the broad perspectives of animal form, function and systems of relationship. The history of zoology nevertheless shows that at all stages of the historical enquiry it is possible to identify a relatively few general circumstances that have been especially important in determining the tempo and changing patterns of the growth of understanding of the invertebrate animals.

The first hinges on the nature of the animals themselves and, in particular, on the way in which the invertebrates present themselves for discovery and interpretation by virtue, for example, of their size, the places in which they live and whether when seen they can readily be recognized for what they are. Secondly, account must be taken of the various means and devices which have been evolved in the course of time as aids to the understanding of animals. These include not only the many kinds of instruments, types of apparatus and techniques that have assisted their discovery and examination and made possible the detailed analysis of their structure and functional machinery, but all the benefits to scientific enquiry that are associated with the cultural and technological accoutrements of a developing civilization.

Among the most important of these developments are the means of travel and communication, the recording and storage of information, and the opportunities which advanced societies afford for learning and research. Finally, and most important of all, we need to consider the historical emergence of the individual and corporate attitudes of mind, thoughts and ideas which, in their attention to the study of animals, have been exploratory and inventive of new knowledge.

None of the many circumstances contributory to the knowledge of invertebrate animals can of course properly be considered in isolation when assessing the pace and progress of the invertebrate revelation. The smallest animals were unknown until microscopes were invented; Darwin's evidences for his theory of evolution relied largely on the experiences of the *Beagle* voyage and on opportunities for study which would not have been available to him a hundred years earlier; and a young zoologist of the present day has at his command a body of knowledge and an armoury of instruments which put within his range discoveries which Darwin could not have made. We may, however, be better able to judge the effects of the confluence of circumstances in shaping the history of invertebrate zoology by considering some aspects of the three main determinants of their history referred to in the previous paragraph, namely the manner of presentation of the invertebrate animals, the means afforded for their discovery and interpretation, and the emergence of the thoughts and ideas which have been creative of their better understanding.

The presentation and the discovery of the invertebrates

An animal is 'discovered' when it is seen and recognized as an animal; and both 'seeing' and 'recognizing' can present their difficulties. If an animal is to be seen it must either be visible to the naked eye or made visible by instruments of magnification. As it happens only two of the major groups or phyla of animals whose characteristics are briefly outlined in the next chapter are predominantly made up of animals too small to be seen with the naked eye. They include the minute individual or colonial unicellular organisms of the phylum Protozoa, most of which are within a size range of 5–500 microns,* and many of the organisms belonging to the mixed assemblage of creatures that are conveniently grouped together within the phylum of the Aschelminthes. Within this phylum are included many thousands of species of the smallest of the thread-worms (Nematoda) of ubiquitous habit; wheel-animalcules (Rotifera), common both in fresh and

* 1 micron (μ) = $\frac{1}{1000}$ mm.

salt water; and organisms known as gastrotrichs and kinorhynchs which live in the muds of fresh and salt waters (gastrotrichs) or exclusively in the sea (kinorhynchs), and which few zoologists can confess to having seen except perhaps on rare occasions.

Although the means of magnifying objects by viewing them through curved lenses of glass and other transparent materials have been known since classical times, lenses of a ten times or so magnification were first developed in the closing years of the sixteenth century by Dutch spectacle makers, and by the mid-seventeenth century microscopes with compound systems of lenses and capable of a much higher magnification were being constructed. The earliest microscopists ground their own lenses and by trial and error sought for the greatest enlargements they could get with the minimum distortion. Some of the best microscopes of the mid-seventeenth century were almost certainly in the hands of Robert Hooke in London, Antony van Leeuwenhoek in Delft, and Marcello Malpighi in Bologna, each of whom used their microscopes to make notable and original discoveries about the anatomy and tissue construction of animals. It is to Leeuwenhoek, who, like Hooke, communicated his results to the newly founded Royal Society of London (1660), that we owe the first descriptions of microscopic animals (ciliated Protozoa and rotifers) which he found in puddles of rain water. It is therefore rather surprising to find that when, almost a hundred years later the great Swedish systematist Linnaeus listed in the tenth edition of his *Systema Naturae* (1758) the four thousand or so animals then known to science, only about ten were of microscopic size. In the early days of the study of invertebrate animals, smallness of size would seem therefore to have been less limiting to their discovery than the fact that they were little sought for or lived in places where it was difficult to find them.

The habitats of the invertebrates

Invertebrate animals of one kind or another are found in almost every part of the seas, fresh water, and land surfaces of the world as well as in the surrounding air. The high temperatures of the magma-heated waters of hot springs and geysers, the cold and desiccating conditions of exposed ice caps, and a lack of oxygen in, for example, the deep still waters of some seas and lakes are among the circumstances which make life impossible for animals. But except for these rather rarely found limiting conditions, animals are able to occupy, each according to its special adaptations, every kind of niche that is available to them.

Of the three generalized categories of the earth's habitations – the sea, fresh water and the land – the sea is almost certainly the ancestral home of all the present-day phyla of animals. There is no single phylum with a well documented fossil history in which fresh water or terrestrial species have been found to pre-date the earliest marine representatives. Moreover, in the many phyla which for various reasons such as the absence of hard parts are rarely preserved as fossils, the so-called 'primitive' forms of relatively unspecialized structure are invariably found in the sea. Further, when a phylum includes, in addition to marine species, animals of fresh-water or terrestrial habit, the non-marine representatives almost always show in their greater specialization of structure and physiological adaptation that, by comparison with some at least of their marine relatives, they are of later evolutionary origin.

Table 1 sets out in a roughly quantitative way the distribution of the present-day invertebrates within the three major world environments. Of the twenty-three phyla of invertebrate animals which it is customary to recognize, eleven are made up wholly of animals that live in the sea and nowhere else; another four are predominantly marine; and of the eight remaining phyla all include a substantial proportion of marine species. No phylum, in other words, has entirely forsaken the sea, and two-thirds of them have either remained wholly marine or, at the best, have produced only a very few species which have made the passage to fresh waters or to the land.

The eight widely ranging phyla have all produced a variety of fresh-water and terrestrial species, and the success of their incursions into these secondary environments may be judged from the data set out in the table. They show (admittedly as rather rough approximations) the number of known species within each phylum and it will be seen from the number of groups represented in each of the biotopes that, although rather few classes of invertebrate animals have succeeded in occupying non-marine habitats, the immense proliferation of species on the land has caused them greatly to outnumber, in their 80 per cent of all known animals, the 17 per cent or so that live in the sea and the mere 3 per cent inhabiting fresh waters. The overwhelming predominance of terrestrial representation is due almost entirely to the success of one class of animals of the phylum Arthropoda – the insects – and, only to a lesser extent, to the contributions of other classes of arthropods, notably the arachnids (e.g. spiders and mites) and the myriapods (centipedes and millipedes). The only other invertebrates of reasonably large size found on land in considerably variety and numbers are the gastropod molluscs (slugs and snails) and the ringed worms (earthworms and leeches).

Table 1 *The distribution of the animal phyla in the sea, fresh waters and on land*

		Approx. no. of described living species	*Notes*
Phyla that are entirely marine (11)			
Echinodermata	Starfishes; brittle-stars, sea-urchins, sea-lilies etc.	5,500	Almost half of all the animal phyla are confined to the sea. They include, however, only about 0·5% of the world total of animal species. Conservative in their styles of structure and in their restriction to the sea all, except for the few species of parasitic Mesozoa, are free living. Echinoderms were in past times more numerous in species than they now are; Brachiopods (30,000 fossil forms) were much more numerous.
Brachiopoda	Lamp-shells	300	
Sipuncoloida		250	
Pogonophora		100+	
Hemichordata	Acorn-worms etc.	100	
Ctenophora	Comb-jellies	80	
Echiuroida		70	
Entoprocta		60	
Chaetognatha		50	
Phoronida		15	
Mesozoa		7	
Phyla that are almost entirely marine (4)			
Cnidaria	Zoophytes, jelly-fishes, anemones and corals	9,000+	These phyla include some 1–2% of the world species. The Cnidaria, Ectoprocta and Porifera have tentatively invaded fresh waters – some tens of species only. A very few nemertines are terrestrial, and fewer still have become parasitic.
Bryozoa		5,000	
Porifera	Sponges	5,000	
Nemertina	Proboscis worms	600	

Phyla with marine, fresh water and terrestrial representatives (8)

		Approx. no. of described living species	*Notes*
Arthropoda	Insects, arachnids, centipedes, millipedes, crustacea etc.	800,000	These phyla include almost 98% of all known animals; the insects alone account for about 80%. The immense invasions of the land by arthropods and to a lesser extent by molluscs, annelids, nematodes and vertebrates; and the more limited invasion of fresh waters by all of these phyla has resulted in an overall occurrence in the sea, fresh waters and land of approx. 17%, 3% and 80% of known living species. The Chordata, and (with a few exceptions) the Molluscs are free living; the Acanthocephala are parasitic. All other phyla include both free-living and parasitic forms.
Mollusca	Snails, bivalves, octopuses, squids, etc.	80,000	
Aschelminthes	Round worms, rotifers etc.	15,000	
Chordata	Lancelets [fishes, amphibians, reptiles, birds, mammals = vertebrates]	45,000	
Protozoa		15,000	
Platyhelminthes	Flatworms, flukes, tapeworms		
Annelida	Ringed-worms	9,000	
Acanthocephala		500	

We conclude from this brief survey of the world occurrence of the invertebrate animals that, while terrestrial habitats have been and will almost certainly continue to be the most richly productive of new species, neither land habitats nor fresh-water environments can provide more than a poor impression of the breadth and variety of invertebrate form and structure within the earth's major environments. The sea alone yields examples of all the different phyla of animals and holds the key to the understanding of the breadth and meaning of the invertebrate panorama.

The search for the marine invertebrates

Seas cover more than 70 per cent of the earth's surface to an average depth of about two and a half miles. Within this vast volume of water are contained myriads of pelagic animals which either float (as plankton) or swim (as nekton) and, though we cannot be sure of this, they probably occur at all depths from the surface to the deep abyss. Other animals, classified under the general term of the benthos, live on or within the various kinds of materials of the ocean bed.

If the seas were to be withdrawn from the earth's surface we would look down upon a terrain of mountains and deep gorges, hills and valleys, and far extending flat plains similar in their scale and relationship to the features of the dry land. The scenery of the ocean bed would, however, be the more dramatic. The great ocean deeps such as the six thousand fathom abyss of the Marianas trench some thirteen hundred miles east of the Philippines could accommodate Mt Everest with a good six thousand feet to spare; many submarine canyons match in their form and scale the Grand Canyon of the Colorado river; and the average depth of the oceans – all the great oceans of the world are of comparable average depth – is substantially greater than the average height of the land.

As the sea floor approaches the continents it ascends steeply from the depths as a continental slope, until it comes within a hundred fathoms or so of the surface of the sea. It then flattens into a more gently graded platform known as the continental shelf. The shelf is usually, though not always, of even gradient, and may vary in width from a few yards to many hundreds of miles. At its landward margin it breaks surface to form shores which in most but not all parts of the world are covered and uncovered daily by the rhythm of the tides.

The refuse dumps of Mesolithic middens, the folk-lore of primitive peoples and the written records of more progressive civilizations all bear witness to the fact that sea shores have long been worked over for fish and other kinds of animals that are good to eat, useful for the making of tools, or valued as body ornaments; and to judge from the life-like representations of marine animals on Minoan vases and later pottery, Mediterranean people of cultivated taste early enjoyed displaying the forms of attractive looking sea creatures such as starfishes, cockle shells and cuttle-fishes as decorations for the home. It required, however, Aristotle's deeper inquisitiveness into the nature of animals to transmute aesthetic appreciation into scientific enquiry. Few men have ranged so widely in their critical examination of the nature of men's actions, ways of thought, moral judgments and political systems, and as the originator of the science of zoology his primacy is unique and undisputed.

Aristotle was probably in his late thirties when he began his studies of marine animals. 'Some of these,' he says, 'live in the open sea, some near the shore, and some on rocks,' and the *Historia animalium* lists animals from all these situations. In many ways Aristotle's narrative is curiously and annoyingly uninformative. He rarely bothers, for example, to name the place where he made his observations, and he often says so little about an animal that it is difficult, except in a general way, to know what he is describing. But when there is something about an animal that really interests him he is admirable in the accuracy of his observations and in the lucidity of his descriptions. And yet, in spite of Aristotle's example, no one was to look at invertebrate animals again with anything approaching the same degree of insight and concentration of purpose until, almost two thousand years later, the Dutchman, Jan Swammerdam (1637–80) made his detailed and beautifully illustrated dissections of the organ systems of the honey-bee and the garden snail among other invertebrate animals.

There is little evidence up to this time of any great interest having been taken in the marine invertebrates. The various accounts of animals that had been written in the sixteenth century, such as the *Historia animalium* of Conrad Gessner (1516–65), borrowed freely from Aristotle and were mainly about vertebrate animals. William Rondelet's *De piscibus marinis* published at Lyons in 1554 is a notable exception. Rondelet, a Huguenot doctor and friend of Rabelais, practised in Montpellier. In his book he describes and figures many shore and near-shore creatures of the Mediterranean coastline, and his drawings of animals, though simple, are often sufficiently accurate for their species to be identified with a fair degree of certainty. Among the hundred or so animals noted by Rondelet there are included about thirty different species of bivalve molluscs, almost as many gastropod sea-snails, as well as a variety of crabs, starfishes, brittle-stars, sea-urchins, bristle-worms, anemones and jelly-fishes. Two kinds of animals figured in Rondelet's book, the worm-like *Sipunculus* and a colonial polyzoan of uncertain identity are of especial interest as the founder members of two of the invertebrate phyla, the Sipunculoidea and the Ectoprocta (Bryozoa).

Little attention seems to have been paid to marine invertebrates during the seventeenth and early years of the eighteenth centuries. There are, it is true, occasional glimpses at this time of naturalists foraging on sea shores and making notes on the animals which they found between the tide marks. Their observations are, however, often more quizzical than informative as when John Ray (1627–1705), the most distinguished English naturalist of his day, and author of a *History of Fishes*, mentions finding, on one of his many visits to the Channel coast, 'a Sea-worm, a yard long and living in holes in rocks, a peculiar creature unusual and almost absurd in appearance'. The vagueness of the description makes it difficult, however, to know to what kind of an animal he is referring. The Welshman Edward Llwyd (1670–1709), who was frequently consulted by Ray, is a good deal more informative about marine invertebrates. He is perhaps best known for his observations on starfishes, and his name is commemorated in the genus *Luidia* which includes two of the most handsome of the British sea stars. Other accounts of this period are interesting in reporting unusual occurrences as when Sir Thomas Browne (1605–82), who maintained a Botanical Garden in Norwich, refers to the large number of sea stars at Yarmouth in or about 1640, or are of value as descriptions of the animals of particular localities. One of the best of the regional surveys is given by William Borlase (1695–1772), rector of Ludgvan in west Cornwall who, almost a century later, describes in his *Natural History of Cornwall* (1754) dogwhelks, periwinkles, topshells, bivalve molluscs, sea-urchins and lug-worms as well as other kinds of animals that he found on the rocks and in the muddy sands of the Mount's Bay shore. 'Of shells,' he says, 'we have great quantities, but rather more varieties than sorts on our Cornish coasts,' though, 'we cannot boast of the rich colourings which the shells of the Mediterranean and Indian shores afford the collections of the curious.'

When Borlase refers to 'the collections of the curious' he has in mind the many collections, or cabinets of shells and other miscellaneous objects, animal, vegetable and mineral, that were being assembled by (mainly) well-to-do amateur collectors in western Europe, and which were largely made up of specimens brought back by seamen engaged in the rapidly growing seventeenth-century trade with the Americas, India and the East Indies. John Tradescant, who for a time was employed as agent to the

Duke of Buckingham, explains the sources of the specimens with which he furnished the Duke's cabinet when he says that 'it was the Duke's pleasure for him to deal with all the merchants from all places, but specially from Virginia, Bermuda, New Foundland, Guinea, the Amazon and the East Indies, for all manner of rare beasts, fowls and birds, shells and stones'. Tradescant himself amassed a considerable collection of shells and other objects which he left to his son the second John Tradescant. The younger Tradescant added greatly to the cabinets which eventually came into the hands of Elias Ashmole and became the nucleus of the Ashmolean Museum collections in Oxford. Although some of the finest of the seventeenth-century cabinets were in France and Holland, several of the British collections were of museum proportions. William Courten's cabinet with its rich collection of shells from the West Indies was valued in 1702 at more than £8,000, a very considerable figure by present-day standards; and Sir Hans Sloane who, following Sir Isaac Newton, was President of the Royal Society for fourteen years (1727–41), is said to have spent £50,000 on the purchase of shells. When he died, in 1753, he bequeathed his possessions to the nation and they were displayed in Montague House as the first collections of the newly founded British Museum. But, without doubt, the largest and most valuable of the shell cabinets of this period belonged to Margaret Bentinck, the Dowager Duchess of Portland, and it is known that Linnaeus and other leading naturalists of the eighteenth century examined and made notes on specimens contained in the collections she maintained at her house at Bulstrode. Many of the leading European scientists of the day enjoyed the hospitality of Bulstrode or of the Duchess's London house in Whitehall. One of her more frequent guests was Abraham Trembley (1710–84) who, though born in Switzerland, spent most of his working life in Holland. Trembley's studies of the minute colonial protozoa of fresh waters and, more particularly, his demonstration that small pieces of the polyp *Hydra* are able to regenerate into whole polyps, excited widespread interest in the academies of Europe. For many years he kept up a lively correspondence with René Réamur (1683–1757), his co-founder of the science of experimental zoology, on the subject of his own work and on Réamur's investigations of the calcification of the shells of molluscs, and while in London his letters are usually addressed from the Duchess's house in Whitehall.

The diaries and journals of eighteenth-century naturalists show that visits to the owners of cabinets were almost always included in the itinerary of their journeys, and when, in the middle years of the century, Linnaeus (1707–78) began to prepare in the successive editions of the *Systema Naturae* (1–12; 1737–68), a catalogue of all the then known species of animals, the cabinet collections were to prove invaluable centres of reference for his enumeration of animals. The tenth edition of the *Systema* (1758) is accepted as the starting point and rule book for the naming of animal species. Each species is referred to by Linnaeus by a latinized binomial name. It will be recalled that Aristotle had also used a binomial system for naming animals, but whereas Aristotle's eidos and genos had said in effect that a thrush and a blackbird (the eidos) are both birds (the genos), Linnaeus' binomial of genus and species says that a thrush (*Turdus philomelus*) is a bird like the blackbird (*Turdus merula*). In other words the so-called trivial names *philomelus* and *merula* establish the species of the two birds and the generic name *Turdus* indicates the more general category to which the two species belong. Different genera were sorted into a number of broader categories (or orders) of animals and orders arranged within different classes. Linnaeus recognised only two classes of invertebrate animals, the Insecta (crustaceans, insects, arachnids, centipedes, millipedes etc.) and the Vermes within which were included the five orders of the Intestina ('worms' of all kinds), the Mollusca (soft-bodied invertebrates of all manner of unrelated kinds from slugs and snails to jelly-fishes), the Testacea (a mixed bag of animals with shelly coverings), the Lithophyta (stony skeletoned corals) and the Zoophyta (mainly softer-bodied polyps and polyp colonies as well as a larger miscellany of creatures including, for example, tapeworms).

The contents of Linnaeus' classes and orders are no more related than the ingredients of a haggis, but his failure to extend the systems of relationship of animals to the higher categories of classificatory groupings is not surprising. For the fact of the matter is that while it is usually possible for a competent systematist (such as Linnaeus) to define correctly the species status of an animal by noting some clearly visible external features that are unique to the species and different from those of other species, the working out of the extended systems of relationship implicit in the grouping of species into genera, genera into orders and so on through the hierarchy of categories is a much more difficult and sophisticated exercise. The slow process of creating a meaningful classification presupposes not only the examination by many people of a wide range of animals but a knowledge of the nature and development of the internal as well as external structure of animals. So, while a modern classification accepts Linnaeus' categories as useful names for inclusion within the classificatory index, it puts a wholly different interpretation on their meaning and content. Two categories have since been added to Linnaeus' list. The family, as a grouping of genera, was added in Linnaeus' lifetime while the phylum, a primary subdivision of the animal kingdom containing animals of a distinctive general style of architecture, was introduced by Ernst Haeckel in 1866. The now accepted hierarchy of the major categories within the animal kingdom therefore runs: phyla, classes, orders, families and genera to the ultimate taxon of the species made up of animals of an essentially like kind, and reproductively self-contained.

Of the four thousand or so animals listed in the *Systema Naturae* about half are included in the Insecta and the remaining half are almost equally divided between the Vertebrates and the Vermes. Marine invertebrates (most of the Vermes and the crustaceans of Linnaeus' Insecta) number around nine hundred, and more than two-thirds of them are molluscs. Almost all live between tide marks or on the sea floor in shallow water, and it is notable that while many were collected from places ten thousand miles and more from western Europe there are few if any from water more than a hundred feet deep. Moreover, although a few of the larger and more conspicuous of the animals that float on the surface of the sea, namely goose barnacles of the genus *Lepas*, some jelly-fishes and a few jelly-fish-like siphonophores such as the By-the-wind-sailor, *Velella*, and the related *Porpita* are noted in Linnaeus' catalogue, not one of the many smaller kinds of pelagic animals that swim or float in the surface and deeper waters of the sea is included in his list.

In the fifty years following the publication of the *Systema Naturae* the influence of Linnaeus is clearly discernible in a more extended search for animals, in the emergence of a deeper and more critical enquiry into the nature and habits of some of the more puzzling kinds of creatures which Linnaeus had examined as dead specimens, and in a growing realization of the inadequacy of the Linnaean classification of the invertebrates.

Thus, when on 26 August 1768 Captain James Cook set sail from Plymouth in the *Endeavour* on the first of his great circumnavigating voyages of exploration a ship's master was for the first time accompanied by naturalists whose purpose was 'to collect specimens of natural history and to study the manners and customs of the natives with whom they came into contact'. The botanist on the voyage was Joseph Banks (1743–1820), later President of the Royal Society for the record period of forty-two years; the zoologist was the Swedish naturalist Daniel Solander (1736–82), a pupil of Linnaeus who, on his return from the voyage, took up an appointment at the British Museum in London. Three artists were included in the ship's company and the drawings and paintings of John Reynolds and Sidney Parkinson provide an excellent illustrated record of the plants and animals collected during the voyage.

Nearer home a less dramatic but none the less important event in the history of the discovery of invertebrate animals was enacted in 1765 when another of Linnaeus' scientific colleagues, Johan Ernst Gunnerus (1718–1773), Evangelical Bishop of Trondhjem, dipped a glass phial into the clear waters of Hammerfest in Finmark, the northernmost province of Norway, to draw from them a specimen of the copepod crustacean *Calanus finmarchicus*. This, the first planktonic organism to be described, is now known to be one of the most numerous and important sources of food for herring and other pelagic fishes.

Meanwhile, in England, John Ellis (1710–76), a merchant of the City of London who for many years regularly corresponded with Linnaeus, was making some of the earliest scientific observations and experiments designed to demonstrate the animal nature of the plant-like growths of hydroid zoophytes and of the colonial polyzoa of which there are many kinds in coastal waters, and was trying his best to persuade Linnaeus that they are animals and not plants. Ellis' favourite collecting grounds were in the Thames estuary and many of his specimens were examined at Whitstable on the north Kent coast. The fishermen there are described as bringing to him 'several irregular Pieces of a fleshy substance sticking to shells' called by them 'Deadman's Toes' (*Alcyonium digitatum*); 'Pudding Weed' or 'Pipe Weed' (*Alcyonidium gelatinosum*), a name which even now the Whitstable oystermen use to describe the soft, light brown cylinders of this locally common polyzoan; 'sea water wash balls' (the egg cases of the whelk *Buccinum*); and various kinds of hydroids such as the handsome foot-long squirrel tail or sea-cypress (*Sertularia cupressina*) which has been described as 'waving in the sea as Swedish junipers wave in the breeze'. Ellis put his specimens in wooden buckets filled with sea water and carefully recorded (as evidence of their animal nature) the movements of the expanded polyps of the colony and of their circlets of tentacles. In order to preserve his specimens in an expanded state for further study Ellis surrounded the tubs in which they were contained with warm water: 'we left them undisturbed for some time till the Polyps had extended themselves out of their starry Cells in which their Tails were fixed: and then suddenly took them out of the Salt-water and instantly plunged them into Brandy, whereby many of their Bodies were kept from shrivelling up'. Save for the gross misuse of brandy Ellis' account of his investigations would not be out of place in a modern scientific journal.

In France, Georges Buffon (1707–88), a brilliant and accomplished man of many parts whose *Histoire naturelle* was widely read and appreciated as much for its liveliness of style as for the interest of its accounts of the natural history of vertebrate animals, was scathing in his criticism of Linnaeus' ragbag classes of the invertebrates and especially of the hotch-potch collection of the Vermes. Buffon himself offered no improvements but he indirectly prepared the way for Lamarck's later clarification of the invertebrate groupings. Buffon early noted Lamarck's abilities as a scientist and teacher, and appointed him as tutor to his son. Largely through Buffon's influence, Lamarck was elected as a member of the Academie des Sciences, a position which enhanced Lamarck's eligibility when in later years he was elected to the Chair of Invertebrate Zoology (as we would call it nowadays) in the Museum of Natural History in Paris.

The immense surge of interest in natural history at the turn of the eighteenth and nineteenth centuries, which Lamarck's appointment as curator of the collections of invertebrate animals in Paris exemplifies in a particular way, was but one aspect of an increasing awareness of the nature, meaning and implications of scientific enquiry which, in the course of a hundred years, was to extend from the continent of Europe to all parts of the world.

By the beginning of the nineteenth century, physics and chemistry had entered upon a period of experimentation during which Priestley had prepared oxygen and demonstrated its elemental character and power to support combustion and respiration: 'fixed air' or carbon dioxide had been shown to exist in combination with certain alkalies, and Cavendish had demonstrated the compound nature of water. In France, Lavoisier's proof that a chemical compound is always made up of the same constituent parts in the same proportions, and that the masses of the constituent parts persist through a series of chemical reactions argued a common sense view that matter is real and indestructible, and invited renewed confidence in the supposition that the key to the interpretation of physical phenomena lies in observation, measurement and experiment. Men of an enquiring mind began to feel that natural philosophy and scientific enquiry were not the narrow province of professional savants but open to every man. Philosophical Societies sprang up in towns and provinces throughout Europe as active centres for the exchange of information and discussion. The preface to the first volume of the *Philosophical Magazine*, the journal of the London Society, sums up succinctly the inspiration and intent of the Philosophical Societies at their inception. A portrait of Lavoisier occupies the first page and the preface proclaims of the new society that '. . . the grand Object of it is to diffuse Philosophical Knowledge among every Class of Society and to give to the Public as early an Account as possible of everything new and curious in the scientific world, both at Home and on the Continent'. Here is made manifest the intellectual climate of the opening years of the nineteenth century which saw the birth and active working life of a galaxy of talented naturalists who were to create a new era in the development of invertebrate zoology and to set it on its way towards the larger developments of the twentieth century. The list is long and in naming but a few of the many distinguished naturalists of the day who made notable contributions to the knowledge of the invertebrate animals mention will be made only of zoologists who are best known for their studies on marine invertebrates.

There were born between 1800 and 1825, Henri Milne-Edwards (1800–1884), a pupil of the great Georges Cuvier, a Belgian of British descent and Professor of Zoology in Paris. He is remembered particularly for his studies of crustaceans and for the first recognition (with his French colleague V. Audouin) of the nature of the zonation of plants and animals between tide marks. Johannes Müller (1801–58), born at Coblenz and Professor in the University of Bonn, was one of the first zoologists to make systematic studies of the marine plankton and to describe the free-swimming larval forms of a variety of marine invertebrates including echinoderms, molluscs and polychaete worms. Michael Sars (1805–69) of Bergen, priest and sea-going zoologist, made many observations on the faunas of northern waters and helped to work out the complicated life histories of the floating barrel-shaped sea-squirts, the salps. Jean Louis Agassiz (1807–73), Swiss born Professor at Neuchâtel, was later, as Professor of Zoology at Harvard, to be the pioneer of marine biology in North America. Sven Løven (1809–1895), born in Stockholm was distinguished for his researches on the life histories of marine animals and especially for his work on the development and adult anatomy of sea-urchins. Although curator of the State Museum he was essentially a field naturalist and is the honoured founder of the Swedish marine laboratory at Kristineberg. Charles Darwin (1809–82), pre-eminent among naturalists for the formulation of his theory of Organic Evolution which transformed the thinking of biologists and gave in the later years of the century new direction to their research, began his scientific life as naturalist on the *Beagle*; and among the many groups of animals on which he was expert his work on the cirripede crustaceans (barnacles) is outstanding among the monographs of all time. Johannes Steenstrup (1813–97), born in Jutland and Professor in Copenhagen, botanist, archaeologist (he was the earliest investigator of Mesolithic middens) and zoologist, like many others of his time, helped to disentangle the complicated life histories of invertebrates and made a broad comprehensive survey of the alternation of generations which characterises the life histories of many free-living and parasitic animals. Edward Forbes (1815–54), Manx born, was for a time Professor of Botany in King's College London and towards the end of his short life Regius Professor of Natural History in the University of Edinburgh. A splendid marine naturalist and palaeontologist, his numerous publications include three large monographs on different groups of marine invertebrates, the *Naked-eye Medusae*, the *History of British Starfishes* and (with Sylvanus Hanley) the *History of British Mollusca.* Félix Lacaze-Duthiers (1821–1901), Professor at Lille and later in Paris, who worked on the anatomy and evolution of the molluscs was the founder of the marine stations at Roscoff and Banyuls. And finally, Thomas Henry Huxley (1825–95), a great teacher, a marine biologist who made many notably original investigations of the habits and life histories of medusae, sea-squirts and crustaceans, the foremost expositer of Darwinism was the founder, in Britain, of the scientific investigation of sea fisheries. Huxley was the leading spirit in the foundation of the Marine Biological Association of the United Kingdom and in the building, in 1888, of the Plymouth Marine Laboratory.

It is evident when we look through this list of naturalists born in the early years of the nineteenth century and note the character of their zoological researches that zoology is entering a new phase in its historical development; and one of the main reasons for the change is apparent from the recital of the posts to which these naturalists were appointed. It will be noted that, except for Darwin, all of them held chairs of zoology either in the newly formed universities or in older universities which for the first time were including natural history among the subjects of their curricula. Moreover, Aggasiz' appointment at Harvard had extended the boundaries of a European tradition of zoological teaching and research to another continent. The establishment of a rapidly growing number of teaching centres under the leadership of a professional class of research-minded zoologists was henceforth to ensure a strong succession of younger zoologists eager to profit from the example of their teachers. And since, as it happens, many of the more distinguished naturalists of the time lived long enough to acquire influential positions in the scientific academies and as advisers to the governments of their countries, they were able to provide, as in the setting up of marine laboratories, conditions of research that were to be of inestimable benefit to succeeding generations of zoologists and to the advancement of invertebrate zoology in the later years of the century and to the present time.

Concurrent with the emergence of a professional class of university-based zoologists the early years of the nineteenth century witnessed a remarkable flowering of natural history studies in the hands of talented naturalists in every way as competent as their university colleagues but representative of other professions. In Britain alone there were born in the twenty-five years between 1795 and 1815 (in chronological order) W.E.Leach, Joshua Alder, J.S.Bowerbank, George Johnston, Albany Hancock, J.G.Jeffreys and Philip Henry Gosse, each of whom, besides his wider contributions to zoology, was to become an authority on one or more groups of the marine invertebrates and by skill and patient labour to compile detailed and beautifully illustrated monographs which even to this day are used by naturalists as valuable works of reference. Johnston's (1838) *The British Zoophytes*, Alder and Hancock's (1845–55) *Monograph of the British Nudibranch Mollusca*, Forbes and Hanley's (1848–53) *History of British Mollusca* and Gosse's (1860) *History of the British Sea Anemones and Corals*, all of which were published within a short space of twenty-two years, bear witness not only to scholarship of a high order but to a careful and systematic searching of shores around almost the entire coastline of Britain.

More and more of the new species described in these monographs were being collected from offshore waters by the naturalists using the small and easily worked iron-framed dredge which in Edward Forbe's words was rapidly proving to be 'an instrument as valuable to the naturalist as a chronometer to the natural philosopher'. An even more important event in the opening up of new horizons of discovery was the invention of the tow-net. A tapering cone of fine-meshed silk attached at its front and wider end to an iron ring and with its narrow end lashed around a canister when towed slowly through the water was found to collect an abundance of the many kinds of small organisms which make up the marine plankton. Many of these such as copepods, arrow worms and pteropod molluscs were recognised as adult animals of small size. Others which were described under generic names such as *Zooea*, *Pluteus*, *Bipinnaria*, *Pilidium*, *Actinotrocha* and *Tornaria* were also for a time thought to be adult animals and named as such.

So far as is known tow-nets were first used by J.Vaughan Thompson, while serving as an army surgeon in Ireland. In 1822, Thompson found in the plankton of the Cork river several small comma-shaped and spine-bearing crustaceans which had previously been named *Zooea taurus*. He kept them in bowls and found that after some days, they lost their cuticular coverings and metamorphosed into small crab-like creatures with an extended abdomen. These were very similar to organisms which Leach had already described as of the genus *Megalopa*. Thompson managed to keep the megalopas until they changed into small shore crabs to which Linnaeus had given the name *Carcinus maenas*. Later, Thompson hatched the eggs of the edible crab *Cancer pagurus* into zooea larvae, thus establishing a life history common to all crabs. In showing that the larval stages of invertebrates – of which there may be many in a life-cycle – are strikingly different in form and habit from the adult animal into which they ultimately metamorphose, Thompson made the search for the larval forms of invertebrates and the elucidation of their life histories a major subject of research which contributed more directly than any other field of study to the early nineteenth-century growth of understanding of invertebrate animals and of their classificatory relationships.

The publication in 1859 of the *Origin of Species* and of Darwin's evidence for the transmutation of species and of organic evolution drawn from the fossil history, anatomy, embryology, adaptations, and the past and present distributions of animals gave a new urgency to zoological research both in the field and in the laboratory. It became important to extend the search for animals to regions and environments which had previously been unexplored and the three-year expedition of H.M.S. *Challenger* (1873–6) had as its main object the collection of animals from the unexplored depths of the deep oceans. The dredge hauls of *Challenger*, taken as deep as 2,690 fathoms, were so greatly successful in bringing to light a new world of animals of unusual adaptations and of outstanding systematic importance that within a few years the challenge of deep-sea exploration had been taken up by almost all the major maritime nations of the world.

Another consequence of Darwin's presentation of the broad vistas of evolutionary change was to stimulate the tracing of evolutionary trends within the phyla of animals and to encourage speculation on the lines of descent and relatedness of the phyla themselves. The study of phylogeny depends on the recognition of the nature and development of the main features of the architectural design of animals and the determination both of their persistence in time and of their adaptiveness to environmental requirements. These studies were greatly aided by the development in the 1870s and 1880s of the microtome, an instrument which cuts animals or pieces of tissue suitably preserved and impregnated with wax into a series of very thin sections which can be mounted serially on slides and examined under the microscope. The contemporary development of fixatives designed to preserve tissues in a tolerably life-like state, of stains to make them more visible and capable of differentiation by colour variations, and of microscopes of improved lens design, became the necessary tools of trade of anatomical and embryological studies which came to dominate the research of the later years of the century.

Our tale of the historical development of invertebrate zoology is but half told for the great and expanding revelation of the invertebrate animals belongs to the twentieth century. Zoology was henceforth and of necessity to become as we noted in the opening paragraphs of this chapter 'increasingly specialized and to be concerned with particular aspects of the lives of invertebrates rather than with the broad perspectives of animal form, functions and systems of relationship'. But, if zoological research has become progressively the province of professionally trained scientists skilled in the use of specialized techniques and complicated equipment, zoology remains at heart the science of looking at animals and learning from the looking; and this is something which we all may do.

Chapter 3

The personae of the invertebrates: the major phyla

AN ANIMAL exists by virtue of the structural materials of which it is made and is recognized as being an animal of a particular kind both by the nature of its constituent materials and by the way in which they are arranged in composition of the general body design. Related species may be so alike in their structure and design that the differences between them are only apparent after a careful and detailed examination of all their parts. Animals may, on the other hand, vary so greatly in their form and construction that they are recognised as being of distinctively different styles of organization and styles of architecture. Differences of this magnitude are used as the criteria for subdividing the animal kingdom into the major categories of the phyla.

In a two-chapter survey of the invertebrate phyla we will follow Hyman (1940) in recognizing twenty-three distinctive categories of animals; and, since it will not be possible to give more than an impression of the immense variety of animals contained within the various groupings, it may be well at the outset to explain the intentions of the survey and to make clear its limitations.

The phyla are reviewed in the order of their size, as measured by the number of their known present-day species, and are considered in groupings approximating to 10^5, 10^4, 10^3, 10^2, and 10^1 species dimensions. The main purpose of this method of presentation is to establish the perspectives of the invertebrate panorama by displaying with especial prominence the kinds of animals (of the 10^5, 10^4 and 10^3 groupings) which by sheer force of numbers and a widespread distribution dominate the environments of the world, and by making only brief mention of the many phyla of minor role, namely the 10^2 and 10^1 species groupings.

Of the twenty-three phyla, one, the Arthropoda (crustaceans, insects, spiders etc.), is of colossal proportions. This phylum with almost a million (8×10^5) species of animals with segmented, skeleton-encased bodies, spills its representatives into innumerable habitats within the sea, fresh waters, land and air. There follow four phyla of immense size, the Mollusca, Nematoda (thread- and roundworms), the Chordata (including the vertebrate classes), and the minute unicell Protista, each running into tens of thousands ($1–8 \times 10^4$) of species. Six phyla, the Echinodermata (star-fishes, sea-urchins and their allies), Annelida (ringed worms and leeches), Platyhelminthes (free-living and parasitic flatworms), Cnidaria (hydroid zoophytes, jelly-fishes, anemones, corals and other coelenterates), the Porifera (sponges) and the Ectoprocta (colonies of box- or vase-contained polyzoan polyps) are of large ($5–9 \times 10^3$) dimensions. The twelve remaining categories, representing more than half of the total number of animal phyla, share little more than two thousand known living species and are together less than half the size of the smallest of the phyla we have so far listed. Ten of them, the Nemertea, Sipunculida, Acanthocephala, Brachiopoda, Hemichordata, Pogonophora, Ctenophora, Echiuroida, Entoprocta and the Chaetognatha range from a few hundreds to less than a hundred species apiece, and few have earned the familiarity of common names. And finally there are two minute phyla, the Phoronida and the Mesozoa, each containing a mere handful of ten or so different kinds of animals.

This present chapter reviews the five largest phyla of the animal kingdom which provide between them no less than 95 per cent of all the known living animals. In briefly characterizing each phylum mention is made of the constructional features that are diagnostic of its particular style of architecture or are indicative of its relationships with other phyla. Reference is then made to the organ systems and bits of structural gadgetry that have been of particular service to the phylum by providing, through the versatility of their form and structural adaptations, the means of exploiting different kinds of environmental conditions and ecological situations. These considerations lead naturally to the identification of the characters that can best be used in classifying animals into subsidiary categories and groupings, and some – necessarily a few – examples are given of animals representative of the various major subdivisions. In concluding the brief review of each phylum some notes are added, when it is appropriate to do so, on matters of especial interest such as the fossil history of the phylum.

Phylum Arthropoda. A colossal assemblage of animal species, ten times larger than the next phylum (the Mollusca) in order of size, and one hundred thousand times as big as the smallest phyla. Of the eight-hundred thousand and more arthropodan species (80 per cent of the entire animal kingdom), some 750,000 are insects. Crustaceans and arachnids (spiders, mites, etc.), each with 20,000–25,000 species, are the other major classes of arthropods.

All arthropods, at some stage during their lives and usually in the adult, are bilaterally symmetrical; have segmented, ringed bodies retained within an encasement of tough exoskeletal material; paired jointed appendages attached to some or all of the body segments; three body layers – ectoderm, mesoderm and endoderm – of differing embryological origin and structure-forming capacity; and have a body cavity formed, for the most part, of enlarged blood spaces. Some of these basic arthropodan features are illustrated in Fig. 3.1, a transverse section through a body segment of the abdomen of a crayfish.

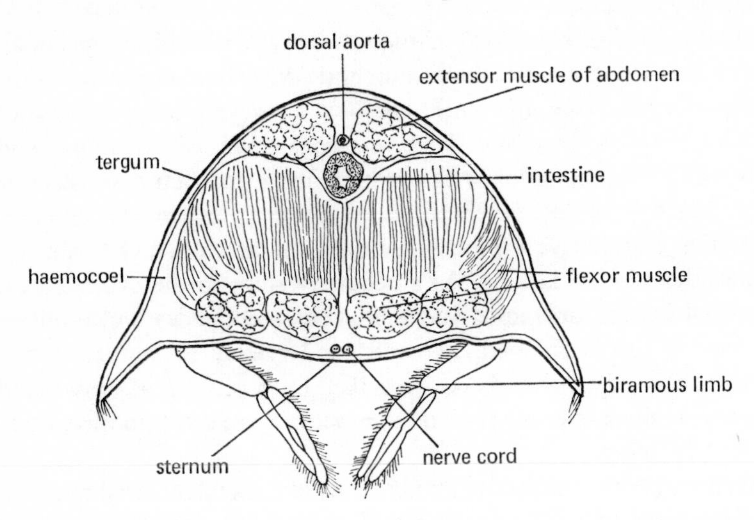

Fig. 3.1 Transverse section through an abdominal segment of a crayfish (after T.H. Huxley, 1880).

The uniqueness of the arthropodan type of body and the justification for regarding the group as a phylum lies mainly in the combination of these features. The nature of the various materials used to toughen the skeleton, some parts of which remain flexible to allow for the hinging of the body segments or of the successive pieces of a multi-jointed limb, and the universal occurrence within the group of an enlarged perivisceral blood sinus are, however, diagnostic of the phylum.

Body segmentation involving the repetition of similar kinds of structures in successive segments (metameric segmentation) is not unique to the arthropods, being found also in the Annelida (p. 51) and Chordata (p. 40).

The chordate segmentation is an unrelated, independent development, but the segmentation of annelids and arthropods is indicative of a near relatedness and of common ancestry. The relationship of the two groups is further supported by the occurrence in the less specialized annelids of paired parapodia having some equivalence to the arthropodan limbs, and by the presence in the two phyla of a central nervous system of a similar type of construction and arrangement within the body.

There is little doubt that the structures and organ systems that have most substantially contributed by their variations of form and function to the immense diversity of the arthropods and to the evolution of the several arthropodan classes are the exoskeleton, the paired appendages, the sensory organs and the central nervous system.

The foundation substance of the exoskeleton is chitin, a material resembling cellulose in its composition and properties, and in the different groups of arthropods, the chitin becomes in varying measure invaded or overlain by other materials. In crustaceans much of the exoskeleton is hardened by impregnation with chalk grains. Lipid (fatty) substances are incorporated into the outer layers, thereby giving to the exoskeleton properties of selective permeability to water and dissolved salts which would otherwise be lacking. As a necessary stage in the evolution of the terrestrial insects, arachnids and myriapods (centipedes and millipedes), the exoskeleton is thinned, lightened and greatly strengthened by the impregnation of proteins bound together by a process of cross-bonding (tanning) to form sclerotin; and over the surface are laid thin waxy coatings for waterproofing.

In all parts of the arthropod body struts of skeletal material project into the haemocoel to form attaching surfaces for muscles. Except in a few organ systems (e.g. the legs of spiders) the muscles attach in antagonistic pairs to moving parts, promoting by their reciprocal contraction and relaxation a two-way movement such as is seen for example in the rapid flexion and extension of the abdomen of a jumping prawn; and in the jointed limbs of arthropods the successive pieces of the limb move in different planes to give the limb great flexibility of movement and positioning. Arthropodan limbs are used as sensory feelers; for walking, swimming, digging and burrowing; for collecting, manipulating and chewing food; for respiration; as organs of offence and defence; for reproduction as sex symbols, copulatory organs, brood pouches and egg carriers; and for a variety of other purposes. Limbs of differing function are positioned on the body in the regions where their functions are most efficiently carried out, and are usually arranged in mutually assisting groups.

No other assemblage of invertebrate animals, except the cephalopod molluscs (octopuses and squids), show by their manner of viewing, feeling and savouring their surrounds, and by their reactions to the situations in which they find themselves, a more obvious awareness of their environment. The many-faceted eyes of arthropods and the development, in for example the insects and arachnids, of a brain which incorporates regions of great neuron complexity have been particularly important agencies in the creation of complex inbuilt patterns of behaviour that are perhaps best seen in the business of food collecting, breeding behaviour and care of the young.

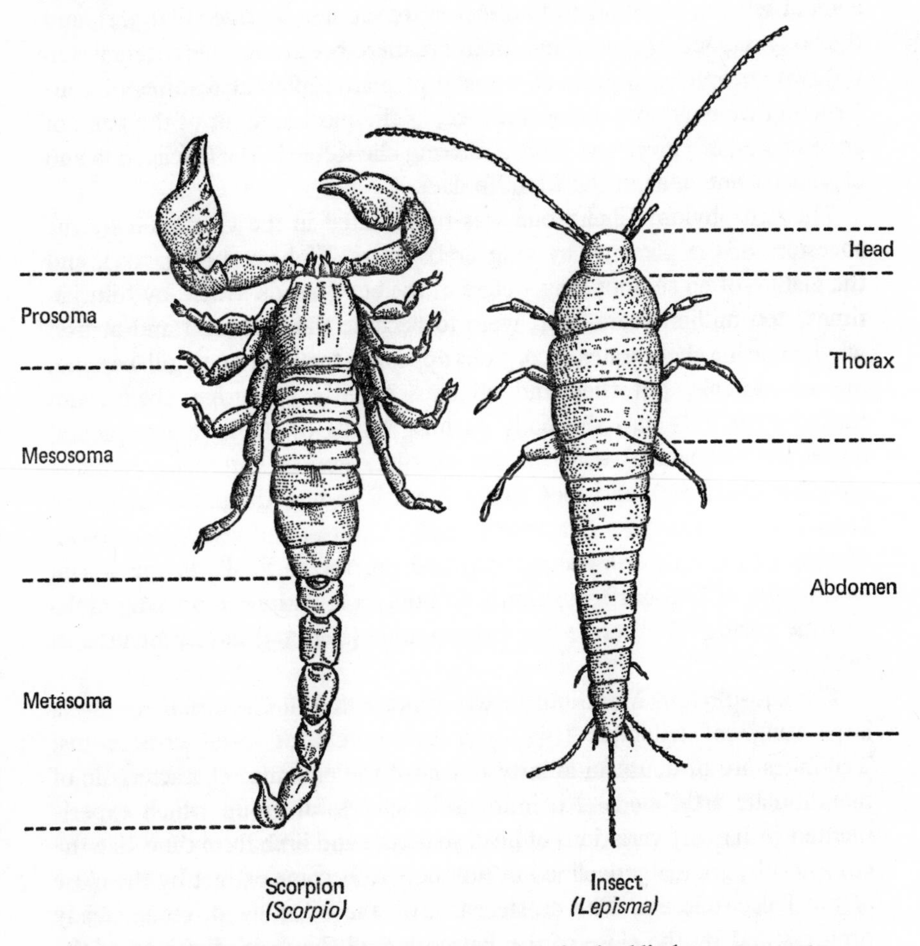

Fig. 3.2 The different body divisions (tagmata) and paired limb arrangements of the chelicerate and mandibulate arthropods as seen in dorsal view of a scorpion and a wingless insect

Fossil arthropods are found in the 600-million-year-old sedimentary rocks of the Cambrian era and a few are found in Pre-Cambrian deposits. Arthropods early showed a disposition to experiment widely with different segmentation patterns, limb form and limb arrangements. The subsequent exploitation of these differing patterns is, however, based on the two essentially different types of segment and limb seriation of the sub-phyla of the Chelicerata and the Mandibulata. Scorpions and the many kinds of arachnids (spiders etc.) are the most numerous of present-day chelicerates; insects and crustaceans are the dominant modern mandibulates. The contrasting form of the body regions and limb series is evident when a scorpion and an insect are set side by side (Fig. 3.2), and the body regions (tagmata) and limbs are therefore given different names in the two sub-phyla. Merely as a token of many different features of construction we may note the occurrence, as the most anterior of the pairs of appendages, of pincer-like food-gathering chelicerae in the Chelicerata and of sensory antennae in the Mandibulata.

The sub-phylum Chelicerata was represented in the Cambrian by the ancestors of the present-day king crabs, class Xiphosura (Fig. 3.3), and the giant – often six foot long – class of the eurypterids which, by Silurian times, 200 million years later, were to become the dominant and at first wholly marine chelicerates. Scorpions appear in the Silurian as gill-bearing, marine animals, but they and all subsequently appearing chelicerates tucked their gills into the body as lung books or (as in most spiders) developed, in parallel with the mandibulate insects, internally coursing air tubes (tracheae), and took to the land. The now dominant terrestrial chelicerates of the class Arachnida (spiders, mites, and a variety of other smaller and mainly tropical groups) first emerge as fossils in the warm, wet jungles of Carboniferous times, and only a few xiphosuran king crabs and the curious chelicerate-like pycnogonids (Fig. 3.3) have remained in the sea.

The sub-phylum Mandibulata was represented in Cambrian seas by a large range of trilobites (Fig. 3.3) and numerous small crustaceans. Trilobites are of doubtful affinity but have the antennae characteristic of mandibulate arthropods. An immensely successful group which experimented with many variations of head structure and limb form they, like the eurypterids, gradually declined in numbers to become extinct by the close of the Palaeozoic era. The crustaceans, on the other hand, made steady progress and by the close of the Palaeozoic all the main divisions of the group which are known to us at the present time and are recorded as fossils were in existence, though the larger, more specialized and most familiar

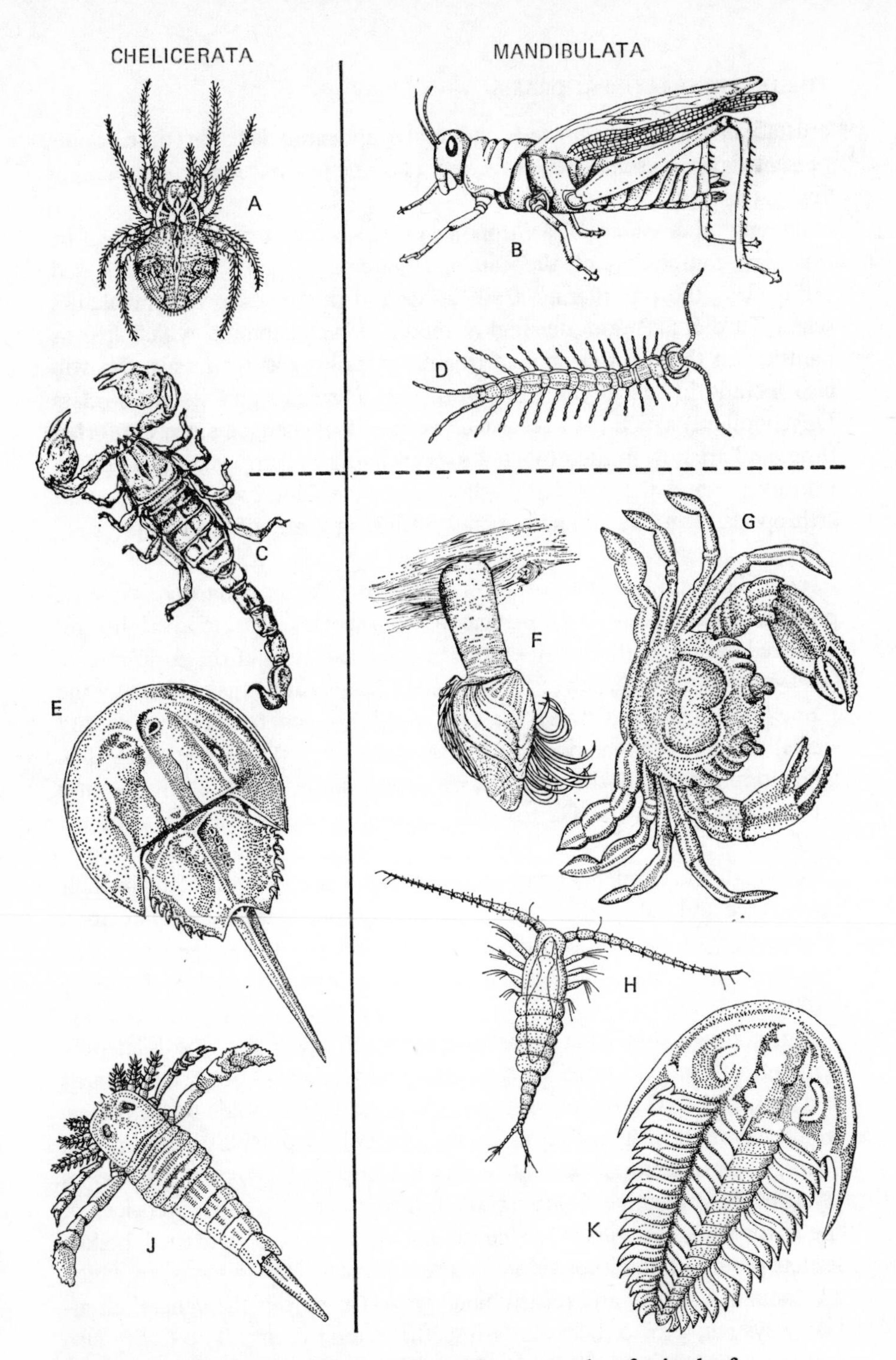

Fig. 3.3 Phylum Arthropoda. A, C, E, and J are examples of animals of different classes of the subphylum Chelicerata; B, D, F, G, H, and K of classes of the subphylum Mandibulata. A, a spider (*Epeira*); B, an insect (*Locusta*); C, a scorpion; D, a centipede (*Lithobius*); E, a king-crab (*Limulus*); F, a goose-barnacle (*Lepas*); G, a swimming crab (*Macropipus*); H, a copepod crustacean (*Calanus*); J, a eurypterid; K, a trilobite (*Olenellus*)

crustaceans, such as lobsters and crabs appeared later in time. Some present-day representatives of the orders of the Crustacea are illustrated in Fig. 3.3.

Among other classes of arthropods, of which only brief mention will be made in completion of the catalogue, are myriapods (centipedes and millipedes), the tiny marine, fresh-water and damp soil bear-animalcules (class Tardigrada) and the highly modified pentastomids which live as parasites in the nasal cavities of (mainly) reptiles and mammals. We will also include in the phylum the worm-like *Peripatus* and its allies (class Onychophora) which have persisted practically unchanged since Cambrian times and are now limited to a few species living in damp vegetation in the warmer parts of the world. Onychophorans combine both annelidan and arthropodan features and are an ancient linking group of the two phyla.

Phylum Mollusca. Next in order of size to the Arthropoda this immense phylum contains some 80,000–100,000 living species. Marine in origin, and still predominantly marine, it embraces the five classes of the Amphineura (coat-of-mail shells), the Scaphopoda (elephant's tusk shells), the Lamellibranchiata (bivalves), the Gastropoda (in general terms, slugs and snails) and the Cephalopoda (octopuses and squids) (see Fig. 3.4). The lamellibranchs and the gastropods have both evolved fresh-water representatives but the Gastropoda alone have achieved invasion of the land.

Although the animals of the different classes are very different in their appearance and habits all are of an essentially similar type of architecture. The molluscan construction is perhaps best seen in the veliger larva, the free-swimming and independently feeding juvenile phase of many molluscs.

Fig. 3.5 is a somewhat stylised drawing of a veliger. The bilaterally symmetrical, unsegmented body projects forwards as a head, downwards as a foot, and carries on its back (postero-dorsally) side folds of the mantle which surround without fully enclosing a part of the outside environmental water. Cells of the mantle wall secrete the organic matrix and calcareous layers of the protective molluscan shell. A pair of thin-walled extensions of the body wall, the gills, each consisting of a series of flattened hollow leaflets mounted on either side of a seam of tissue (the gill axis) in which are contained supply and return blood vessels as part of the general circulatory system, project backwards into the mantle cavity. The cavity also serves as a repository for waste materials voided from the rectum and through the openings of the paired tubular excretory organs. The inner

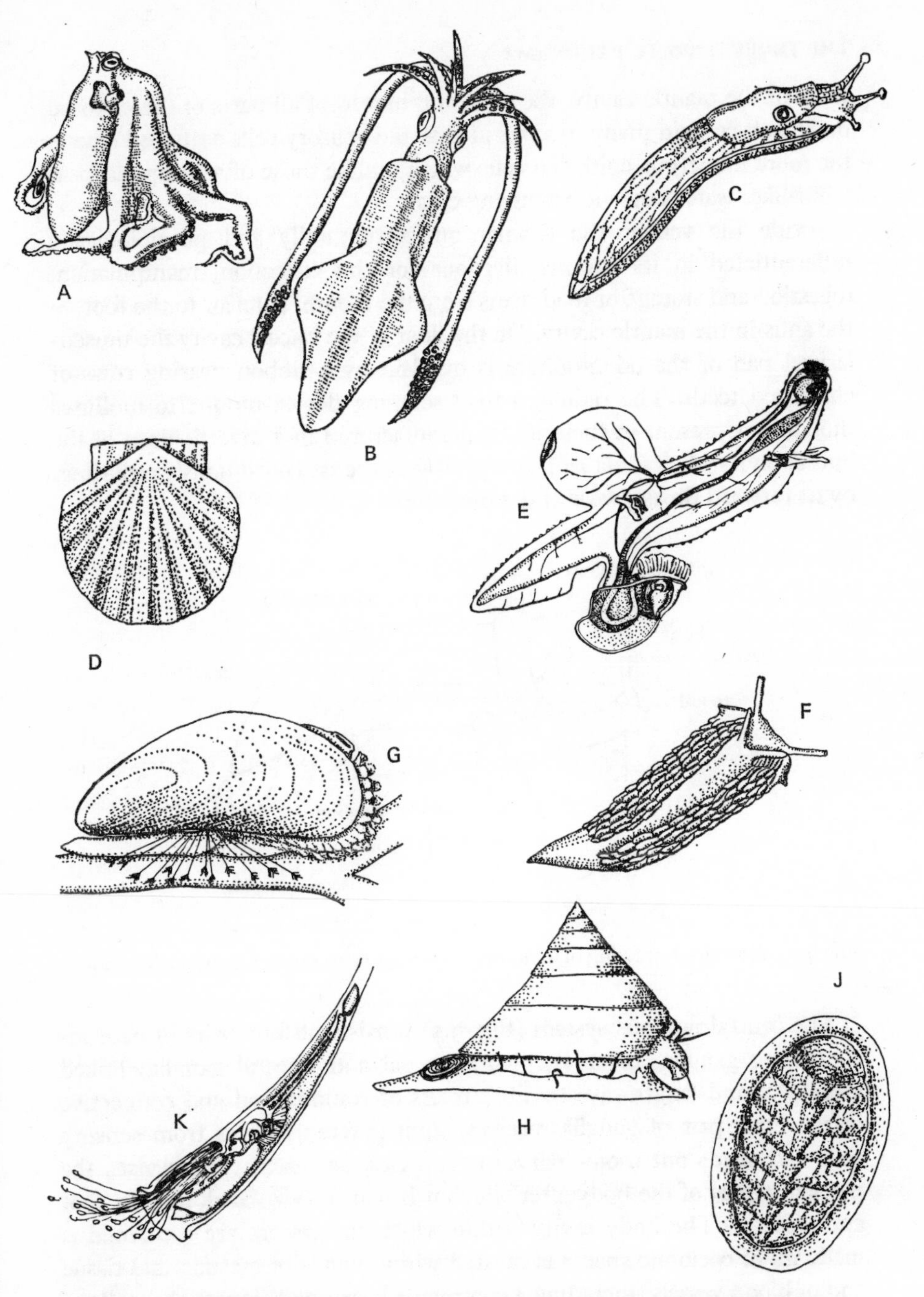

Fig. 3.4 Phylum Mollusca. Some examples of animals of the five classes of the Mollusca. Class Cephalopoda: A, *Octopus*; B, a squid (*Loligo*). Class Gastropoda: C, a slug (*Arion*); E, a heteropod (*Carinaria*); F, a nudibranch (*Aeolis*); H, a top-shell (*Calliostomam*). Class Lamellibranchiata: D, a scallop (*Pecten*); G, a mussel (*Mytilus*). Class Amphineura: J, a coat-of-mail shell (*Chiton*). K, Class Scaphopoda: an elephant-tusk shell (*Dentalium*)

walls of the mantle cavity, the gills, and indeed of all parts of the exposed body wall contain many mucous glands and sensory cells scattered among the more numerous epithelial cells which, unlike those of arthropods, bear whip-like, water current producing cilia.

Inside the veliger the tubular gut, structurally and physiologically differentiated in its various divisions for the collection, manipulation, digestion and storage of food, runs from the mouth, anterior to the foot, to the anus in the mantle cavity. On the floor of the buccal cavity the muscularized pad of the odontophore is overlain by a ribbon bearing rows of chitinized teeth. The radula, a food scraping device unique to molluscs though not present in them all, is manufactured in a diverticulum of the buccal cavity (the radula sac), new growth in the sac continuously replacing, by its forward thrust, wear at the front end.

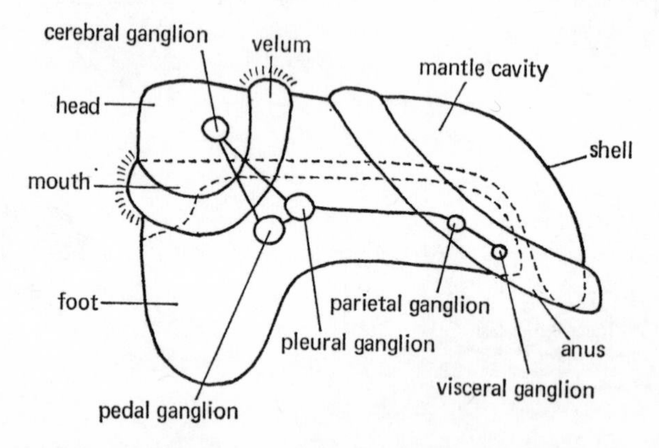

Fig. 3.5 A stylized drawing of a molluscan veliger larva viewed from the left side.

The central nervous system (Fig. 3.5) consists of four pairs of neuron-containing ganglia (cerebral, pleural, pedal and visceral ganglia) linked crossways and lengthways by fibre tracts of commissural and connective cords. Each pair of ganglia receives input (afferent) fibres from sensory cells and sends out motor nerves to muscles, and each pair services the particular part of the body after which it is named (viz. head, mantle, foot and viscera). The body cavity within which the viscera are contained is made up of coelomic spaces excavated within bands of mesodermal tissue and of blood vessels (including a contractile heart) and sinuses the walls of which are formed by embryologically less coherent mesenchymal cells. It is characteristic of molluscs that the coelom takes the form of four inter-connected sacs. The heart surround (pericardial coelom) and the germ cell producing cavity (the gonocoel) are single cavities; the excretory coelom (renocoel) consists of a pair of sacs.

The fundamental shapes, body orientations and body axes of animals of the different classes of mulluscs can all be derived from the veliger type of construction (Fig. 3.6). Amphineurans flatten the foot into a ventral adhesive sole and, in their most characteristic form of the coat-of-mail shells or chitons, cover the upper (dorsal) side with eight movably articulated cross plates. The mantle, enclosing several pairs of gills, stretches along the length of the two sides of the body, the rectum and the excretory tubes opening into it at the hind end. The veliger organization is thus transmuted without radical change into an adult composition. Gastropods have a like body orientation and primitively, at least, have a flattened adhesive foot. The head is large and equipped with sensory tentacles, eyes, and other environment-testing organs; and the shell (unless it is secondarily lost) is in one piece as a cone or a twisted spiral of various forms. All gastropods, at least in their early stages of development, twist the mantle cavity through 180° to bring its opening to a position behind the head which, on retraction, can be drawn into the protective shelter of the shell. Lamellibranchs, characteristically of a sedentary habit, flatten the body from side to side, reduce the head to a structurally indistinguishable bit of the front end, and plaster side plates on the enlarged mantle flaps. The shell valves, hinged dorsally, can gape widely and are drawn together by the contraction of adductor muscles to give complete cover to the withdrawn animal. The most elaborate and functionally important organs of lamellibranchs are the elongate gills made up (in the more advanced forms) of complexly folded filaments palisaded into lamellae which extend along the length of the mantle cavity as cilia-bearing curtains which draw in, collect and transport to the mouth the minute particles on which the animals feed.

The elaborate cephalopods fuse the head and foot and make of the two systems a complicated head-arm complex furnished with eyes – superficially similar to the eyes of vertebrates – and drawn out into highly tactile food-seizing tentacles of which there may be many (*Nautilus*), ten (squids) or eight (octopuses). The mantle cavity is curved downwards and forwards and its strongly muscularized wall dilates and contracts to draw in and to expel in a funnel-directed stream the jets of water which thrust the animal, according to the direction of pointing of the funnel, backwards or forwards through the water.

The main lines of evolution and of the adaptive radiation of molluscan structure has depended, in the main, on the ability of molluscs to develop

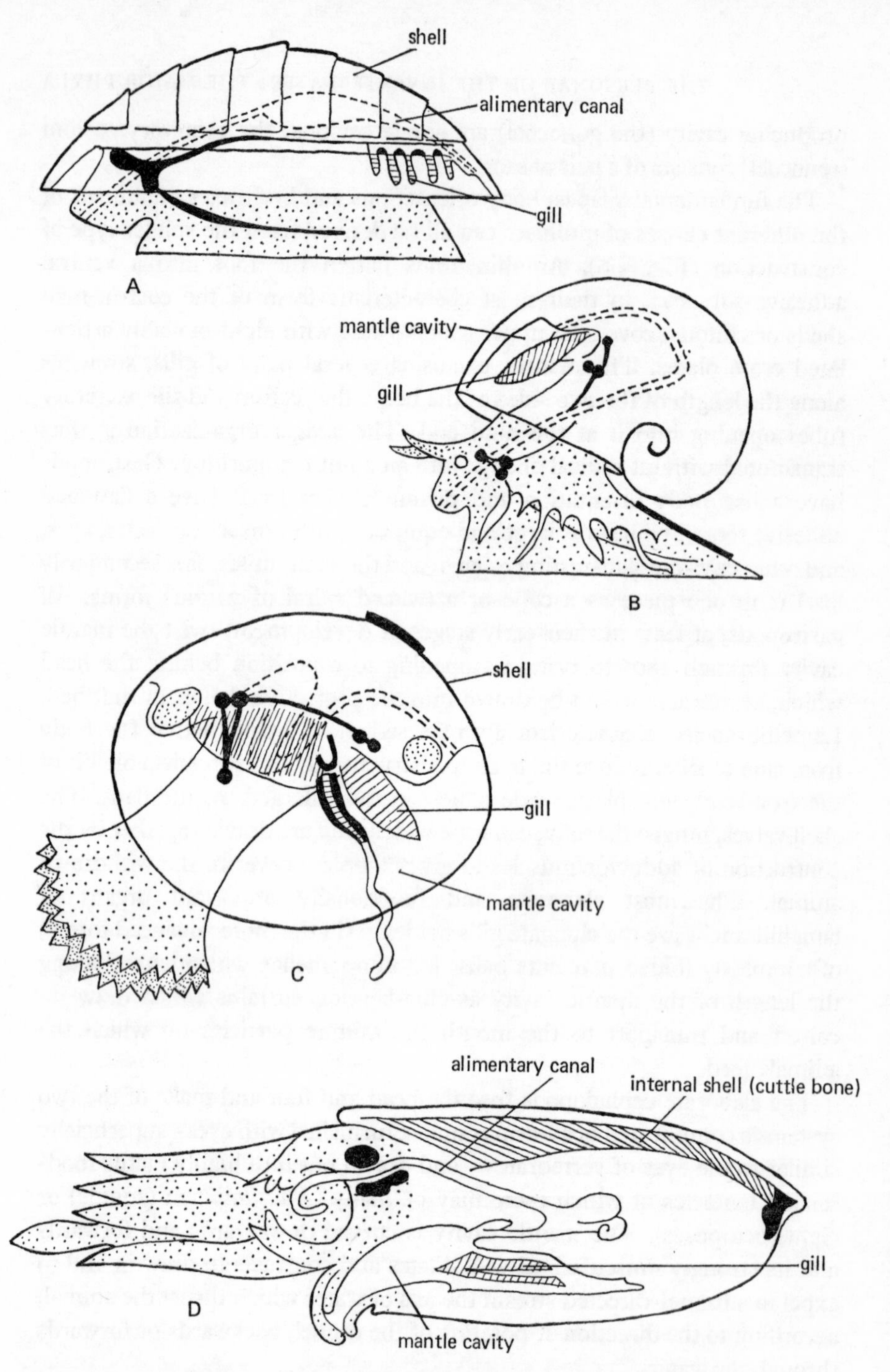

Fig. 3.6 Phylum Mollusca. The body form of different classes illustrated schematically in side view. A, an amphineuran; B, a gastropod; C, a primitive lamellibranch; D, a cephalopod. The head and foot regions are stippled, the mantle and shell regions are unshaded. The principal nerve ganglia and nerve connectives of one side are shown as black circles and thickened lines.

and exploit, in innumerable ways and for a variety of functions, the special varying arrangements of the mantle complex of organs, the foot and the shell. In addition to these capabilities, cephalopods, by concentrating largely on head development, have acquired a perfection of eyes and of organs of touch and balance which, combined with the development of a brain of extraordinary neuron complexity, and in some regions dissociated from the traffic of afferent and efferent discharge, has given to these animals a memory system and capacity for learning that is matched only by the vertebrates.

There is little in the anatomy of adult molluscs to suggest that the phylum Mollusca finds its closest relationships with the segmented animals of the phylum Annelida (p. 51). The two phyla, so divergent in their expressions of adult form, exhibit, however, a remarkable similarity in the pattern of cleavage in their eggs and in the identity of the cells and cell groups (blastomeres) which give rise to similar embryonic structures. These similarities suggest a common ancestry dating far back into Pre-Cambrian times, but stemming from animals of unknown form and history.

Phylum Aschelminthes (mainly class Nematoda). The members of this phylum of bilaterally symmetrical, unsegmented animals are mostly worm-like and barely visible to the naked eye. All have a body cavity which is the first formed space (the blastocoel) of the embryo – a condition known as pseudo-coelomate. While this common feature probably justifies the inclusion of the five classes: Nematoda (thread-worms), Rotifera (wheel animalcules), Gastrotricha, Kinorhyncha and Nematomorpha within one phylum (see Fig. 3.7), the classes are more diverse in structure than is usually allowed for the subdivisions of a phylum. The species content of the five classes is approximately: nematodes (10,000), rotifers and gastrotrichs (each 1,500), kinorhynchs (100) and nematomorphs (<50) but, since nematodes are ubiquitous in habitat, have not been throughly searched for, and are very difficult to identify to species, their real number may safely be put as more than 100,000 and may be many times greater. For this reason the Aschelminthes are promoted to a place among the immense phyla.

Nematoda. No other group of animals has made so much of so little. The cells of nematodes, apart from the germ cells of the gonad, are few in number and are individually very large. There are probably fewer than five hundred cells inside the body wall in the smallest (< 1 mm) and the largest (more than one metre long) thread-worms. No nematode cell is ever ciliated. Nematodes are somewhat conservative in their range of structural variations, though the sucking foregut takes on many forms including at times an armament of piercing stylets. Economy of structure is matched by a physiological toughness which enables most nematodes to survive extremes of cold, heat, desiccation, oxygen lack and a variety of bathing fluids. Highly productive egg producers, nematodes hatch as wriggling larvae which make their home in marine, fresh-water and damp terrestrial soils, and enter as parasites into almost every kind of suitably sized plant or animal host.

Rhabditis, a tiny nematode common in soils containing dead animals, exemplifies the basic nematode organization (Fig. 3.5). The body wall consists of a middle layer (hypodermis) with few nuclei and no discernible cell boundaries (a syncytium), coated on the outside with a fibrous, elastic keratin cuticle, and bounded internally by four quadrants of longitudinal muscles which, with their large proportion of non-contractile substance, are peculiarly nematode in character. The muscles of each side of the body alternate in their contractions to produce side to side sickle flexures or waves rippling along the body. The deformations give locomotory thrust when pressed against surrounding objects. The tubular gut opening anteriorly at the mouth and subterminally at the postero-ventral anus is regionally divided into a relatively thick-walled muscularized sucking fore-gut with a circlet of six lips surrounding the mouth, a long thin-walled mid-gut and a short capillary hind-gut. Most of the remaining part of the pseudocoel is filled with the gonads. In the female the ovaries are paired and open by a vagina about half way along the under side of the body. The testis of the male is single and the amoeboid spermatozoa pass to the outside down a discharge duct (the vas deferens) into a common rectum-vas deferens chamber (the cloaca) which opens within the shallow bowl of the bursa copulatrix. During copulation, spicules are thrust into the vagina of the female and spermatozoa pass into a receptaculum within the ovary to fertilize eggs which, before extrusion, are coated with a cuticle. The nervous system consists of a ganglionated collar encircling the fore-gut from which dorsal and ventral cords and other finer nerves run backwards along the body in the skin. A pair of excretory cells open by a common duct and pore on the under side a little way behind the mouth.

The tiny thread-like larvae of nematodes grow in jerks by moulting the tightly fitting cuticle, swelling their cells and forming a new cuticle. The third of the larval stages is characteristically motile and foraging, and in parasitic forms is usually infective of the host. The parasitic forms often have secondary (vector) hosts and nematodes cause a variety of diseases disabling or fatal to man; for example, hookworm, or miner's disease (*Ancylostoma*), elephantiasis (*Wuchereria*) and, guinea worm disease

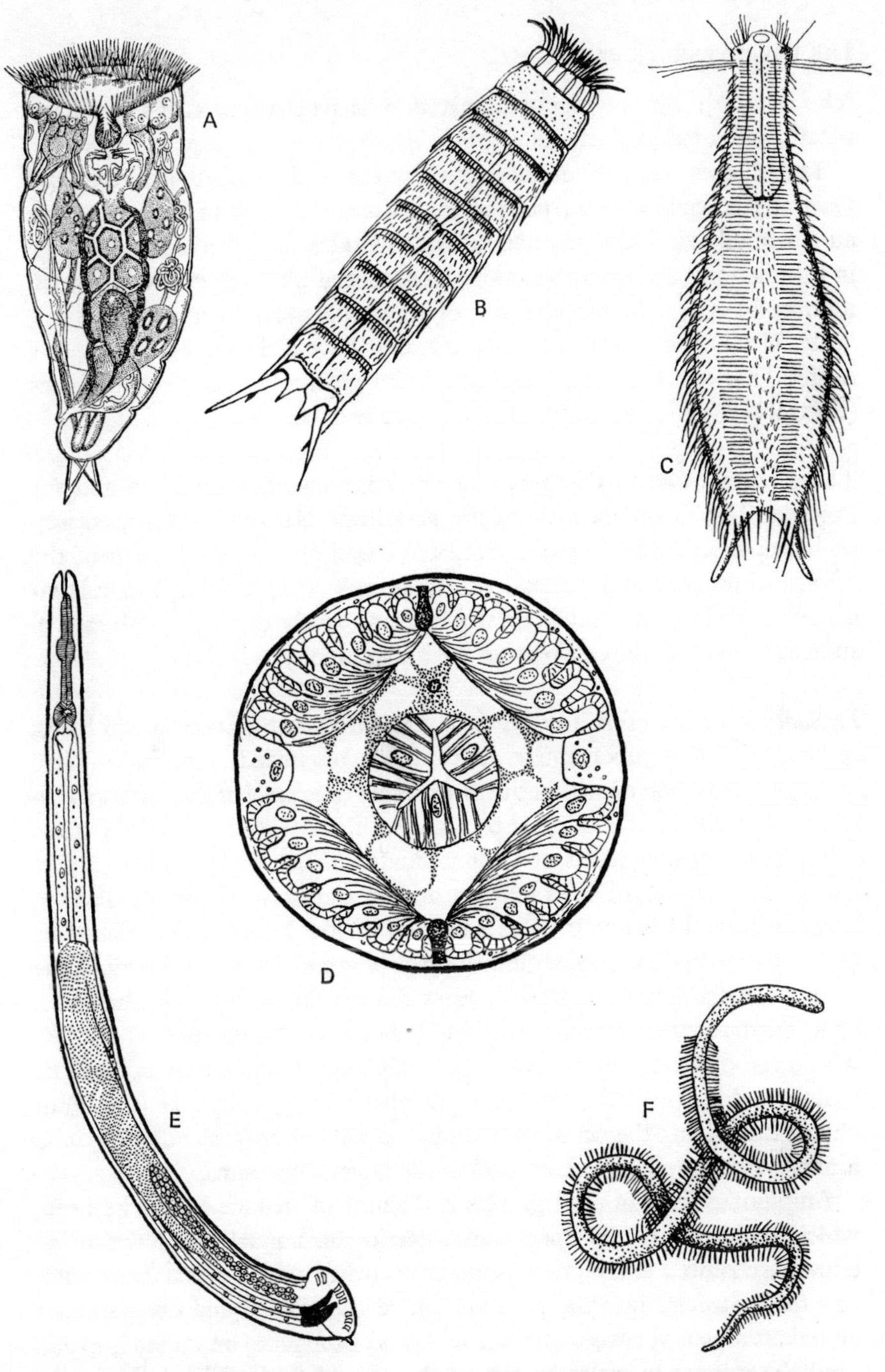

Fig. 3.7 Phylum Aschelminthes. Some examples of animals of different classes of the Aschelminthes. A, a rotifer (*Hydatina*); B, a kinorhynch; C, a gastrotrich (*Chaetonotus*); E, a nematode worm (*Rhabditis*); F, a nematomorph worm (*Emplectonema*); D, T.S. of a nematode worm (*Ascaris*) showing the cuticular body wall, the longitudinal muscles of the body wall, the central unmuscularized intestine, the pseudocoel body cavity, the dorsal and ventral nerves and the laterally placed excretory tubes

(*Dracunculus*). Among many nematodes destructive to crops are beet, potato, and tomato eel worms.

The classes of the minute fresh-water and marine Rotifera and Gastrotricha can be seen from the generalized diagrams of figures to have an essentially similar construction to nematodes, and features not shown in the diagrams such as the pseudocoel nature of the body cavity, the small number of cells, and the presence of an elastic cuticle heighten the similarity. Unlike the nematodes, however, they have a ciliated body wall, and rotifers (Fig. 3.7) have elaborate front-facing girdles of cilia of various forms (the trochus) which they use for locomotion and feeding.

There remain two further phyla of immense size, the Chordata and the Protista. The Chordata include the vertebrate classes of the fishes, amphibians, reptiles, birds and mammals as well as the sub-phylum of the Cephalochordata (the lancelets) which, because they lack bony or cartilagenous vertebrae, are invertebrates. A few words on each phylum will suffice to complete the catalogue of the immense phyla.

Phylum Chordata. Fourth in order of size with some 45,000 named living species, the large, perceptive and highly organized vertebrates have originated and evolved within the measurable time scale of the Cambrian to Recent succession of geological periods, with now extinct types of jawless fishes giving rise to fishes of various kinds, ancient fishes to the earliest amphibia, early amphibia to reptiles, and reptile stocks of former times to both birds and mammals. The small spear-like lancelet, *Branchiostoma* (amphioxus) and its allies within the sub-phylum Cephalochordata, alone among the chordates, are invertebrates though they possess, in their own style, all the more generally identifying chordate features such as gill slits within the pharynx wall to serve as exit channels for water taken in at the mouth, a dorsal tubular nerve cord, a pliable supporting rod (the notochord) underlying the nerve cord, a pharyngeal gutter (the endostyle), and a muscularized tail projecting backwards beyond the anus.

Amphioxus lives in shell gravels and sands of shallow seas. The head, which projects into the open water, carries on tentacles and within an entrance chamber to the mouth, numerous tracts of cilia which draw water and food particles into the pharynx where the ciliated endostyle spreads abundant mucus sideways over the gill-bars of the pharynx wall to trap and bind the minute particles on which the animal feeds. The delicate slit-fenestrated pharynx is protected by a surrounding atrium, an enclosed sea-water chamber into which the pharyngeal water is expelled. Segmentally arranged body-wall muscles, on contraction, produce strong front to back undulations used in swimming and burrowing, the elastic notochord opposing the stresses of muscle contraction.

There is little doubt that the Cephalochordata are the survivors of the first pre-fish essays in chordate design and link, far back in time, with the Memichordata (p. 67) and with the early ancestors of the echinoderms, a phylum (p. 55) now represented by starfishes, sea-lilies and allied kinds of organisms.

The phylum *Protista* with its ten thousand and more living species grouped within five classes of different forms of unicells completes the series of the immense phyla). Some (within the class Flagellata) are plant-like in their photosynthetic type of nutrition and in having, in some instances, cellulose walls; others feed like animals on large organic molecules. Here, among the flagellated Protista, is the present-day memorial of the primeval emergence of animals from plant ancestors.

The arthropods, molluscs, nematodes, chordates and the protistans are pre-eminent among the animal phyla not only because they are the largest in the number of their species but because they are also the most successful. Size and success, though interdependent, are not, however, synonymous, and 'success' is a term fraught with philosophical difficulties. Some explanation of meaning is therefore needed in justification of the opening statement.

The basic need of an animal is to keep alive and its species responsibility (if we may put it that way) is to provide from its substance a new generation of animals of a fundamentally like kind. An animal that comes to terms with its environment and stays alive is a successful animal, and one that contributes from its substance to a new generation has contributed to the success of its species.

In assessing the success of a phylum we need, however, to extend our horizons of judgement beyond the needs and purposes of individual animals and species. A phylum is, at any moment of time, an assemblage of species occupying a range of environments and habitats; it is also a generalized architectural style capable, by the transmutation of its individual species, of producing new forms of varying capacity for extending the environmental representation of the phylum into previously unexplored situations, and even in the course of time into new kinds of environments: for example, from sea water to fresh water. We have, therefore, on the one hand, phyla which, through the genetic transformation of their individual species and the development in consequence of these changes of appropriate larval and adult adaptations, that are able to exploit new ecological situations and, on the other, a world made up of environments and habitats of differing opportunity and hazard that await occupation. The success of a phylum must therefore be judged by the extent to which it has availed itself of these opportunities and, by overcoming hazards, has taken possession of a wide variety of world habitats. By these standards of judgement the arthropods, molluscs, nematodes, chordates and protista, quite apart from their very large size, have been immeasurably more successful than the smaller phyla we are to review in the next chapter. For they are not only present in marine, fresh-water and terrestrial situations but occur in each of a wide variety of habitats and in great abundance.

By contrast, as we shall see later, of the eighteen smaller phyla eleven have remained wholly confined to their ancestral environment, the sea; and, although three of these remaining assemblages have both free-living and parasitic representatives in all of the major environments, none has exploited any of them outside a restricted range of habitats.

By what means, we may then ask, have the five wide-ranging phyla achieved their success? To this question there must be many answers, for in respect both of its physiological and morphological adaptations each phylum has framed its innovations within the resources of its own particular style of architecture and material composition. But if we are to speak in general terms of the conditions of success we may perhaps regard as the most important (a) a high fecundity or, if egg production is low, the development of compensating methods of brood care by which to ensure the production of successive generations in quantities adequate for the maintenance of populations of large or, at the very least, of viable size, and (b) a high capacity for genetic variation by which to make trial of new expressions of physiological function and morphological adaptation in the exploration of new ecological situations and previously unfamiliar environments.

Although we know rather little about the conditions which determine the fecundity and rates of reproduction of the very many organisms contained within the five dominant phyla of the animal kingdom and have, except for a very few insects and mammals, practically no knowledge of the rate and adaptive value of their genetic mutations, the high productivity of their reproductive processes and their considerable capacity for variation can hardly be doubted. Arthropods, nematodes, protistans and, to a somewhat lesser extent, chordates and molluscs all include species that are known to occur in immense numbers of individuals. It is true that this is not an attribute exclusive to the largest phyla but it contrasts sharply with the relatively small number of individuals produced by animals belonging to many of the smaller phyla. Phoronids, sipunculids, nemerteans, and brachiopods, for example, are rarely plentiful even in the habitats best suited to them. And, as evidence of the genetic variability of the large phyla, we need only point to the great number of species that they have produced within a time period comparable to that which the smaller phyla have had at their disposal for experimentation but which they have used to far less effect.

Chapter 4

The personae of the invertebrates: the minor phyla

THIS CHAPTER includes short descriptions of six large phyla of $1\text{–}9 \times 10^3$ species dimensions and briefly surveys the more numerous smaller categories of invertebrate animals.

The size of a phylum is of course in no way a measure of its zoological interest. Some of them have earned the special attention of zoologists because they furnish important clues to the evolutionary relationships of the larger groupings; for example, the Hemichordata exhibit in their structure glimmerings of the vertebrate style of organization and show, by the manner of their development, features that relate them to the starfishes and other animals of the phylum Echinodermata. Others, like the arrow-worms (phylum Chaetognatha), are recognized by marine biologists as sensitive planktonic indicators of water masses of differing chemical and physical properties, and of differing fisheries potential. And, in another context, the Pogonophora, the most recent of the phyla to be discovered, serve as a perpetual reminder to zoologists that animals may assume unexpected forms (the Pogonophora look like pieces of thin thread) which may easily be overlooked when searching through sea-dredged samples or other kinds of collections. Some of the smaller phyla, for these or other reasons, are referred to in later chapters. However, since our main purpose in this, as in the preceding chapter, is to present, so far as is possible, a picture of the present-day representation of the different kinds of animals, the smaller phyla receive but scant attention and, in many instances, are passed over without mentioning more than a few of the many things that are interesting about them.

The large phyla

Phylum Cnidaria (Coelenterata) (Fig. 4.1). The 9,000–10,000 species of the almost entirely marine hydroid polyps and medusae (class Hydrozoa: *c.* 3,000 species), the exclusively marine thick-walled jelly-fishes (class

Scyphozoa: *c.* 250 species), and the anemones and corals (class Anthozoa: 6,000 species), are the only animals which form tissues of specialized function from two embryonic layers – the ectoderm and endoderm.

The coelenterate body, in its simplest form, is a hollow sac clothed externally by an almost invariably ciliated epithelium and with a single (mouth) opening leading into the central food cavity of the coelenteron lined by a flagellated food-digesting sheet of endodermal cells. A ring of food-capturing tentacles surrounds the mouth. The tentaculated sac is bilaterally symmetrical about the oral-aboral axis and radially symmetrical in the transverse plane.

Coelenterate diversity is founded mainly in the varying extent of the cellular and tissue differentiation of the two body layers, the degree of development and structural organization of an ectodermally secreted jelly-like mesogloea sandwiched between the ectoderm and endoderm, and by the differing (polymorphic) forms and functional adaptations of their solitary or colonially joined individuals of which there are two basic types, the cylindrical (usually sessile) polyp, and the bell- or saucer-shaped (usually free-swimming) medusa.

Both ectoderm and endoderm contain some cells that remain essentially individual and do not bond into tissues. They include groups of potentially multipurpose interstitial cells which may differentiate into single cell mucous glands, multicellular gonads, or secrete mesogloea and form within it collagen fibres and skeletons of inorganic or organic materials. Many migrate in great numbers to the tentacles or (in the Anthozoa) to thread-like filaments (acontia) in the coelenteron where, within an enlarged cell vacuole, they fashion bulb and thread structures (the cnidae) peculiar to cnidarians and which, when irritated by contacts and/or chemical substances, explosively evaginate the thread to lasso or pierce contacting organisms, often with the injection into them of highly toxic fluids. In addition to the epithelial sheets of tissue there are differentiated from both ectoderm and endoderm fibrous strings of contacting, excitation-conducting nerve cells, and layers of contractile muscle fibres. The nervous system consists of sensory cells in fibre to fibre contact with deeper-lying intra-epithelial fibre systems made up of short multidirectional fibre nets of short excitation reach, to which are added in the larger more highly organized coelenterates long-fibred linearly arranged through-conducting pathways which serve to bring distantly placed groups of muscles into co-ordinated patterns of contraction. The muscles of the simplest types of hydroid polyps are developed as extensions of the epithelial cells laid along and across the mesogloeal surface as body-contracting and body-lengthening

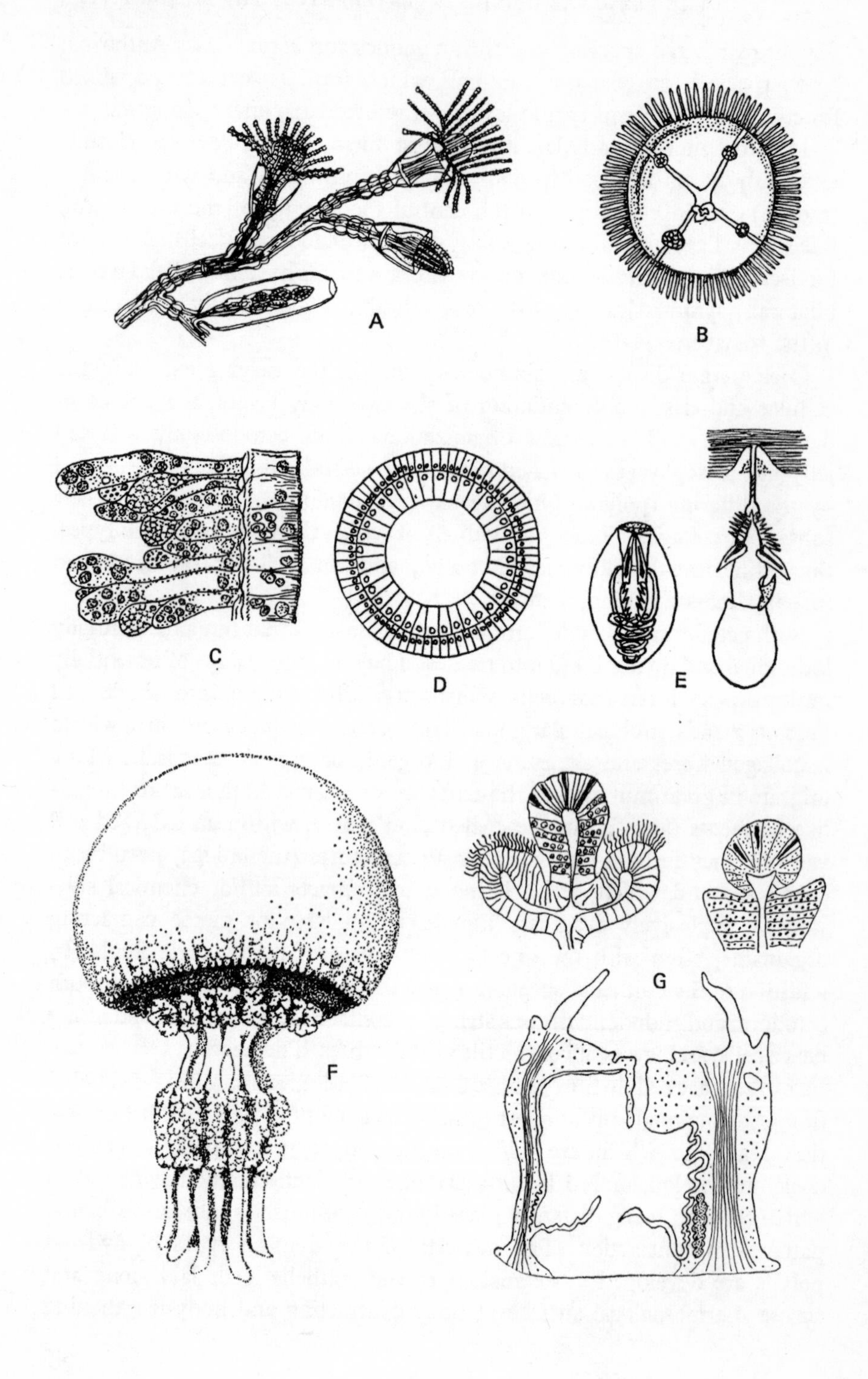
A
B
C
D
E
F
G

fibres. In larger polyps and medusae the muscles are separated from the epithelium and are often intucked into the mesogloea where, in attachment to collagen fibres, they form powerful contraction systems; in medusae they produce the pulsating swimming movements.

Class Hydrozoa. Throughout this class the polyp and medusoid individuals, which may be solitary or joined in colonies, are always of a simple type of construction with a thin non-cellular mesogloea, a network arrangement of nerve fibres and poorly developed muscle. Hydrozoans show, however, a remarkable capacity to vary the form of the individual zooids throughout the various phases of their life history. These progress through a cycle of changes involving a sexual medusoid stage that produces eggs or spermatozoa, a fertilized egg which develops into a motile two-layered, mouthless ovoid planula larva and which (usually) attaches to the substratum to grow into a solitary polyp or, by continued asexual budding, into a colony of polyps. Sprigs of the polyp or colony (blastostyles) bud off medusae to complete the alternation of generations. Some hydrozoans are represented by a long persistent polyp phase, others by the medusoid generation. Examples of polyp-constructed hydrozoans are the many sorts of chitin-tube encased hydroids commonly found on stones and weeds on shores and in shallow seas, and the coralline millepores of tropical reefs which secrete solid and often massive calcareous skeletons overlain and penetrated by the living tissue. By contrast, many of the trachyline medusae, a few of which live in fresh water, have no known polyp phase in their life history. And finally, the beautiful pelagic siphonophores, such as

Fig. 4.1 Phylum Coelenterata. The classes Hydrozoa, Scyphozoa and Anthozoa. Class Hydrozoa: A, a small piece of the sessile, colonial hydroid *Obelia* showing tentaculate feeding polyps (gastrozoids) in various states of extension, and a medusa-budding blastostyle; B, the free-swimming tentacle-encircled medusoid phase of *Obelia*, with a central mouth, four radial canals of the coelenteron, and gonads; C, body wall of *Obelia* in longitudinal section with elongate digestive endodermal cells bordering the coelenteron, outer epithelial cells of the ectoderm, and the thin intermediate mesogloea; D, schematic transverse section of *Obelia* showing the two-layered body wall and the central coelenteron; E, undischarged and discharged cnidae of *Obelia*. Class Scyphozoa: F, the jelly-fish *Rhizostoma* with a convex exumbrella surface and the subumbrella surface extended centrally as a manubrium bearing multiple mouth openings arranged in two zonal rings. Class Anthozoa: G, a diagrammatic longitudinal section through an anemone. Circlets of tentacles surround an oral disc and the mouth is intucked into the coelenteron as a stomodaeum. The coelenteron is partitioned by vertically hanging, radially arranged sheets (mesenteries) in which are developed longitudinal muscles and gonads. The mesenteries are clothed with ciliated and digestive cells (see smaller figure) and bear cnida-laden filaments (acontia)

the 'Portuguese man-of-war' (*Physalia*) and the 'By-the-wind-sailor' (*Velella*) which are summer visitors to British waters, are floating colonies compounded of numbers of feeding, stinging and reproductive polyps and of medusoid floats and swimming bells.

Class Scyphozoa. The large thick-walled jelly-fishes such as the translucent *Aurelia*, the blue *Cyanea* and the brown *Chrysaora* that are often found stranded on British shores are more complicated than the hydrozoan medusae. The mouth, mounted on a mobile food-catching stalk (the manubrium), leads into a food cavity into which project groups of filaments loaded with digestive cells: and from the central cavity there extend into the jelly radiating canals through which pass slow respiratory and food-carrying streams of water. Sense organs at the base of the tentacles which surround the margin of the medusa register changes in the animal's orientation and, by a combination of localized bell contraction and excitation of the through-conduction tracts, cause the circular muscles to contract and produce pulsations which are both locomotory and corrective of abnormal orientations.

The gonads of the two sexes are formed from cells of the gut wall. Eggs when fertilized develop into planula larvae which, in many jelly-fishes, attach to stones, grow into a small polyp and bud off successions of small star-shaped ephyrulae which, by growth and filling in of the indentations, produce little jelly-fishes so completing the life-cycle. In oceanic species the planula changes directly into a small medusa. Jelly-fishes are of many shapes, sizes and colours, and one group, the Stauromedusae, are unusual in attaching by the convex (aboral) surface of the umbrella to weeds and stones.

Class Anthozoa. This class includes the soft-bodied anemones and many kinds of skeleton-forming corals, including the reef-building madreporarians. Anthozoans are solitary or colonial polyps and are never, at any stage in their life history, medusoid. The polyps show all the more advanced features of nerve, muscle and mesogloea organization referred to earlier in the description of the phylum. They also show elaborations of body structure not seen in other coelenterates such as an intucked tube (the stomodaeum) leading from the mouth into the food cavity, and a pleating of the internal wall of the column to form muscularized septa coated with digestive endoderm. The variations in the number and sequence of development of the septa, in the arrangement of the septal muscles, and in the nature and placement of either a calcareous or a horny skeleton, when combined with the various patterns of growth of the colonies, give to anthozoans an immense variety of form and styles of construction. Corals of a different type of organization from the present-day forms were among the commonest animals of warm Palaeozoic seas and many limestones are largely composed of their skeletons.

Phylum Platyhelminthes (*c.* 9,000 species) (Fig. 4.2). This phylum comprises animals that are bilaterally symmetrical, soft bodied, longer than broad, and flattened dorso-ventrally into leaves or wedges. The containing layer of the body (a ciliated epidermis or a firm cuticle) surrounds a solid pack of cells. The three classes of the Platyhelminthes, the Turbellaria, Trematoda and Cestoda, are readily distinguished by the shapes of the body, the presence or absence of adhesive suckers, the nature of the bounding layer, the degree of elaboration of the packing cells into (mainly) digestive and reproductive organs, and by their life histories and habits. All have marine, fresh-water and terrestrial representatives, but while turbellarians are, with a few exceptions, free-living, trematodes (flukes) and cestodes (tapeworms) are parasitic on or in other animals (flukes), or always live within their hosts (cestodes).

The most simply organized platyhelminths are found in the group Acoela of the class Turbellaria. Small, usually 1–3 mm long, marine organisms which glide or wriggle over the sea bed and among weeds, the thin ciliated epidermis contains many mucous glands, sensory cells and nerve fibres. Two intucks of the epidermis form a ventral mouth and a posterior germ cell discharge duct. Only a few of the interior cells are visibly differentiated for specialized functions. Some are elongated into muscle fibres around and along the body under the epidermis and a few traverse the packing tissue. Others form cords of egg and spermatozoa producing cells arranged in side by side rows on either side of the midline. The worms copulate to transfer spermatozoa. There is no gut; organisms taken in at the mouth are digested by the packing cells; and there are no excretory organs.

The muscularized ectodermal intucks associated with feeding and reproduction, the prominence of germ cells, and the brain rudiments foreshadow in the Acoela the main lines of organ development and evolution in the Turbellaria and in the Platyhelminthes generally. All other platyhelminths have mucus-like rods (rhabdites) in the skin and develop within the packing tissue flagellated (flame) cells which draw waste materials from the parenchyma spaces to discharge the fluids through collecting ducts opening by a few or many excretory pores. The different classes of the Turbellaria are conveniently categorized by successive phases of elaboration of the gut through a solid rod (rhabdocoels), a tubular sac

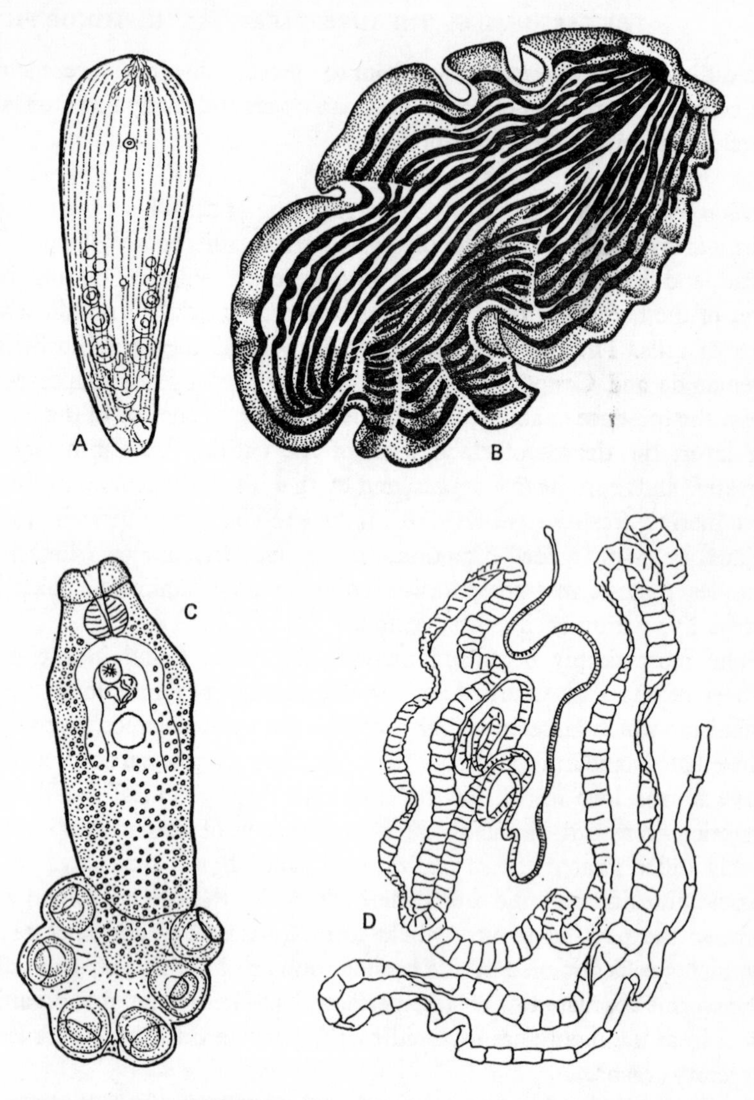

Fig. 4.2 Phylum Platyhelminthes. Class Turbellaria (free living flatworms): A, a simply organized turbellarian (Order Acoela) in which the alimentary canal, reproductive and excretory systems show little trace of structural elaboration. B, *Prostheceraeus vittatus* (Order Polycladida), an example of a turbellarian of complex internal organization which includes a much branched alimentary canal, a reproductive system embodying glands and ducts, and a nervous system comprising a ganglion and longitudinal cords. Class Trematoda (parasitic flukes): C, *Polystomoides*, an example of the Order Monogenea in which the flukes have only one host during their life cycle and usually possess a large posterior attachment organ subdivided into suckers. Class Cestoda (parasitic tapeworms): D, *Taenia saginata*, not uncommonly found in the alimentary canal of cattle and of man

(alloiocoels), a much branched three-lobed cavity (triclads), to the multiple and extensive ramifications of the polyclad system. Parallel and even more far-reaching developments of the reproductive system include, on the female side, the formation of special yolk-producing cells, spermatozoa stores, fertilization chambers, egg-shell glands and shell-shaping organs and, on the male side, a muscularized evaginable penis as well as numerous connecting tubes within the system, and arrangements for ensuring cross-fertilization.

In the Trematoda a ciliated epidermis is present only in some types of larvae; the adult fluke has a cuticle resting directly on the interior tissue. Ectoparasitic flukes, most of which attach to the gills of fishes, have suckers at the mouth and, often of a more elaborate form, at the hind end of the body. The eggs of these flukes develop into ciliated free-swimming larvae which seek out and attach to the host. The single host and simple life history characterize the Monogenea. The Digenea are, with rare exceptions, internal parasites of vertebrates. They have a single (anterior) sucker and their complicated life histories involve first a motile ciliated larva (the miracidium) which penetrates the tissues of a second host – usually a sea, fresh-water, or land snail – to give a succession of self-multiplying larvae culminating in larvae which seek out and enter the original host species. There are numerous variations of the life history with sometimes two or three intermediate hosts, but the minimum two-host cycle gives these flukes the entitlement of Digenea. The sheep liver fluke *Fasciola* and the human blood fluke *Schistosoma* (*Bilharzia*), a scourge of the tropics, are examples of this large and troublesome group of parasites.

Adult Cestodes nearly always live in the intestines of vertebrates. The elongate body with an attaching scolex at one end is, in most tapeworms, divided into a number of structurally repetitive blocks (proglottides) budded off in succession from behind the scolex and gaining in sexual maturity as they enlarge and become pushed further along the length of the body ribbon. There is no gut, liquid food from the host passing across the cuticle into the interior tissues. Folds (bothria) and sometimes hooks are present as attaching organs on the scolex. The life histories are complicated and there is at least one intermediate host. The damaging tapeworm of man, *Diphyllobothrium latum*, for example, gains entry by the eating of larva-containing raw or partially cooked fish.

Phylum Annelida (*c.* 9,000 species) (Fig. 4.3). Worms with segmented and ringed bodies. Except for a very few simply organized worms (class Archiannelida) and aberrant parasitic forms (class Myzostomaria) annelids

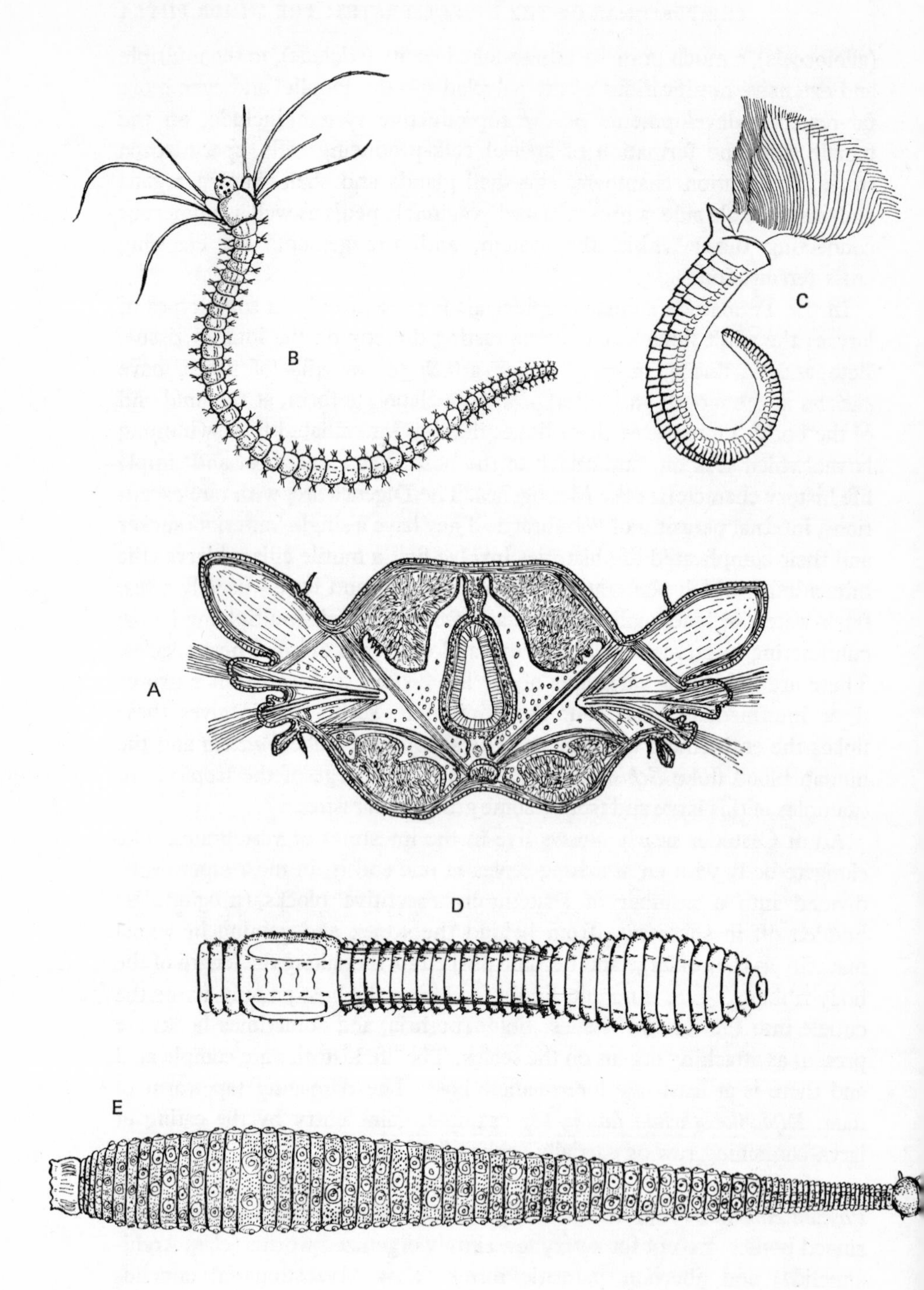
B
C
A
D
E

belong to the classes of the Polychaeta: a marine group of worms carrying, in segmental series, paired side appendages (parapodia) furnished with strong projecting bristles; Oligochaeta: sparsely bristled, limbless marine, fresh-water and terrestrial (earth-) worms; and the Hirudinea: marine, fresh-water and terrestrial parasitic leeches, limbless, almost always bristle-free, and equipped with adhesive suckers. The classes number respectively some 5,500, 3,000 and 300 species.

Annelids have a soft elastic covering cuticle secreted by a thin cellular, sensitive epithelium in which there are numerous mucous glands from which slime exudes through cuticular pores on to the surface of the body. The body wall is strongly muscularized with circular and longitudinal fibres which, by different combinations of contraction and relaxation, produce wave-like side to side locomotory flexures of the body (as in many polychaetes), travelling concertina extensions and contractions (some polychaetes and most oligochaetes), or dorso-ventral undulations and loops (leeches). Muscle contractions are made effective in locomotion by the turgidity of the fluid-filled coelomic body cavity. In each segment the coelom develops as two sacs, each arising as a cavitation of the embryonic mesoderm on the two sides of the central tube of the alimentary canal. In the midline of the segment the walls of the sacs provide the dorsal and ventral mesenteries of the gut, while the end walls join with the coelomic walls of neighbouring segments to make the diaphragms of the inter-segmental septa. Small spaces – all that remains of the first formed body

---

Fig. 4.3 Phylum Annelida. The classes Polychaeta, Oligochaeta and Hirudinea. Class Polychaeta: A, Transverse section through a body segment of the rag-worm *Nereis*. The body wall, clothed by ciliated ectodermal cells, is underlain by circular muscles and by four bands of strongly developed longitudinal locomotory muscles. The alimentary canal is slung between dorsal and ventral mesenteries each enclosing a blood vessel, and is surrounded by a coelomic body cavity. A longitudinal nerve cord lies in the mid-ventral line. Paired parapodia are present as lateral extensions of the body segments and have lobes variously developed as swimming and bristle (cheta) bearing appendages. The parapodia are moved by obliquely set muscles. B, a free-swimming nereid worm (*Platynereis*). C, a tubicolous worm (*Bispira*) with a tentacular crown developed for fine particle food collection and respiration. Class Oligochaeta: D, anterior end of an earthworm (*Lumbricus*) in ventral view. Parapodia are absent, but each segment bears paired groups of chaetae. Glands of the swollen clitellum, to the left of the figure, secrete the cocoon in which eggs are fertilized and the embryos developed. Class Hirudinea: E, a marine leech (*Pontobdella*). The body, which lacks parapodia and chaetae bears anterior (to the right of the figure) and posterior attachment suckers. Each body segment is subdivided externally by annular grooves

cavity (the blastocoel) – between the adjoining membranes of the mesenteries and septa become walled with contractile tissue to form a system of dorsal and ventral blood vessels connected by intersegmental loops. Vessels connect to this system to distribute blood to and from the skin and the various internal organs. Haemoglobin and, in some polychaetes, other respiratory pigments serve as oxygen stores and carriers.

Segmentation is displayed externally by the parapodia (if present) and by the body rings, and internally by the blood vessel arrangement, the presence of paired excretory tubes with closed or open internal ends, by the paired ridges of mesodermal tissue from which the gonads develop, and by the ganglionic swellings of the double ventral nerve cord which loops round the fore part of the gut to the dorsal ganglia of the 'brain'. Only the first and last of the body divisions, the prostomium and the pygidium, are not true segments, having neither coelomic cavities nor the normal complement of segmental organs. The gut in annelids is a straight tube, regionally differentiated for food catching, storage and digestion. It opens at the mouth behind and below the prostomium and terminally at the anus.

Polychaetes, like arthropods, owe their diversity of form mainly to the differing character of their limbs which, among other uses, are locomotory, respiratory and current-producing; to the various forms of the head (prostomium and peristomium) region and of its tentacular embellishments; and to the differing patterns of regionalization of the body into regions of distinctive styles of construction. The group of the Errantia, freely moving and often predatory, usually have a well developed head bearing sensory tentacles and eyes. Many have protrusible biting jaws and most of them have substantially constructed parapodia of various shapes and functional subdivisions. The Sedentaria, on the other hand, often live within tubes which they secrete around them, have barely visible parapodia, and carry on their head respiratory and food catching devices that are both elaborate and beautiful in their range of colour and design.

Oligochaetes are outwardly much less complicated but carry within them the marks of a greater specialization which is particularly evident in the closing off of a part of the coelom within which reproductive cells are collected, and by the development of reproductive ducts for the discharge of eggs or spermatozoa. Oligochaetes are always hermaphrodite and associate in pairs to ensure cross-fertilization. Some oligochaetes, like some polychaetes, may reproduce asexually by fission and budding.

Leeches are the most specialized of annelids. Groups of segments at the two ends of the body combine to form anterior and posterior suckers used for locomotory and host attachments. The gut has a muscularized blood-sucking pharynx often armed with piercing jaws and an enlarged, much-lobed crop stores the blood taken when feeding and kept liquid by an anticoagulant salivary secretion. The coelom of leeches is largely obliterated by packing cells which are both excretory and haemoglobin producing.

Many annelids have yolky eggs which develop directly into young worms without producing intermediate larval stages. Some polychaetes, however, produce free-swimming cilia-girdled trochophore larvae prior to metamorphosis. In this type of development the alternating shift to right and left of the successive quartets of cells (blastomeres) of the growing embryo (spiral cleavage); the specificity of the blastomeres in forming particular types of tissue and organs; the formation of the coelom by a splitting of the mesodermal bands; and the production of a trochophore type of larva are characteristics reminiscent, at least in some measure, of the Platyhelminthes and the Mollusca, and are indicative of a common bond of relationship which would not be suspected were we merely to compare their adult forms.

Phylum Echinodermata (*c.* 5,500 species) (Fig. 4.4). The five present-day classes are the Crinoidea (sea-lilies and feather stars), Asteroidea (sea-stars), Ophiuroidea (brittle-stars), Echinoidea (sea-urchins) and Holothuroidea (sea-cucumbers). All are marine and few can withstand brackish water. They live on or burrow into the sea floor. A few holothurians are pelagic, and feather stars (stalkless crinoids), which normally attach to stones and weeds, can swim.

The basic form of an echinoderm – though no echinoderm adheres exactly to this pattern – is a sphere with a superimposed pentaradiate symmetry of its body-wall structures and interior organs. The body wall is divided into oral (upper) and aboral (lower) hemispheres and contains calcareous plates which are unique among invertebrate skeletons in being formed by and lodged in the intermediate mesoderm layer. Almost all the muscles of echnoderms are formed from mesoderm of the body wall and attach to ossicles or connective tissues. The mesoderm is overlain by a thin cuticle-covered ectoderm made up of ciliated epithelial cells, mucous glands, and containing numerous scattered sensory cells which connect with nervous pathways of fibres threaded intra-epithelially between the skin cells. Eruptions from the skin form spines and (in asteroids and echinoids) pincer-like food-grasping and skin-cleansing pedicellariae.

The body cavity is unusual in consisting of three compartments, a spacious perivisceral cavity surrounding the gut – a tube aligned along the vertical axis of the sphere from the mouth below to the anus above – and

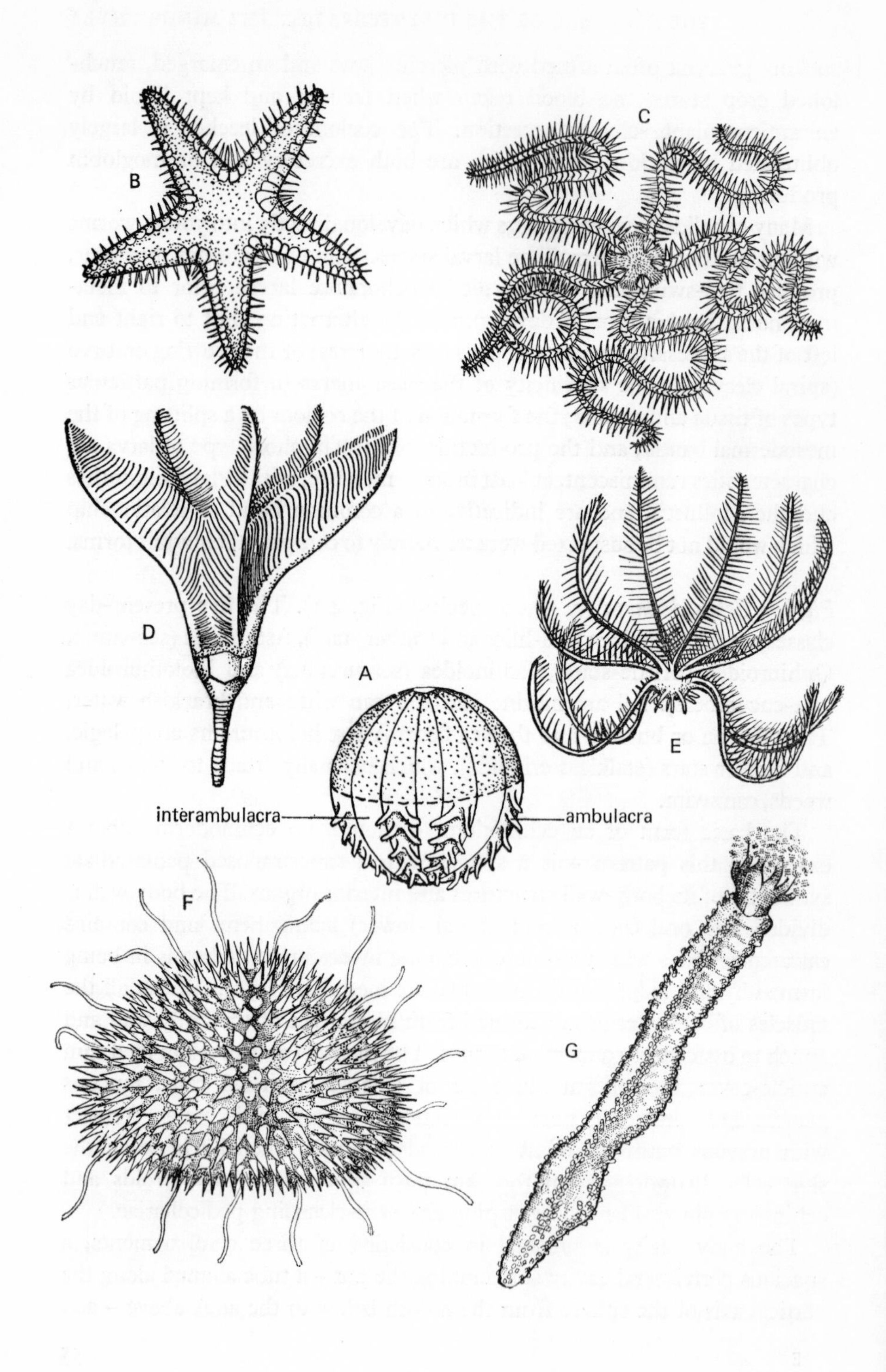
B
C
D
E
A
interambulacra
ambulacra
F
G

two much smaller enclosed circumoral rings of the water-vascular and perihaemal cavities, both of which send an ascending tube towards the aboral pole. The calcified stone canal of the water-vascular system lies within the perihaemal axial sinus and opens externally through a sieve plate – the madreporite. In the oral hemisphere the pentaradiate symmetry is expressed in the structural arrangements of the five radii (ambulacra) and the five inter-radii (interambulacra). Each radius has a midline thickening of the skin nervous pathways (the radial cord), the cords being connected round the mouth by a nerve ring. The cords may be external to the skeleton (asteroids and crinoids) or become enclosed within the body (other echinoderms). Below the nerve cords are the radial canals of the water-vascular and perihaemal system and they too issue from central rings. Side branches of the water-vascular canal project into finger-like projections of the body wall, the tube-feet, which in many echinoderms have internal bulbs, the ampullae.

The shapes and body orientations of the five classes of echinoderms can all be derived as modifications of this plan. In the pelmatozoan crinoids the animal is inverted with the mouth uppermost and the gut curves to bring the anus to the water facing surface. The aboral pole extends into an attaching stalk or, if the stalk is secondarily lost as in the feather star *Antedon*, temporary attachments are made by flexible cirri. Arms, branched at the base, grow out radially from the disc and develop side extensions – the pinnules. Small-particle food is trapped by mucus on the oral surface and is transported by cilia along the midline of the arms to the mouth. All other living echinoderms are stalkless and freely ranging (Eleutherozoa) and usually have a mouth to the ground orientation. In asteroids and ophiuroids short, or more often long, arms extend radially from the disc, the oral and aboral regions of the body wall being equally developed in both disc and arms. In echinoids there are no arms, and in both the subspherical (regular) urchins and the flattened (irregular) heart-urchins and cake-urchins with a partial bilateral symmetry, the aboral hemisphere is

Fig. 4.4 Phylum Echinodermata. A, generalized external features of an echinoderm showing the pentaradiate symmetry of the body with alternating meridional interambulacra and ambulacra. The aboral hemisphere is stippled, the oral hemisphere with two rows of projecting tube feet in each ambulacrum is unshaded. The mouth is at the oral pole. B–G, schematic representations of animals of the five living classes of echinoderms. B, Class Asteroidea: a starfish in aboral view. C, Class Ophiuroidea: a brittle star in aboral view. D & E, Class Crinoidea: side views of a stalked sea-lily and of a feather-star. F, Class Echinoidea: side view of a sub-spherical (regular) sea-urchin. G, Class Holothuroidea: a sea-cucumber.

reduced to a small apical cap. The ambulacra and interambulacra of the enlarged oral hemisphere thus come to encircle the rigid plate-encased body almost completely along latitudinal lines of the test. Holothurians are similarly constituted, but have a pliable, highly muscularized and almost skeleton free body wall and most of them lie on their sides.

Echinoderms are in many respects of primitive design. They have an intra-epithelial skin plexus of nerve fibres and lack, for example, a blood system and excretory organs. Opportunist and unconventional in their varied devices for performing essential functions such as respiration they have utilized to the full the potentialities of the structures that are peculiar to them. The tube-feet, variously adapted for locomotion, food gathering and respiration, the skeleton and the muscularized body wall, spines and pedicellariae are the main devices which have given to echinoderms a great variety of form and functional adaptations.

An ancient group, all the present-day classes are represented in Palaeozoic strata. In the Palaeozoic, crinoids and other now extinct Pelmatozoa were dominant. Echinoderms show in their development and in the character of their larval phases a body cavity construction similar to that of the Hemichordates (p. 67) to which phylum they are probably related.

Phylum Porifera (the sponges: *c.* 5,000 species) (Fig. 4.5). Sponges, like coelenterates, may be described as two-layered animals, basically of sac-like form, and with the central cavity (the spongocoel) opening to the outside water through a pore (the osculum) that is visible to the naked eye. But at every stage of the description the features described are zoologically so very different from the analogous features of coelenterates that the sponges are placed not only in a different phylum but are assigned to the subkingdom of the Parazoa in order to distinguish them from the coelenterates and all other multicellular animals of the subkingdom Metazoa.

A lens inspection of a simple tubular sponge such as *Sycon*, commonly found attached to shaded overhangs of rock on most shores of temperate seas, shows the sac to be perforated by innumerable tiny pores (prosopyles) through which water is drawn by flagellated cells into the spongocoel and out by the osculum. The folded body wall consists of two layers of cells, an outer pinacoderm and an inner choanoderm, and between them are numerous isolated amoeboid cells and large numbers of three-rayed skeletal spicules of calcite. The pinacoderm is a tissue in that its constituent cells are welded into a continuous sheet but it gives rise neither to nerve nor muscle. Such capacity as a sponge has for responding to (shall we say) injurious contacts or noxious chemical substances is dependent on the limited sensitivity of the pinacoderm tissue and the considerable power of contraction of individual pinacocytes. The choanoderm is composed of flagellated cells each with a membranous collar surrounding the root of the flagellum, a type of cell found elsewhere only in some of the single-celled Protista. The choanocytes, which are current producing and engulf and digest food particles, are separate entities not bonded into a continuous sheet of lining tissue. They may lose their collars and move into the deeper recesses of the sponge wall. Some differentiate into egg and spermatozoa producing cells, and it is possible that others transform into amoebocytes having other functions. Most of the amoebocytes, however, are probably not engaged at any time as pinacocytes or choanocytes. Some form the sponge skeleton which, in different groups of sponges, may be made up of calcareous or siliceous spicules or (as in the bath sponge) be composed of a soft organic spongin. Other amoebocytes are slime producing, and others are food storers or pigment carriers.

In the more specialized sponges the body wall becomes greatly folded and the choanocytes leave the spongocoel border and become confined within spherical chambers which have pinacoderm lined entrance and exit canals for the through-ducting of water.

There are three or, according to some authorities, four classes of sponges. The large and beautiful hexactinellid sponges such as Venus's flower basket (*Euplectella*), *Hyalonema* and *Rossella*, are composed for the most part of a skeleton made up of six-, five- and four-rayed siliceous spicules bound together into a rigid tubular or vase-shaped lattice frame. The hexactinellids are probably the most primitive of sponges in that they lack pinacoderm and consist only of amoebocytes and choanocytes. Most of them live in deep water and they have a fossil history dating back to Cambrian times.

The calcareous sponges (class Calcarea) are small (1–10 cm) growths forming little erect vases or prostrate systems of tubes. They are mostly confined to shallow waters and are common between tide marks. Their three-rayed spicules are found in rocks from the mid-Mesozoic period onwards, but there is little doubt that the Calcarea originated much earlier.

All of the more massive sponges and the few fresh-water species have either siliceous spicules of various forms other than the six-rayed variety, or a skeleton of spongin, or they may have both spicules and spongin. Some (e.g. *Halisarca*) lack skeletal material. Although a somewhat mixed assemblage they are usually all grouped within the class of the Demospongia in that they have rather similar types of choanocyte chambers and canal

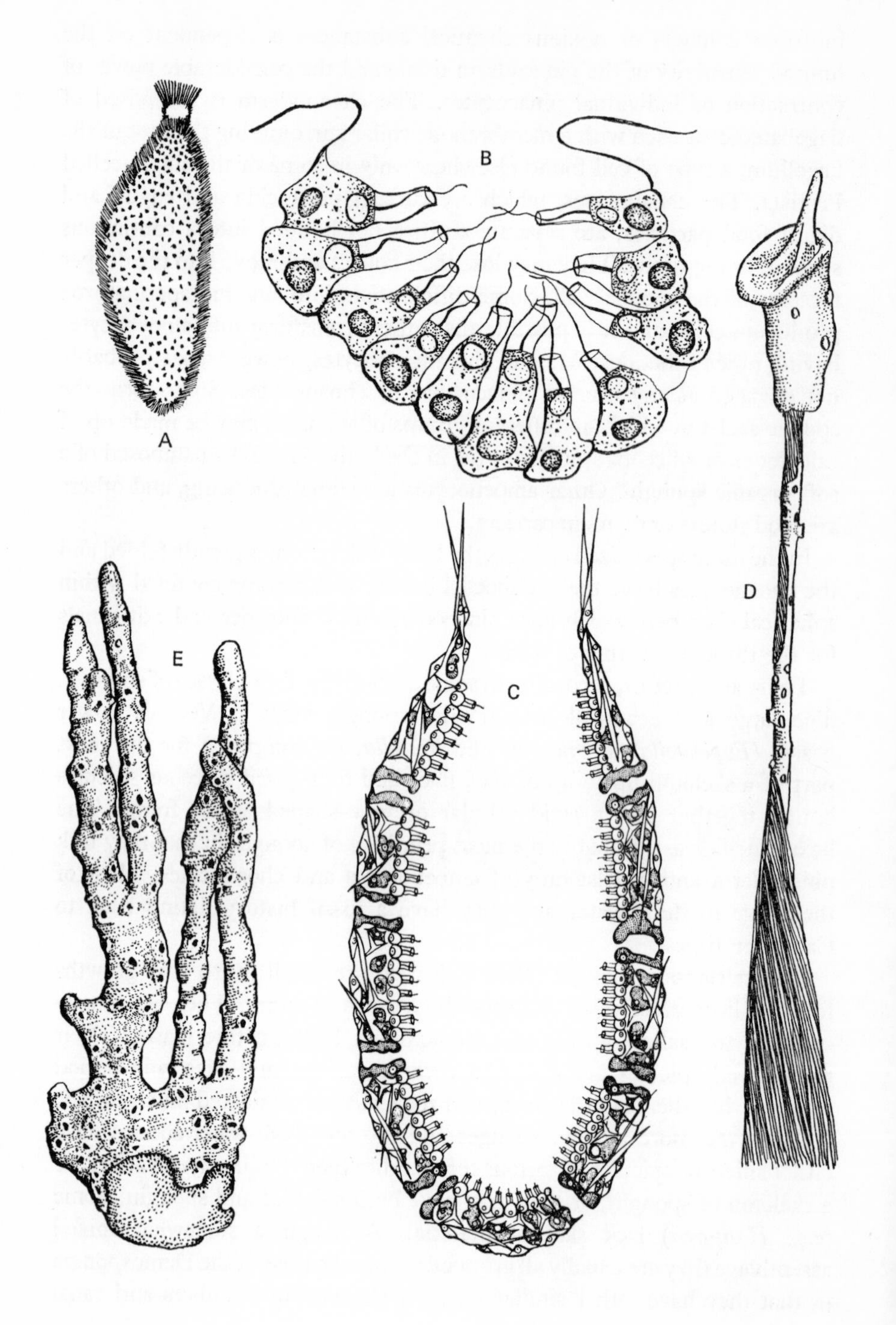
A
B
C
D
E

system arrangements; and their choanocytes are small. Many, such as the red *Hymeniacidon* and the yellowy-green *Halichondria*, are familiar as encrusting growths on intertidal rocks. The bath sponges, *Euspongia* and *Hippospongia*, occur only in warm seas and are cemented to rocks by their spongin which, when first secreted, is viscous.

All sponges have considerable powers of regeneration even when reduced to a relatively few but mixed assemblage of cells. Some, including all the fresh-water species, produce from time to time balls of skeleton-coated cells (gemmules) which germinate into a new sponge growth; and all reproduce sexually. The fertilized eggs cleave into motile flagellated larvae, attach to a rock surface, and by a uniquely poriferan process of cell inversion envelop the flagellated cells (potential choanocytes) with cells (potential pinacocytes) that were previously in the interior of the larva.

Although most sponges are rather drab objects, many are vividly coloured with red, blue, yellow, purple and other tints, and sponges often harbour crustaceans and other animals within their cavities and tissues.

Phylum Ectoprocta (sea-mosses and sea-mats: almost 5,000 species) (Fig. 4.6). The colonial ectoprocts occur in great numbers and variety of species on stones, rocks and weed of seashores and shallow offshore waters. Many of them look like hydroid colonies or algal tufts, others form encrusting mats or sticks of coral, and a few grow into large contorted calcareous sheets a foot or more in diameter. Each of the many ectoderm-, mesoderm- and endoderm-constituted zooids of a colony has a box- or vase-like body wall

Fig. 4.5 Phylum Porifera. Some examples of animals of different classes of sponges. Class Calcarea: A, *Sycon*, a sponge of simple vase-like form with a skeleton of calcareous spicules, many of which protrude from the surface. Water enters the central cavity (paragaster) through numerous pores (prosopyles) in the body wall and is expelled through the terminal aperture (the osculum).
B, flagellated collar cells (choanocytes) which line the paragastral cavity. Water is drawn through the prosopyles and expelled from the osculum by movements of the choanocyte flagella. Trapped food particles are ingested by the choanocytes.
C, a diagrammatic longitudinal section of *Sycon*. Flattened cells (pinacocytes) form the outer covering; choanocytes line the paragaster. Spicule-secreting and other types of cells occupy the intermediate space. Porocytes (stippled) each have an excavated cavity for the inward passage of water. Class Hexactinellida:
D, *Hyalonema*, the swollen body of the sponge consists of choanocytes and amoebocytes laced within a rigid skeleton of (mainly) six-rayed siliceous spicules. The long rooting stalk is composed of greatly elongated spicules. E, Class Demospongia: *Chalina*, a branching sponge with a skeleton composed of siliceous spicules and horny, ceratose material. The numerous prosopyles are minute; the larger apertures are oscula

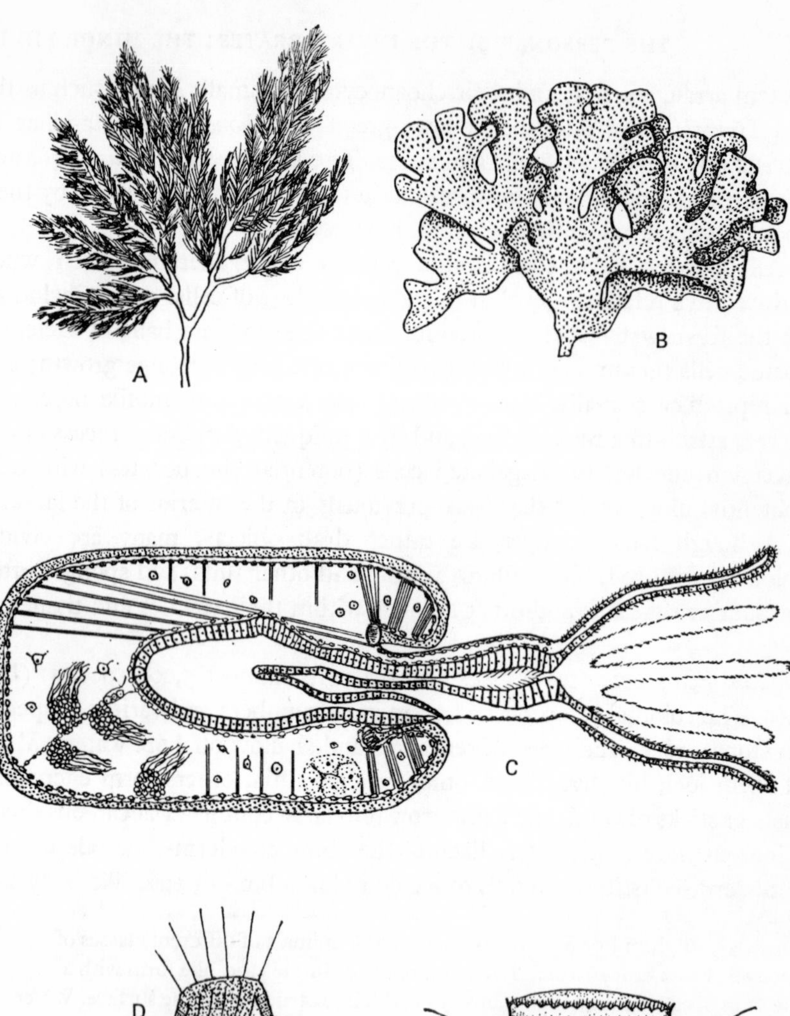

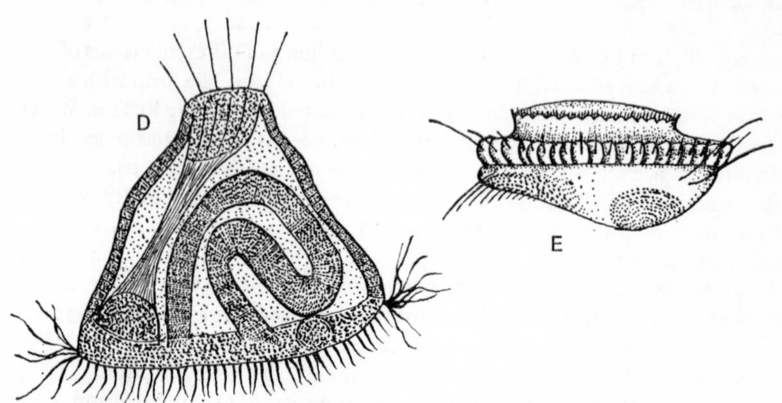

Fig. 4.6 Phylum Ectoprocta. A, *Bugula*, a much branched tufted polyzoan colony. B, *Flustra*, a colony with flattened round-ended fronds. C, an extruded polyzoan polyp showing among other features, the mouth, tentacles, recurved alimentary canal, retractor muscles and gonads. D and E, larval forms of the Ectoprocta

enclosing a coelomic cavity traversed by muscles but otherwise mainly occupied by a U-shaped alimentary canal open at either end. The mouth is encircled by a ring of ciliated food-capturing tentacles (the lophophore), which can be opened out umbrella-like into the sea water when feeding, or be withdrawn into the body-wall box. The anus opens near the mouth at the side of the lophophore.

All ectoprocts secrete a skeleton of chitinous material and many add to this a calcareous layer between the chitin and the soft body wall. A part of the upper (sea-facing) wall of the box remains flexible. Muscles, attached at one end to this wall and at the other to the fixed side of the box, on contraction, pull the wall downwards, compress the fluid-filled body cavity, and push the lophophore outwards through a forwardly directed pore in the box. The lophophore is drawn back by retractor muscles.

Ectoprocts have a nerve ganglion at the base of the lophophore from which nerve fibres extend throughout the body wall and innervate the muscles. Probably because of their small size the zooids lack a circulatory system, and the gathering of excreted material seems to be effected mainly by wandering cells and by cells within the lophophore and gut wall. Reproduction is both asexual and sexual. A colony grows by the extrusion from the body wall of existing zooids of stolons of living tissue from which new zooids are differentiated. Ectoprocts are hermaphrodite, the ovaries and testes being formed from strings of germinal tissue within the coelom. In some species eggs escape directly into the sea through pores in a mouth-encircling membrane and, after fertilization, cleave to form ciliated larvae of various forms, one of which, the cyphonautes, when first discovered in 1836, was thought to be a rotifer. But most ectoprocts brood their eggs to a late stage of development within a brood chamber, an enlarged bulbous growth which is shown to be a modified zooid by the presence within it of an (often much modified) alimentary canal. Polymorphism of the zooids is further seen in many ectoprocts in the production of parrot-beak aviculariae which, like the pedicellariae of echinoderms, snap on organisms which might otherwise settle on and overgrow the colony; of bristle-like vibracula which sweep away debris; and of other types of zooids of unknown function, including dwarf forms. And, in addition, the normal zooids regenerate from time to time by breaking down into 'brown bodies' the fragmented material of which is incorporated within an internally regenerated new polyp.

The classification of the ectoprocts is based primarily on the shape of the lophophore. The wholly marine and most numerous class of the Gymnolaemata have a circular lophophore and the few fresh-water Phylactolaemata have a horseshoe-shaped circlet of tentacles. Subdivisions of the Gymnolaemata are conveniently made on the shape, lip moulding and spine ornamentation of the mouth opening, though there are many other features distinctive of the several orders.

Ectoprocts with calcified walls are common as fossils from Ordovician (early Palaeozoic) times and some fifteen thousand species of fossils have been described.

The small phyla (Fig. 4.7)

Phylum Nemertina (or *Rhynchocoela*: 600 species). The soft 'proboscis worms' have ribbon-shaped, cilia-covered and mucus-coated bodies that glide or move by peristaltic contractions over the surfaces of rocks and stones or among weeds. Some can swim by undulations of the body. Almost all are marine, though a few live under logs or on soils on land, and a very few shelter as commensals with other animals, mainly bivalve molluscs. They vary in length from a few millimetres to 80 cm and more, and the coiled black bootlace worm, *Lineus longissimus*, sometimes to be found under stones on Atlantic shores, attains an unravelled length of 20–30 metres; and many nemerteans are vividly coloured. The body, fashioned from three embryonic layers, is highly muscularized, with a well-developed central nervous system consisting of a brain and longitudinal cords, and containing an alimentary canal opening by a mouth and an anus. Some nemerteans are armed at the mouth end with piercing stylets. Like the Platyhelminthes the proboscis worms are acoelomate and are stuffed with mesodermal packing tissue. Flame cells drain excretory fluids from the mesodermal spaces and duct them to the outside water through minute pores in the body wall. The feature unique to the nemerteans is an eversible proboscis developed in a sheath which lies above the front part of the gut. Some nemerteans use the proboscis for capturing and passing small animals to the mouth; others feed mainly on minute plants and animals. The nemerteans are of separate sexes and the fertilized eggs grow into freely swimming larvae, of which one form is the planktonic pilidium, discovered in 1847 but for long not known to be the young stage of a proboscis worm.

Phylum Brachiopoda (lamp shells: 300 living species). These wholly marine animals are superficially rather like bivalve molluscs but in almost every detail of their body construction are different from them. The two valves of the shell are made of horny material impregnated with calcium carbonate or calcium phosphate and are both closed and opened by

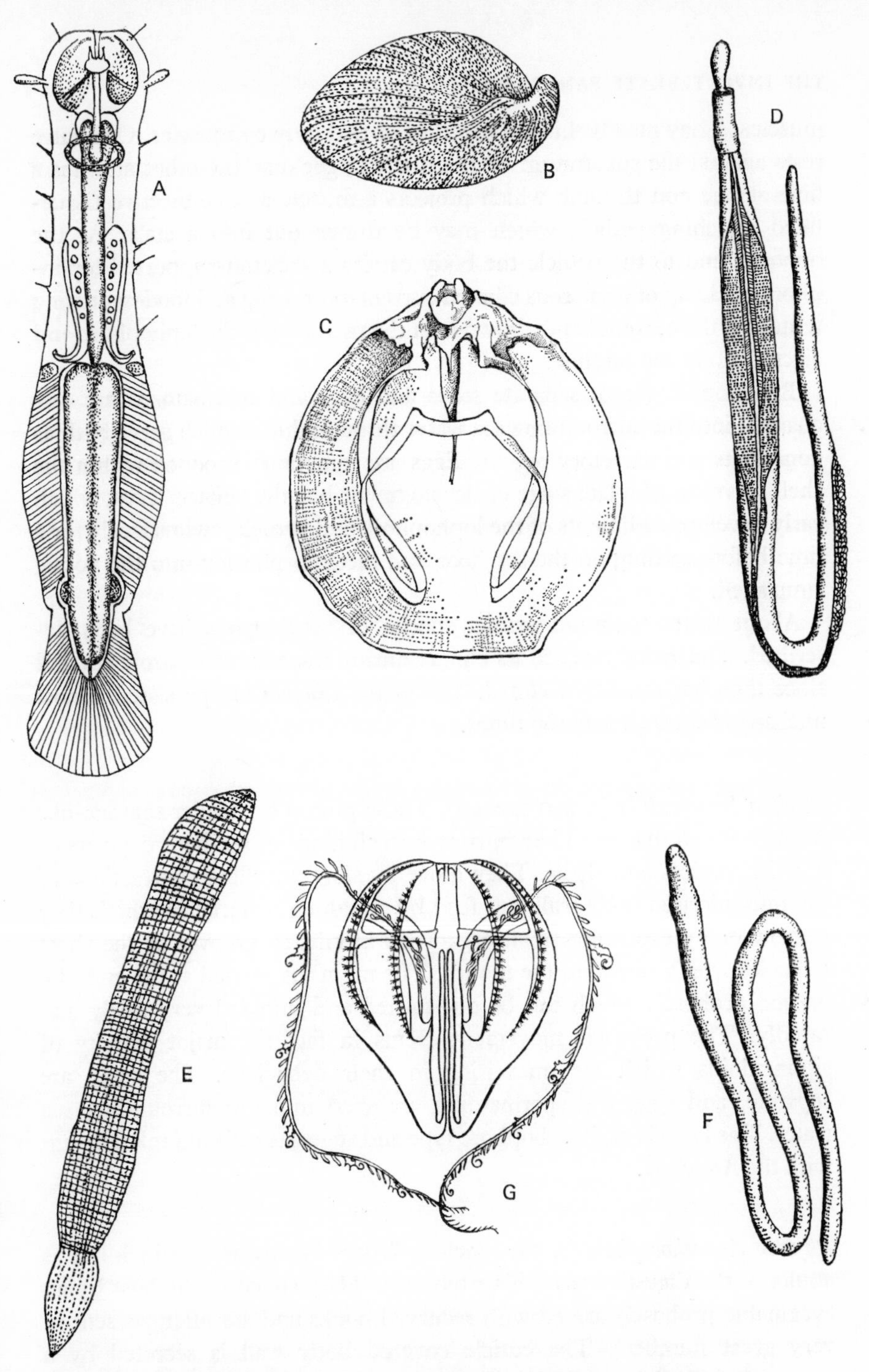

Fig. 4.7 The small phyla. A, Chaetognatha (arrow-worms): *Spadella*. B, Brachiopoda (lampshells): *Neothyris*. C, interior of upper (dorsal) valve of *Neothyris* showing the skeleton which supports the ciliated feeding organ, the lophophore. D, Hemichordata: the acorn-worm *Balanoglossus*. E, Sipunculida: a sipunculid worm. F, Nemertina (proboscis worms): *Lineus*. G, Ctenophora (comb-jellies): *Pleurobrachia*

muscles. They may be hinged together or tied only by muscles. One valve rests against the substratum. It is usually larger than the other and has a hole at one end through which projects a muscle and connective tissue-filled attaching pedicle which may be drawn out into a stalk. At the opposite end to the pedicle the body carries a skeleton-supported lophophore made up of numerous ciliated current-producing and food-capturing tentacles, the current being drawn in at the two sides of the lophophore and ejected along the midline.

Brachiopods are of separate sexes and eggs and spermatozoa are discharged into the surrounding sea water through ducts which serve both as gonoducts and excretory organs. Eggs are sometimes brooded within the shell valves until a late stage of development but the ciliated larva, which early develops rudiments of the lophophore and pedicle, swims freely for a time before settling on the sea floor and metamorphosing into the young lamp shell.

About thirty thousand extinct species of brachiopods have been described. The group reached its climax during the early Palaeozoic era and since then has steadily declined. The genus *Lingula* has persisted almost unchanged since Ordovician times.

Phylum Sipunculida (*c.* 250 species). These plump or slender sausage-like animals are all marine. They burrow into all kinds of sea-floor deposits or hide in crevices and shells. They move by slow travelling contractions of the muscularised body wall and feed by engulfing material in which they burrow or by capturing small organisms in ciliated grooves of the short tentacles which surround the mouth. The mouth is carried at the end of a barbed proboscis which can be out-turned and infolded very easily and rapidly. The proboscis movement seems in fact the major activity of sipunculoids which are unexciting in their behaviour. The sexes are separate, and eggs and spermatozoa are shed into the surrounding sea water. The larva is of a trochophore type and suggests a distant relationship with the Annelida.

Phylum Acanthocephala (*c.* 500 species). Worm-like parasites which live as adults in the digestive tract of vertebrates. They attach to the host by an evaginable proboscis armed with recurved hooks and are often present in very great numbers. The cuticle covered body wall is secreted by a syncytial epithelium and the body cavity, which is not coelomic and lacks a lining layer bordering the weak musculature, is almost entirely filled by the gonads. The sexes are separate and the females are almost always larger than the males. There is no trace of an alimentary canal. The only elaborate organ systems of the Acanthocephala are the proboscis and the copulatory apparatus. As befits parasites they are very effective egg-producing machines and the larval stages are carried in intermediate hosts, usually insects, isopods or amphipods. Acanthocephalans are sufficiently different from all other kinds of animals to rank as a phylum but have some affinities with nematodes.

Phylum Hemichordata (*c.* 100 species). The phylum includes two classes of marine animals, the acorn worms (enteropneusts) and the tube-enclosed pterobranch colonies made up of many minute zooids, of very different outward appearance but of a like kind of anatomy. All hemichordates have a body divided into three regions, an anterior protosome, a mesosome and a posterior metasome. The protosome of enteropneusts is a burrowing proboscis (sometimes acorn shaped), and in pterobranchs is a shield which secretes the encasing tube and bears a crown of ciliated food-gathering tentacles. The mesosome (or collar) has the mouth opening at its front end, but the main part of the alimentary canal is carried within the metasome which in enteropneusts is a long cylinder and in pterobranchs a rounded sac. The gut, with a terminal anus, is a straight-through tube in enteropneusts, and U-shaped in pterobranchs. The few features of the internal anatomy to be mentioned are selected because of their importance in relating the group to other phyla.

Hemichordates have a body cavity made up of a single coelomic sac in the protosome and bilateral pairs of sacs in the mesosome and metasome. The protosome and mesosome cavities open to the outside by pores. Along the mid-dorsal line of the collar the intra-epithelial fibres of the skin-contained nervous system are inrolled to form a thickened ridge or a hollow tube. A short diverticulum of the gut wall (the stomatochord) underlies the front of the nerve tube and projects forward to support the proboscis. In enteropneusts the side walls of the gut in the front region of the metasome are perforated by gill slits which open to the outside via pores in the body wall to release water taken into the gut. In pterobranchs gill slits are absent or only slightly etched in as a rudimentary single pair.

The rudimentary dorsal nerve tube of the collar; the possibility that the stomatochord may foreshadow the vertebrate notochord – a rod of tissue below the nerve cord; and the presence of gill slits are vertebrate features unknown in other invertebrates. On the other hand the 1,2,2 coelom arrangement and the coelomic pores of the protosome and mesosome recall, especially in their arrangement within the tornaria larva of the enteropneusts, developmental sequences that are strikingly similar to those of echinoderms.

Phylum Pogonophora (100 or more species). These fine thread-like, wholly marine animals up to a foot or so in length (10–35 cm) and usually no more than 0·5–1·0 mm across live in chitinous tubes partially sunk within fine oozes of the deep sea bed. They are in outward appearance and internal anatomy oddly reminiscent of a mixture of enteropneust and pterobranch features though they lack any chordate characters, and most surprising of all have no trace of a gut.

The body has protosome, mesosome and metasome divisions. The protosome projects forwards as a cephalic lobe and bears from 1–200 partially ciliated tentacles. The protosome is often joined into one piece with the mesosome and neither is notable for external markings. The metasome, on the other hand, is warted with numerous papillae which attach to the tube. Nerve tracts run intra-epithelially within the skin and there is a circulatory system containing haemoglobin. Capillary loops within the tentacles are presumed to be respiratory. It is possible that food is absorbed as dissolved organic matter and that some external digestion of solid food may occur at the surface of the skin. There is, however, little direct evidence of this. The tripartite body and the 1,2,2 coelomic sac arrangement in the three divisions strongly suggests an echinoderm-hemichordate relationship for these strange and little-known animals.

Phylum Ctenophora (comb-jellies: *c.* 80 species). Wholly marine and usually pelagic these medusa-like animals differ from coelenterates in many ways. The most obvious difference is the presence on the surface of the bell of eight latitudinal lines of locomotory comb plates made up of transversely fused rows of cilia. Other peculiarly ctenophoran features are an apical (aboral) sense organ of elaborate structure; the presence, in some forms on the equator of the bell, of a pair of bilaterally placed tentacles rooted in and retractible into deep pits; and a gut made up of an entry chamber (stomodaeum) flattened in the plane of the tentacles, a stomach flattened at right angles to this plane and, in termination of the system, eight meridional canals which underlie the comb plates.

Some ctenophores flatten the bell from side to side and draw it out along the plane of the tentacles to produce a long undulating ribbon (e.g. the beautiful Venus's girdle, *Cestum veneris*); others are compressed in the oral-aboral plane and come to look superficially like turbellarian flatworms.

Like them they creep about on stones. The Ctenophora are more advanced than coelenterates but are related to them; and their platyhelminth affinities are suggested by the occurrence of mesodermal muscle and the development in the Müller's larva of some polyclad turbellarians of eight equatorial ciliated lobes, a feature which is difficult to explain unless it be a recalling of a common ctenophore-platyhelminth ancestry.

Phylum Chaetognatha (arrow worms: 50 species). Marine, and, with few exceptions, planktonic animals up to 4 cm long, arrow worms have bilaterally symmetrical slightly flattened bodies, with a head armed on either side with hard, curved food-seizing spines; paired side fins and a dorso-ventrally flattened tail fin. The fins are membranous non-muscularized flotation aids and body stabilizers. Chaetognaths, though they lack a circulatory system and excretory organs, are very elaborately organized animals. They are furnished with eyes and batteries of touch receptors, and have a well-developed nervous system comprising a brain and associated ganglia. Numerous nervous pathways connect the central ganglia with the sensory organs of the skin and the many muscle systems of the body. All the muscles are striated and their rapid contractions are used for snapping the spinous jaws and in producing the extremely quick shivering swimming flexures of the body. The skin, covered by a thin cuticle and regionally ciliated, encloses a coelomic body cavity of the 1,2,2 arrangement found also in the Hemichordata, Pogonophora and in echinoderm larvae. A tubular gut runs from head to tail and to the side of it lie the paired ovaries and testes. Self-fertilization is probably the rule in planktonic species, though in the benthic *Spadella*, which attaches loosely to stones, copulation occurs, the male-acting animal depositing a ball of spermatozoa near the crescentic opening of the oviduct of its female-acting partner. Eggs hatch into larvae which are, in most respects, small replicas of the parent.

Phylum Entoprocta (*c.* 60 species). These are stalked animals of very small (5 mm) size, solitary or colonial and mostly marine. They are widely distributed in shallow water but are never found in large numbers, and although they usually attach to stones one genus (*Loxosoma*) settles on other animals. All of the more important organ systems are contained within a bulbous swelling (the calyx) mounted on the top of the stalk. A circlet of ciliated, food-collecting tentacles surrounds the upper surface of the calyx and both the mouth and the anus of the U-shaped gut open within the tentacle ring. Entoprocts reproduce asexually by producing buds, and the hermaphrodite zooids release eggs which, when fertilized, give rise to larvae that have an unusually complicated development prior to metamorphosis. This little-known group of rather insignificant animals shows many features that are reminiscent of a rotifer type of organization but its affinities with other phyla are somewhat problematical.

Phylum Echiuroida (*c.* 70 species). Marine sausage-shaped animals usually from 10–15 cm long that live in sandy muds or in crevices inter-tidally or offshore. They are readily distinguishable from other sausage-shaped animals such as sipunculids by their flexible, non-retractible proboscis which is ciliated, grooved and used for collecting small-particle food. The occurrence, in some echiuroids, of a few bristles on the body and of a larva of a trochophore type suggests a relationship with annelids.

The minute phyla

Phylum Phoronida (horseshoe worms: *c.* 15 species). The animals live in tubes and occur singly or in groups attached to stones in mud and clays on shores and in shallow offshore waters. The horseshoe-shaped lophophore made up of many ciliated tentacles extends from the tube as a current-producing food-capturing device and the body within the tube is divided, as in other phyla already noted, into the three regions of a protosome, mesosome and metasome. It is swollen at the hind end and contains, among other systems of organs, a gut that is recurved to bring the anus near to the mouth. The animals are usually hermaphrodite and fertilization is external. The larva of some phoronids is an elegant actinotrocha which, after a period of planktonic life, settles and undergoes a complicated metamorphosis into a young adult worm.

Phylum Mesozoa. The very few of these little-studied animals that live as parasites in various organ systems of a variety of marine invertebrates are essentially tiny ciliated cylinders with a cellular filling. Few zoologists will confess to have seen them and various interpretations have been given of their zoological status. They may be degenerate flatworms but are best placed, for want of a better understanding of them, in a separate phylum.

Most of the phyla that have been reviewed in this chapter play but a minor role in the world panorama of the invertebrates; indeed, so much so, that more than half of the major categories of animals (the small and minute phyla of our listing) contribute only some 0·2 per cent of all the known living kinds of animals. And, even if we were to remove from the scene the vast concourse of the insects, the representation of the underdeveloped half of the animal kingdom would still be less than 1 per cent of the whole.

Before remarking further on this strange disparity of species number in the different phyla we need to be assured that the census which has been made is a reasonably true bill of numbers. In terms of absolute numbers it is certainly not, for new species are discovered every day and there is much more searching and re-assessing to be done. There is, however, no reason to believe that the small and minute phyla will achieve higher places in the hierarchy of size as the census of animal species is progressively perfected; and, indeed, if we are to point to the phyla which are now vastly underestimated, the groups with the greatest potentiality for further species discovery are almost certainly the already great assemblages of the Arthropoda, Aschelminthes and Protista.

In some instances, notably the Brachiopoda, the fossil record shows that they were once much more numerous than they are now, and for all we know – for they have left no fossil record – other small phyla may be but remnants of a former glory. But the question remains: why are these phyla small today? It cannot be due solely to the fact that almost all of them are confined to the sea and have never achieved the physiological, anatomical and developmental adaptations necessary for entry into fresh waters and on to the land; for the large phyla not only have fresh-water and terrestrial representatives but are abundant in the sea. Have, then, the small phyla met with such severe competition from other more successful groups that they have been pruned down in numbers to the point where their breeding populations have been cut to low levels of out-breeding capacity with a consequent diminution in speciation potential? This is a question to which it is impossible to give an informed answer, though it has often been suggested that the decline of a group of animals, for example the brachiopods, may be attributed to the appearance of newly evolved competitors, in this instance the lamellibranchs. But the suggestion is highly speculative and, to say the least, rather improbable.

It seems therefore probable that the failure of some phyla to break up into numerous species is a consequence of something that is intrinsic to them and, though we have no evidence of this, is rooted in a genetic conservatism the mechanics of which might well repay the attention of cytogeneticists.

Chapter 5

The styles of invertebrate architecture

ANIMALS COME in an almost endless variety of forms, shapes and sizes. In fact, so great is the diversity that at first sight it is difficult to discern any consistent styles of animal architecture, yet if the special and peculiar features of the animals are stripped away, the fundamental styles are as obvious as romanesque, gothic or baroque styles in building. In fact, allowing for the differences in terminology, an account of animal architecture could be written in terms as familiar to the art historian as to the biologist. We find major and minor stylistic trends, regional variations, conservatism, the gradual evolution of new styles, and the great innovators in the history of animal structure just as in the history of architecture.

Understanding the basic designs of animal structure is made a good deal easier if two fundamental precepts are borne in mind. First, animals have generally tended to become more complicated in the course of evolution, and second, animals are constructed on sound mechanical principles to do particular jobs.

As with most things, the tendency for animals to become more complicated is not an inviolable law of nature. There are situations in which a progressive simplification of structure takes place, as it commonly does in animals that become internal parasites or for one reason or another become very small. There is no virtue in complexity for its own sake but one feature of animal evolution has been the gradual liberation of the individual from the demands of the environment, permitting it to control more and more its own activities – to swim rather than drift, to seek and select its food rather than accept whatever comes to hand. To achieve this demands the evolution of complicated structures to carry out an ever increasing range of activities, and complexity breeds complexity, for a complicated animal requires a complicated co-ordination and control system so that each organ works in concert with the rest.

One of the great achievements of comparative anatomists in this century has been to show how closely animal structure is related to its functions. Animals are built on sound mechanical principles and might almost have been designed by an engineer. This is equally true of quite simple and very complicated animals, but always the basic structural plan of the animal sets limits to what it can and cannot do and one of the chief interests of anatomists lies in understanding what essential features an animal must possess if it is to live in a certain way, exploit a particular situation, or perform a certain activity. Looking at the problems of animal structure in this light, we find that in most situations there are only a few ways in which a job can be done efficiently. Swimming, for example, can be carried out by using paddles, by throwing the body into waves that pass from head to tail, or by jet propulsion, and nearly all animals swim by one or other of these methods. Very small animals are able to swim by ciliary action, but as soon as they exceed a millimetre or so in length, the inertia of the body is so great compared with the forces generated by the beating cilia, that instead of the animal being driven through the water, water is driven over the animal which remains stationary. Swimming is largely a matter of overcoming a number of physical and mechanical problems which relate only to the physical properties of water and the size, shape and density of the animal. These considerations apply equally to all animals and so it is not surprising that many animals of quite different fundamental structure should have arrived at the same solution to the problem of how to swim and have evolved comparable structures to carry out this activity.

As we shall see, fundamental animal structure does not show endless variation. There are actually only a few basic designs, but each of them has been subject to a riot of experimentation in which its potentialities have been exploited to the full. But however else animals may be elaborated, they must conform with the physical and mechanical requirements of the environment in which they live and in this they have little freedom of choice. Thus it is that we find that the most basic features of animal structure relate principally to mechanical considerations and although the structures concerned with feeding, or breathing, or reproduction may be more spectacular, they are also more superficial.

The simplest animals

The animal body is composed of cells, and differences in the ways the cells are organized in relation to one another are sufficiently important for multicellular animals to be separated into three groups, Parazoa, Mesozoa and Metazoa, on this basis. Of these, the Metazoa have been by far the most successful and it is the only group to have shown any substantial evolutionary advance and to have produced large, complicated animals. In metazoans, the cells are specialized in various ways and assume different specialized functions, but their activities are closely co-ordinated so that the animal behaves not as a collection of cells but as a single integrated whole. More than this, cells of a particular type, sometimes in conjunction with cells of other types, are associated with one another to form tissues and organs which function in the body as units of a higher order than individual cells. The capacity of forming tissues and organs appears to be essential for the development of complicated animals for in the Parazoa and Mesozoa, cells are much less specialized than they are in the Metazoa and have a greater autonomy. As a consequence the Parazoa and Mesozoa remain among the simplest multicellular animals we know.

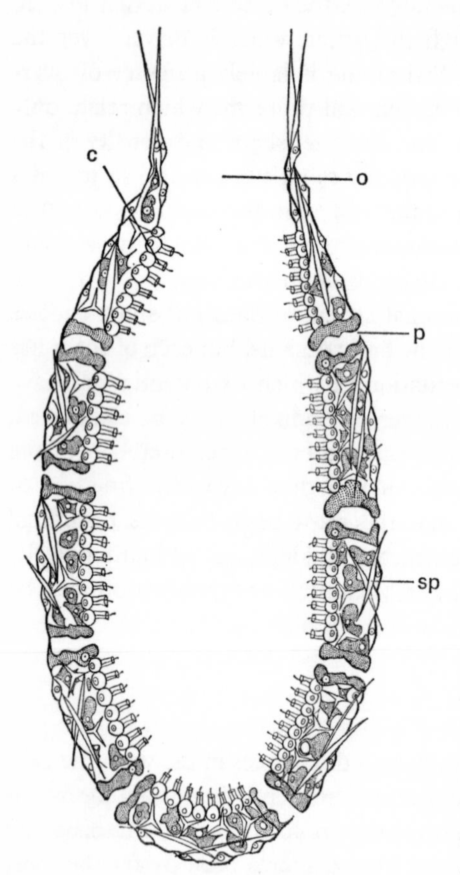

Fig. 5.1 The essential features of the body of a simple sponge. Water is drawn into the central cavity by the beating of flagella on the inner layer of choanocytes or collar cells (c); it enters through the minute pores (p) and leaves via the osculum (o). Spicules (sp) are laid down in the mesogloea between the choanocytes and the thin outer epithelium

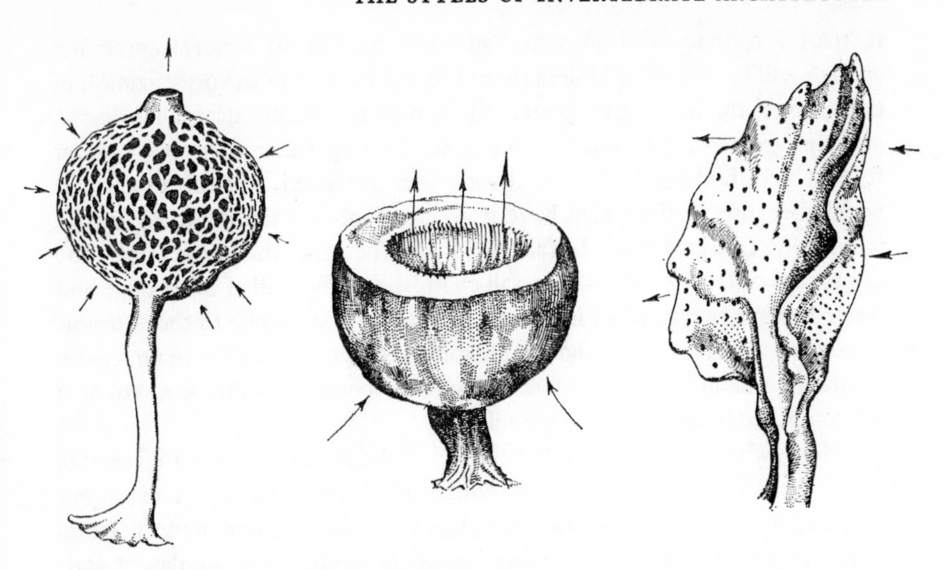

Fig. 5.2 Variable forms of sponges, showing ways in which the inhalant and exhalant currents of water are separated from one another

At their simplest, sponges which constitute the Parazoa, are little vase-shaped animals attached by their base to plants or rocks (Fig. 5.1). They are all aquatic and almost all live in the sea. The surface of the sponge body is peppered with pores through which water is drawn into the central cavity and escapes from there through the neck and mouth of the vase. This is the most conspicuous opening in the sponge body (the pores are microscopic); it is not the mouth of the animal for nothing enters through it, and it is known as the osculum. The water current is produced by the beating of the flagella of the cells lining the central cavity in the sponge, and the same flagella also trap food particles that are carried in the water.

In a delightfully simple series of experiments carried out towards the end of the last century, Dr George Bidder showed that in sponges, as in all animals that filter food particles from the water, water is expelled from the osculum in such a way as to contaminate as little as possible the water that is being drawn into it (Fig. 5.2). The exhalant water is discharged at a considerable velocity from the osculum and this water, from which food particles have already been removed, is carried well away from the areas from which water is being drawn into the animal. Now the flagella which produce this current are only some 30 μ (0·03 mm) long and their beat will not have much effect on water that is out of range of the tip of the flagellum. Consequently, if the cavity within the sponge is enlarged, there will be an increased volume of dead water and the velocity of water leaving the sponge will be reduced. Unless there is some change in the organization of the sponge, the larger the sponge, the slower the water current leaving it, and the greater the contamination of its ingoing water current by water from which food particles have already been removed, and this in animals which because of their size have a greater food requirement. As Euclid would say, this is absurd. In fact, in all but the very smallest sponges, the flagellated cells are confined to small chambers in the wall of the sponge and water leaving them passes into a central cavity and thence to the exterior. The fact that many flagellated chambers empty into this main cavity results in a fast current of water leaving the sponge as many slow streams are combined in one narrow channel.

Bidder, who delighted in rather startling calculations, commented on this apparatus: 'The remarkable achievement of the perfected hydraulic organ in sponges is that from this waving of hairs 1/10,000 of an inch in thickness at a mean speed of seven feet an hour, there is produced an oscular jet with an axial velocity of over half a foot a second (280 times the speed of the flagellum), which in *Leucandra* (one of the sponges he examined) throws to a distance of nine inches five gallons a day or a ton in eight weeks'.

For a sponge to work effectively, particularly if it is a large one, it is essential for it to stand erect, and the construction of the sponge body is so simple that this would be difficult without the aid of supporting structures. The outer covering of the body is of very thin polygonal cells, the lining is of the flagellated cells and between the two there are various kinds of amoeboid cells lying and moving in a gelatinous matrix. The skeleton which supports the body is secreted by some of the amoeboid cells and takes the form of spicules of calcium carbonate or silica (in which case the skeleton is as though composed of spun glass) or of a fibrous material, spongin. A bath sponge is in fact the fibrous skeleton of the sponges *Spongia* and *Hippospongia*; the loofah is not a sponge in the zoological sense but is the fibrous part of a dried gourd.

Sponge spicules take a variety of forms (Fig. 5.3). Single calcareous spicules are secreted by a single amoeboid cell in the gelatinous middle layer of the body wall (Fig. 5.4). The cell divides and one daughter cell remains stationary near the outer edge of the sponge while the inner one migrates inwards, secreting the spicule as it goes. When the spicule has been formed, the outer cell migrates along it, secreting more calcium carbonate and thickening it as it does so. The commonest spicules of calcareous sponges have three rays and these are formed in much the same way by triplets of amoeboid cells.

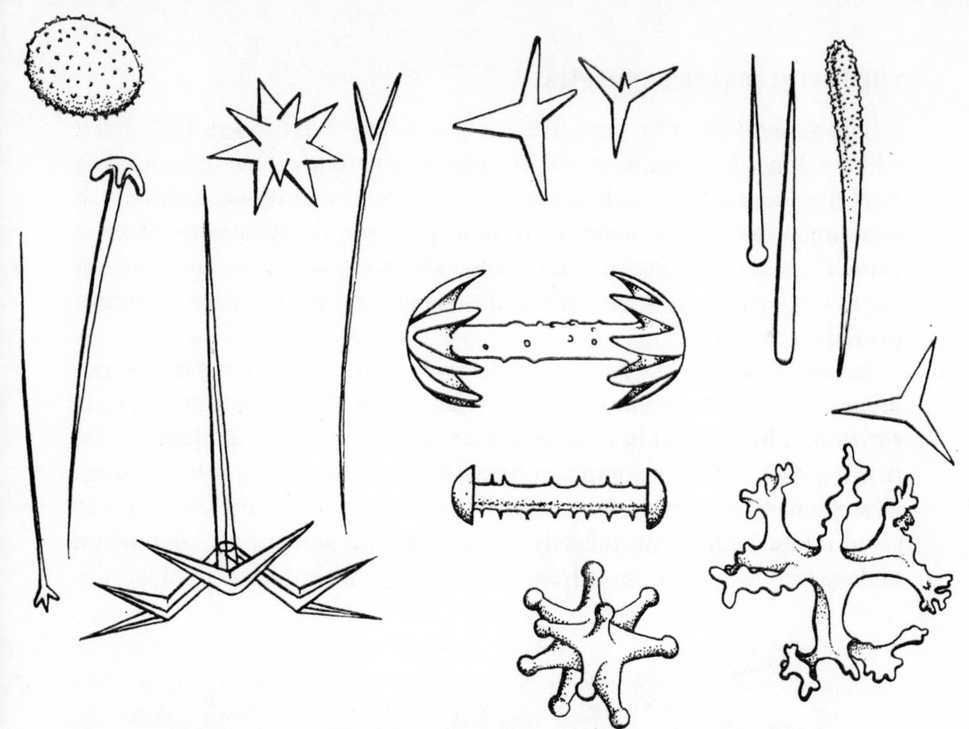

Fig. 5.3 Various forms taken by sponge spicules (different magnifications)

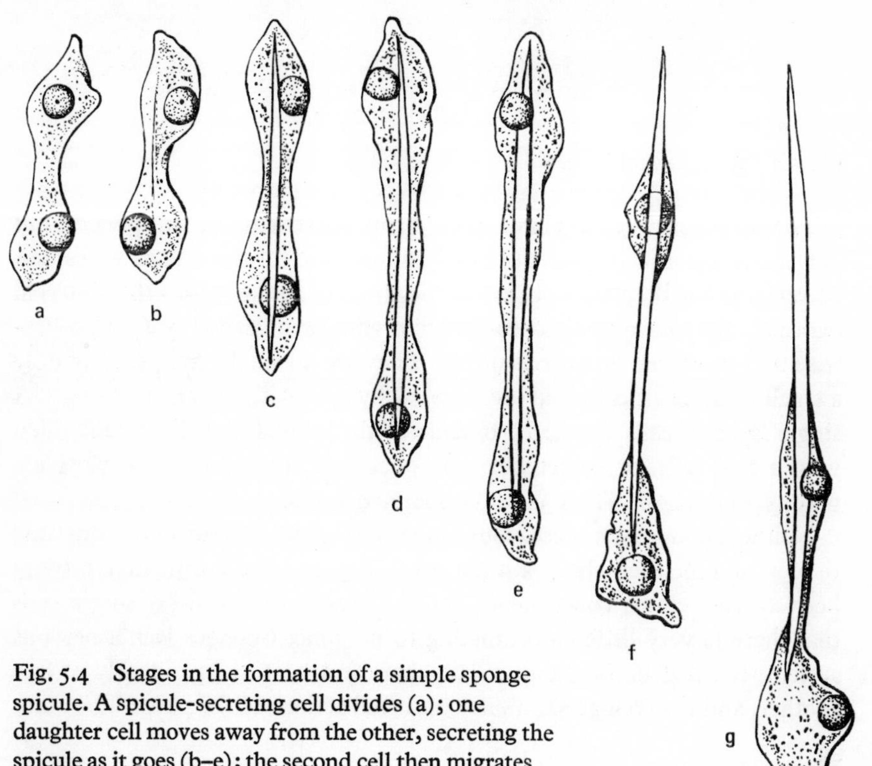

Fig. 5.4 Stages in the formation of a simple sponge spicule. A spicule-secreting cell divides (a); one daughter cell moves away from the other, secreting the spicule as it goes (b–e); the second cell then migrates along the spicule thickening it as it does so (f–g)

The general form of the spicule is genetically determined, but, as Dr Clifford Jones has demonstrated, the precise details are often influenced by local mechanical factors. If a migrating cell meets an obstacle while it is secreting a spicule, it is deflected from its path and the spicule it secretes is bent. Triradiate spicules vary in the angles between the three spines according to the position of the spicule in the sponge and the mechanical stresses prevailing there.

Isolated spicules may strengthen the body wall and, since the sharp ends often project, afford some degree of defence, but they cannot form a rigid skeleton. This is done in glass sponges which have six-rayed spicules that fuse together to form a complete lattice. Because they have a rigid skeleton, glass sponges are the most symmetrical and erect of all sponges and may stand a foot high. Unfortunately, these beautiful animals are commonest in deep waters and our most frequent encounter with them is in museums.

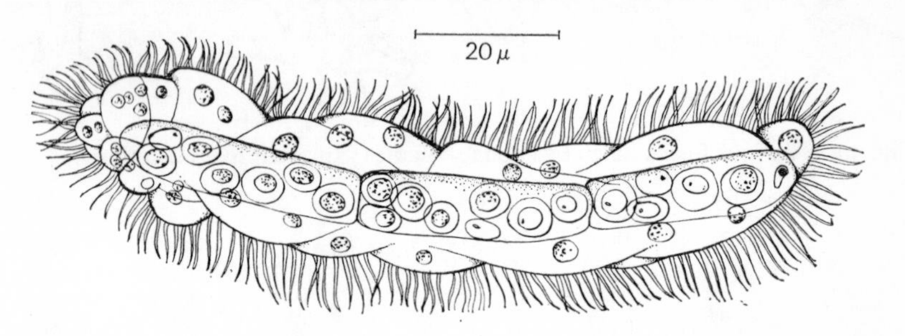

Fig. 5.5 A mesozoan, *Dicyema*

The Mesozoa (Fig. 5.5) have an even simpler structure than sponges, but they are all internal parasites of other animals and for this reason we cannot be certain that they are not simply 'degenerate'. Like most other internal parasites, they have no food-collecting organs or digestive systems; otherwise they resemble no other animals. They are all small, composed of only a small number of cells and are covered by cilia with which they swim. A single layer of cells forms the outside of the animal and the core is filled with a few cells, or sometimes only one, that are concerned exclusively with reproduction. There are no other structures.

Neither sponges nor mesozoans have cells organized to form recognizable tissues and neither show a very great degree of co-ordinated activity between the cellular constituents. Indeed, mesozoans consist of so few cells that there is very little co-ordinating to be done. Sponges lack a nervous system, the flagella in the flagellate chambers beat independently of one another, and the sponge shows no concerted actions of the parts. The outer layer of cells may contract slightly causing the sponge to become a little smaller, and the cells guarding the minute pores into the flagellate chambers may contract, so cutting off the supply of water entering the chamber, but apart from this the sponge's life is very dull. As we shall see, the situation in the Metazoa is much different.

The tissue layers of the body

Metazoan animals differ from sponges in having tissues and organs. At some stage in the development of most metazoans, three distinct and separate layers of cells are formed and these are ultimately responsible for the formation of quite different structures in the adult animal. The outer layer of cells, or ectoderm, produces the epidermis, and it often forms the excretory organs, and sometimes part of the musculature. The innermost layer, or endoderm, produces the greater part of the digestive system. Between the endoderm and ectoderm there is a middle layer of cells, the mesoderm, which contributes the musculature, the reproductive system, and parts of the circulatory system such as the pericardium.

The simplest metazoans are the coelenterates, including sea-anemones, corals and jelly-fish. In some ways they seem little more advanced than

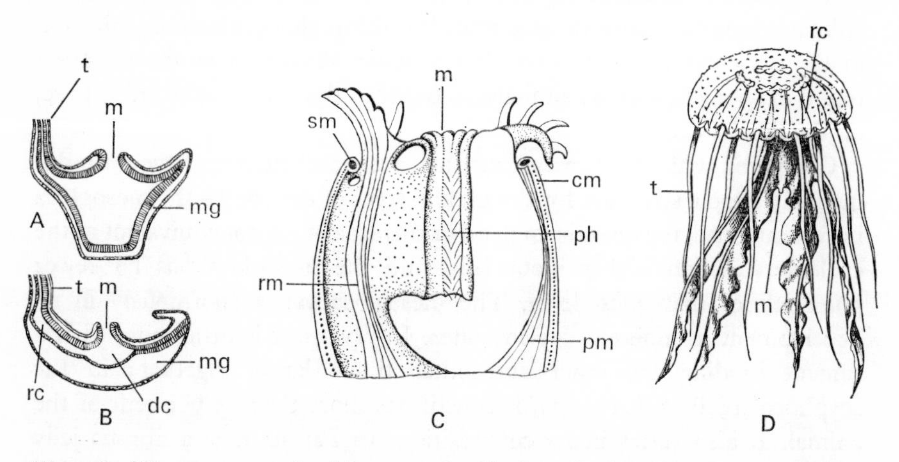

Fig. 5.6 The two types of architecture of cnidarians. A, the polyp; B, the medusa (here shown upside down for ease of comparison with a polyp); C, an advanced polyp, as found in sea-anemones (Anthozoa); D, an advanced medusa, as found in jelly-fish (Scyphozoa). cm, circular muscle; dc, digestive cavity; m, mouth; mg, mesogloea; ph, 'pharynx' with ciliated channel (siphonoglyph); pm, parietal muscle (longitudinal muscle of body wall); rc, radial canal; rm, retractor muscle; sm, sphincter muscle; t, tentacle

sponges and apparently lack a mesodermal tissue layer, but in other respects they show important similarities with the more advanced Metazoa. The coelenterate body has only a single opening, the mouth, and this opens into a single internal cavity which serves as a stomach and circulatory system (Fig. 5.6). The least complicated version of the coelenterate body can be seen in such animals as the familiar *Hydra*. It consists of a simple, more or less cylindrical bag attached at one end to a rock or other solid surface and with the mouth surrounded by a ring of tentacles at the other. The walls of this bag consist of a layer of epidermal cells on the outside and a lining of digestive cells, and the two layers are separated by a middle layer called the mesogloea.

At first sight the structure of a simple polyp like *Hydra* appears comparable to that of a simple sponge, but the differences between a sponge and a coelenterate are profoundly important. The coelenterate has a mouth and a digestive cavity which, although there is no intestine and anus, are essentially the same as in other metazoans. In sponges, the main opening is not a mouth but a route by which water leaves the central cavity and the latter is in no sense a stomach. More important, even cells of the same type behave independently of one another in sponges and the whole animal is incapable of showing any concerted activity. In coelenterates cells of the same type are arranged in tissues, there is a nervous system, and the animal behaves as a co-ordinated unit. It is this higher order of organization in the coelenterates that represents a major advance over the situation found in sponges, and the similarities between the two types of animal are, on the whole, deceptive.

Compared with other metazoans, however, coelenterates have a fundamentally simple structure for the middle layer of the body – the mesogloea in coelenterates, the mesoderm in other metazoans – is not equivalent in the coelenterates and higher metazoans, and the mesogloea has far fewer potentialities than mesoderm. The mesogloea varies enormously in its development in different coelenterates. In *Hydra*, it is little more than a cement binding the inner and outer tissue layers together; in the scyphozoan jelly-fish *Pelagia*, it constitutes more than 95 per cent of the animal. It also varies in its composition. In *Pelagia* it is a fibrous jelly completely devoid of cells; in many sea-anemones it contains wandering amoeboid cells, fibres, and other structures, and amounts almost to a connective tissue which in some corals (though not the reef-building corals) contains closely set calcareous crystals or spicules constituting a skeleton and is not far removed from what we have already seen in sponges. Although it may be structurally quite complicated, mesogloea differs from mesoderm in one crucial respect. It never contributes, and rarely even includes, systems of organs such as are developed from true mesoderm. The formation of the reproductive cells, which is such a constant feature of the mesoderm, takes place in coelenterates either in the epidermis (in *Hydra*) or, more commonly, in the walls of the digestive cavity. Coelenterate muscles are formed at the base of the cells of the inner or outer cell layers, never from cells in the mesogloea although the muscles may separate from their parent cells and migrate into the mesogloea. The development of a mesoderm, with its capacity to form organ systems between the inner and outer layers of the body seems to be an essential prelude to the development of complicated structures and because of its lack, coelenterates are limited in an important way and are invariably constructed on a very elementary fundamental architectural plan.

Rigid and hydrostatic skeletons

The progression in complexity from sponges, in which there is almost no overall co-ordination of activities, to coelenterates in which two tissue layers are formed and the animal behaves in a co-ordinated fashion, and finally to the higher Metazoa with a true, organ-forming mesoderm, is obvious. Now we turn to animals which all have a mesoderm and all have

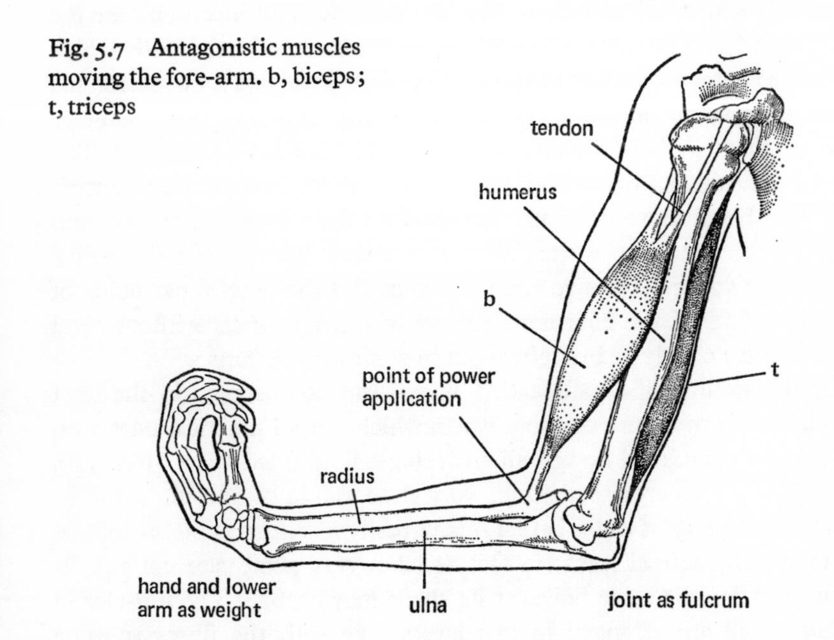

Fig. 5.7 Antagonistic muscles moving the fore-arm. b, biceps; t, triceps

complicated systems of organs. Progress takes a new direction and is concerned in the first place with changes in the way the body is organized to function in different types of environment. Here the physical and mechanical features of the environments become important and inevitably we are led to consider skeletons and muscles by which the animals exert force against the outside world.

Muscles are designed for shortening and exerting a tension, they cannot exert a thrust and cannot, simply by ceasing to contract, restore themselves to their original uncontracted length. A restoring force outside the muscle must extend it after it has finished contracting and before it can contract again. This restoring force is almost always supplied by other muscles. Perhaps the most familiar example of this is the functioning of the biceps and triceps of the upper arm of man (Fig. 5.7). Contraction of the biceps raises the forearm and at the same time extends the triceps. Contraction of the triceps lowers the forearm and restores the biceps to its original length. The biceps and triceps are termed antagonistic muscles, and for such a system to work, one end of each muscle must be fixed and immovable relative to one another. In the human arm, the biceps and triceps are both attached to the head of the humerus and only their insertions on the bones of the forearm are free to move when the muscles contract.

Paired antagonistic muscles form the basis of the skeletal musculature of all animals with hard structures that can provide rigid insertions for the paired muscles. The greater part of the musculature of arthropods which have a rigid outer exoskeleton naturally functions in this way, but occasional paired antagonistic muscles can also be found in many other animals wherever there are hard structures to which muscles can be attached. The body of caterpillars, for example, has no hard parts, but the head capsule is hardened and the important jaw muscles have fixed insertions and behave in the same way as the paired antagonistic muscles of animals like vertebrates with an extensive rigid skeleton. But the body movements of caterpillars and of the enormous variety of other animals without rigid skeletons must clearly be brought about by some other means.

The functioning of a soft-bodied animal can be observed in the least complicated form in an echiuroid worm which, for all practical purposes, consists of a cylindrical body wall enclosing a fluid-filled space (Fig. 5.8). The fluid is sea water, or something very close to it in composition, and an important property of water is that it is virtually incompressible. It follows that unless the animal leaks (in fact, it takes very good care not to), its volume remains the same however its shape may change. The muscles in the body wall are arranged in two layers, one with the fibres running longitudinally, the other with the fibres circumferential, or circular. If all the circular muscles contract, the diameter of the worm is reduced, and it becomes very thin, but because the total volume of the worm remains unchanged, this is possible only if there is a compensatory increase in length. So contraction of the circular muscles makes the worm longer as well as thinner, and in doing so the circular muscles extend the longitudinal muscles. If the longitudinal muscles now contract the worm is shortened with a compensatory increase in girth, and so the circular muscles are restored to their original length.

In these movements, the circular and longitudinal muscles act antagonistically to one another, but more complicated changes of shape are also

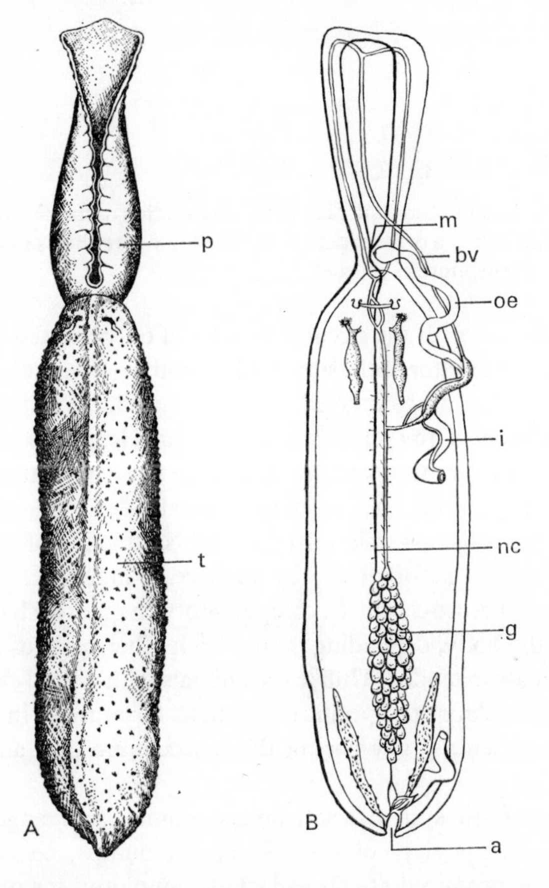

Fig. 5.8 A, external view of an echiuroid worm; B, internal organs. a, anus; bv, blood vessel; g, gonad; i, intestine; nc, nerve cord; m, mouth; oe, oesophagus; p, proboscis; t, trunk

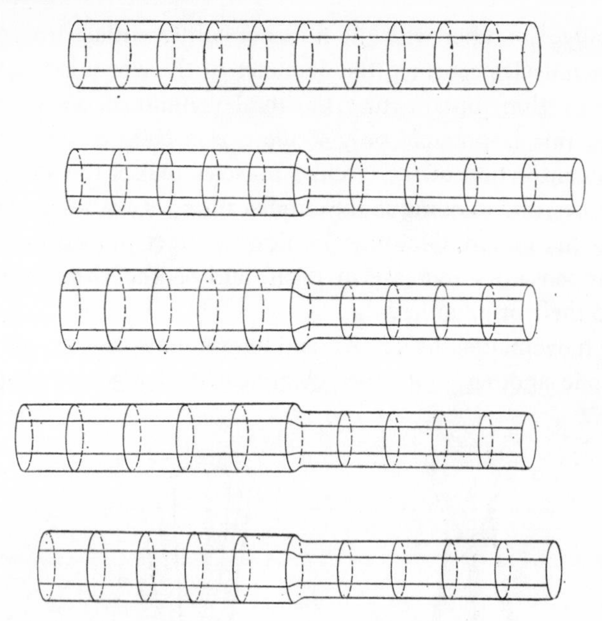

Fig. 5.9 Some possible consequences of the contraction of the circular muscles of the right-hand half of a cylindrical worm. The top figure shows initial condition with all the muscles relaxed

possible in this worm. If the circular muscles of only half the body contract a variety of compensatory adjustments is possible (Fig. 5.9). If the longitudinal muscles remain contracted and resist extension but the remaining circular muscles relax, the worm remains the same length as before, but one half of it becomes very thin and the other very fat. Alternatively, the entire longitudinal musculature may relax and be extended while the remaining circular muscles resist extension. In this case contraction of part of the circular musculature is accompanied by an increase in length. Clearly, if one part of the musculature contracts, compensatory changes of shape may occur anywhere in the body, depending upon which other parts of the musculature relax and are extended. Unlike the animals with a rigid skeleton, there are, generally speaking, no specific antagonists for muscles in animals with a hydrostatic skeleton – any part of the musculature may antagonize any other part.

With a hydrostatic system, a soft-bodied animal can change its shape in an almost endless variety of ways – lumps, bumps, constrictions and dilations can be produced at will and can be employed for many different purposes. An animal with a rigid skeleton is much more constrained and for the most part its movements are restricted to the manipulation of jointed appendages. In this it has an important advantage over soft-bodied animals: it can use the principle of levers to amplify both the speed and extent of the movement of the appendage. In animals with a hydrostatic skeleton, there can be no amplification of the movement; if the longitudinal muscles contract, the body of the animal shortens by precisely the same amount. The muscles of soft-bodied invertebrates accordingly have rather different properties from those of arthropods and vertebrates and are adapted to working over a much greater range of lengths than skeletal muscles are. Except in arthropods, invertebrate muscles are generally much

Fig. 5.10 Changes of shape and size of a sea-anemone, *Metridium*. All the drawings are of the same individual on different occasions and all are to the same scale

slower in their reactions than skeletal muscles and this follows at least in part from the fact that they must change their length so much more.

Some coelenterates show this to a remarkable degree. Changes of shape are produced in exactly the same way as in worms providing the mesogloea is sufficiently thin to permit extensions and contractions, as it very often is. The extent to which the animal changes shape varies enormously, even in sea-anemones in which the phenomenon is particularly conspicuous. *Calliactis parasitica*, an anemone that lives attached to shells inhabited by hermit crabs, has a thick, stiff mesogloea and is therefore capable of only minor shape changes. The plumose anemone, *Metridium senile*, has a thin, diaphanous body wall and undergoes the most spectacular contortions (Fig. 5.10). The same individual may on one occasion be a sad, squalid heap, but on the next a magnificent, inflated, coloured balloon with a great crown of feathery tentacles.

An important difference between coelenterates and other animals that employ a hydrostatic skeleton is the fact that in the former the gastric cavity is used for this purpose and opens to the exterior via the mouth. Unless special precautions are taken, an increase in internal hydrostatic pressure caused by contraction of part of the musculature, instead of extending muscles elsewhere in the body, simply squirts water through the mouth. The anemones, in fact, provide a useful example of the consequences of failing to maintain a constant volume. If leakage of fluid is prevented, completely reversible changes of shape are possible and are practised by sea-anemones so long as they keep their mouths shut. In sea-anemones the mouth leads to the gastric cavity by way of a tubular actinopharynx which acts as a sleeve-valve. If the internal hydrostatic pressure is raised, the actinopharynx is squashed flat and no water escapes through it. When water is lost, the anemone becomes a much smaller animal and a return to its original size and shape depends entirely upon the ability of the animal to re-inflate itself by a steady inward trickle in a ciliated channel at the side of the mouth.

The hydrostatic system of coelenterates is a very unusual one in that it is the gastric cavity and opens to the exterior. In other animals different arrangements are made.

The cavities of the body

Although the digestive cavity may be used as a hydrostatic skeleton, as it is in some coelenterates, it is anything but ideally suited for this purpose. It works in coelenterates only because the body wall is very thin and contraction of the muscles immediately under the epidermis compresses the digestive cavity directly. With the development of a bulky mesoderm and systems of organs between the body wall and the gut, forces generated by the body-wall muscles are dissipated by compressing these tissues and their effect on the gut contents is very slight. The hydrostatic skeleton must be in immediate contact with the muscles that relate to it.

So long as this condition is met, it is not even essential to have a fluid-filled cavity as a hydrostatic skeleton. Worms in the phyla Platyhelminthes and Nemertea are generally described as 'solid-bodied' (Fig. 5.11A) because the space between the body wall and gut is filled with a loosely packed tissue consisting largely of relatively undifferentiated cells that are

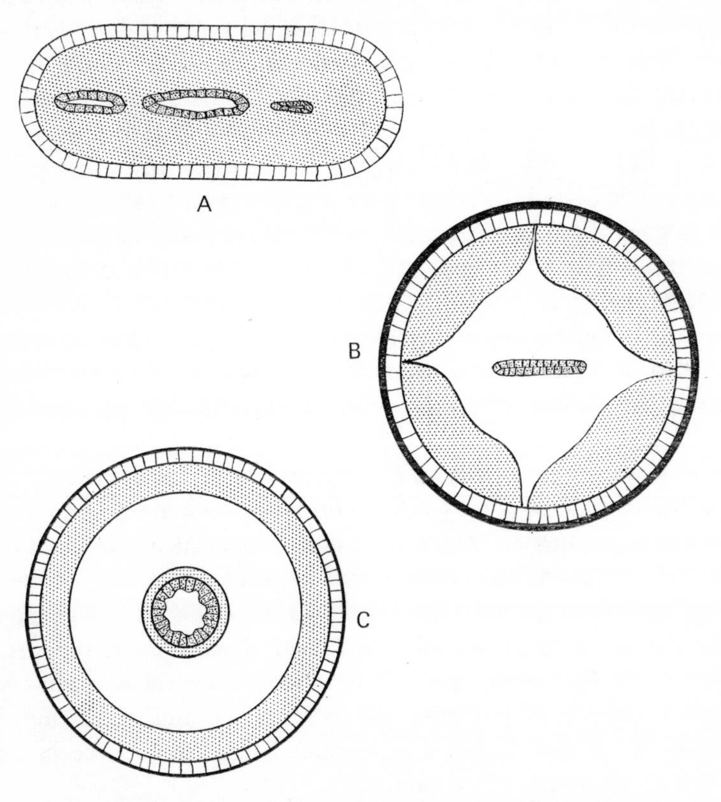

Fig. 5.11 Diagrammatic cross-sections of worms with different types of hydrostatic skeleton. A, a flatworm (e.g. turbellarian) with the space between body-wall and gut filled by tissue; B, a pseudocoelomate (e.g. nematode) with a secondary body cavity (pseudocoel) between the gut and the body-wall muscles; C, a coelomate (e.g. annelid) with a secondary body cavity (coelom) within the mesoderm and muscles on both the gut and the body-wall

almost embryonic in character, some muscle fibres, and a little connective tissue together with the reproductive and excretory organs. Contraction of the body-wall muscles distorts the cellular contents of the body which behave in very much the same way as a liquid would except that, because of the muscles and connective tissue and also simply because the body contents are cellular, the pressure changes are damped down and the changes of shape of the body are slow.

The consequences of using this mechanically inefficient hydrostatic skeleton are immediately obvious in the movements of platyhelminths and nemerteans. The worms elongate and contract in exactly the same way as other worms by the antagonistic contractions of circular and longitudinal muscles in the body wall, but the movements are slow and weak and they are not used to perform work. Locomotion is by cilia or by waves of contraction that pass along the longitudinal muscles of the ventral surface of the body; in neither case is the hydrostatic skeleton implicated. These muscular locomotory waves are rather difficult to observe in turbellarians but they can be seen quite clearly in snails and many other gastropod molluscs. The dotted track left by the common garden snail shows by the small patches of mucus left behind on the garden path where each point on the ventral surface of the body came to rest during the successive waves of muscle contraction.

From an examination of turbellarians, it becomes clear that unless a fluid-filled cavity can be produced within the body to serve as a hydrostatic skeleton, the entire body-wall musculature, with the circular muscles antagonizing the longitudinals, cannot be employed for locomotion or any other task. In the great majority of worms a fluid-filled cavity is formed within the mesoderm (Fig. 5.11C). This is the coelom or secondary body cavity and in its simplest form it is a single compartment running from one end of the body to the other with the gut suspended from the body wall and entirely freed from the influence of contractions of the body-wall muscles.

The development of a coelom is a critical advance in the structural complexity of the animal body. It represents a logical advance over the turbellarian grade of structure and permits the full realization of the potentialities of turbellarian architecture. It also allows the animals to invade and exploit a new environment.

When the coelom is used as a hydrostatic skeleton, the longitudinal muscles antagonize the circulars and when the latter come into play, the worm has a circular cross-section. This has important consequences for locomotion. In turbellarians the circular muscles are completely relaxed with the result that the body assumes a flattened shape. Thus, the greatest possible surface area is presented to the ground and almost half the longitudinal musculature of the body wall can be in contact with it and contribute to the generation of locomotory waves. With a circular cross-section, coelomate worms have a minimal area of contact with the ground and so are hopelessly ill-adapted for living in the same way as a flatworm. But if they are buried, the whole body-wall musculature can contribute to the generation of locomotory forces. This is just as well, for much greater forces are necessary to burrow through the soil than to glide over it. In other words, burrowing in the substratum is an activity for which animals

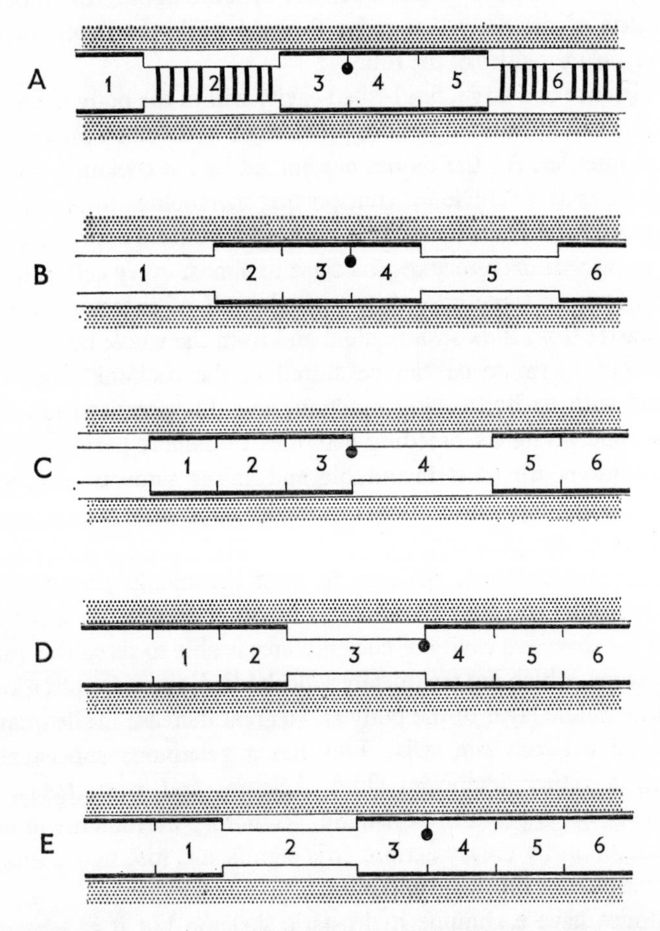

Fig. 5.12 Behaviour of the musculature of a cylindrical worm during successive phases in the passage of a peristaltic locomotory wave along the body. Circular muscles are contracted in regions where the body is thin and elongated, but they are indicated only in A

with a true hydrostatic skeleton in the form of a coelom are designed and flatworms are designed for creeping on surfaces.

Coelomate worms burrow by means of bulges which travel along the body, that is by peristaltic locomotory waves. Waves of contraction pass along the longitudinal muscles, each wave followed by one of circular muscle contraction. If a worm crawls along a tube, as in Figure 5.12, its only points of contact with the sides of the tube are in those parts of the body where the longitudinal muscles are contracted and the circulars relaxed. At intervening points the circulars contract so that the body becomes thinner and fails to make contact with the walls of the tube. These are the parts of the body that move forward while the fatter regions are wedged against the sides of the tube.

The existence of a large, fluid-filled cavity within the body has a number of advantages to the animal besides providing a hydrostatic skeleton for the body-wall muscles. All the tissues are bathed by the coelomic fluid which is able to serve as a vehicle for transporting metabolites to the tissues and excretory products from them. It is no longer necessary for the animal to provide numerous excretory organs close to almost every cell in the body; instead, a few, or sometimes only a single pair of excretory organs can remove wastes from the coelomic fluid and from the whole body. Developing eggs and spermatozoa can be stored in the coelomic fluid without interfering with its hydrostatic function, and this permits the coelomate worm to concentrate its breeding activities to a short period of the year when conditions are most favourable and at the same time produce as many eggs as animals which breed repeatedly but cannot accumulate many eggs for lack of storage space.

Although a hydrostatic skeleton is most commonly provided by the coelom, the same end may be achieved by other means. We have already seen that the digestive cavity of coelenterates is able to serve this function. In nemerteans, which are technically solid-bodied, there are often very few cells in the middle layer of the body and a great increase in the quantity of the material between the cells. This has a gelatinous consistency and represents a rather inefficient fluid skeleton, and nemerteans, unlike turbellarians, often perform peristaltic locomotory movements to complement locomotion by ciliary action. It is significant, too, that a number of nemerteans burrow into sand and mud.

Nematodes have a genuine hydrostatic skeleton but it is a pseudocoel (Fig. 5.11B), a persistent embryonic cavity rather than a secondarily formed cavity as the coelom is. Nematodes are unusual animals. The best known example is *Ascaris*, the common intestinal roundworm, and most investigations have centred on it. Most other nematodes, though usually much smaller than *Ascaris*, appear to resemble it in their basic architecture. It has an extremely thick, complicated cuticle, the body-wall musculature is composed solely of longitudinal muscles, and the internal pressure of the active worm may rise to half an atmosphere; even in a resting worm it is far in excess of that recorded in the body fluids of any other animals. An inevitable result of the high internal pressure is that the animal is distended to the full extent permitted by the thick cuticle and for this reason always has a circular cross-section. Contraction of the longitudinal muscles might normally be expected to cause an increase in the girth of the body as the worm shortened, but this is clearly impossible when the cuticle is already stretched to the limit by the high internal pressure. It is not surprising that overall changes of length of the worm are slight, and when the muscles contract they do not shorten very much but increase the internal pressure instead. As soon as the muscles relax, the tension in the cuticle is sufficient to restore them to their original length. Nematodes are not designed for peristaltic movements and do not use them to any appreciable extent. Their movements generally take the form of lashing and coiling or undulatory waves, and are produced by contraction of one set of longitudinal muscles antagonizing the other.

The structure of nematodes is dominated by the high hydrostatic pressure of the body fluid. The gut, for example, is completely collapsed so that food cannot be transported through it by cilia or by peristalsis, as it usually is in other animals, but must be pumped through by a powerful pharyngeal pump apparatus. All openings to the exterior – mouth, anus, genital and excretory pores – must be closed by sphincter muscles to prevent the contents of the intestine, genital ducts and excretory ducts being expelled from the body. In their basic structure nematodes are very stereotyped and monotonous animals, and all because of their high internal pressure, but they are no less successful for all that. It has been claimed that were nothing but nematodes visible, the outline of the earth and everything on it would still be discernible. This may be something of an overstatement, but gives some idea of how numerous and ubiquitous these animals are.

Segments

Coelomate worms have the great advantage of being able to use antagonistic contractions of the body-wall muscles to perform work. They can burrow into the substratum and crawl by peristaltic locomotory waves,

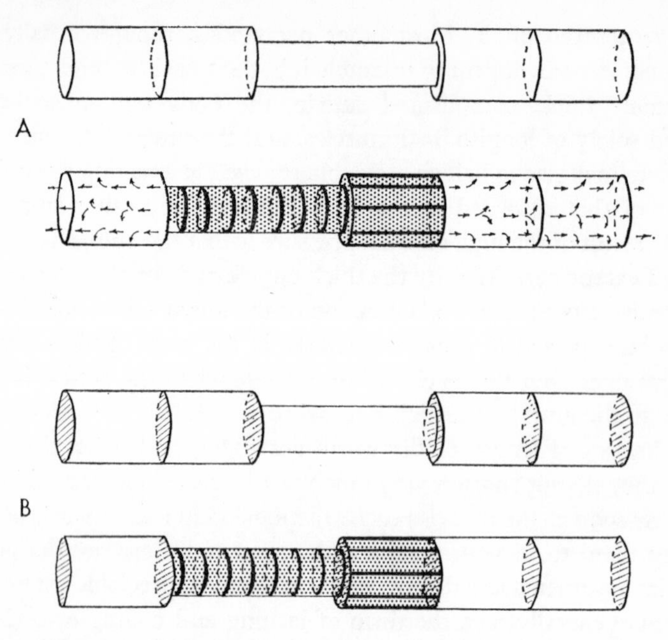

Fig. 5.13 During the passage of a peristaltic locomotory wave along the body of a non-septate worm (A) pressure changes affect all parts of the body wall. In a worm with septa (B) pressure changes are confined to segments in which muscle contraction occurs

but although in these respects they show a great advance over the acoelomate worms, their fundamental architecture can still be improved from the point of view of mechanical efficiency.

When a peristaltic wave travels along the body of a crawling worm, different parts of the body are subjected to quite contrary stresses. Where the circular muscles are contracting, the body contents are under radial compression, and in neighbouring regions where the longitudinal muscles are contracting, the body contents are under longitudinal compression. Because the coelom is undivided and extends from one end of the body to the other, pressure differences generated by the contracting muscles are transmitted freely to all parts of the body wall (Fig. 5.13A). As a result, in a region where the circular muscles are contracting two things are happening: the longitudinal muscles of that region are being extended, but at the same time the circular muscles themselves are exposed to entirely opposite forces produced by the contraction of the longitudinal muscles in the adjacent parts of the body, which is causing circular muscle extension. The conflict between opposing forces can be abolished if regions of the body which are engaged in different activities can be isolated from one another. This can be done by separating each region from its neighbours by rigid, water-tight bulkheads; then pressure changes in one region act only on the body wall of that part of the body and have no influence elsewhere (Fig. 5.13B). We find precisely this structure in the segmented annelid worms, more particularly the earthworms, in which the coelom is subdivided by a succession of muscular diaphragms, the septa.

The septa do not provide complete mechanical isolation of each part of the body they delimit. To do so it would be necessary for them to be completely rigid and this is very difficult to achieve in a flexible, muscular diaphragm. The septa bulge slightly as a pressure difference develops between the two adjacent regions they separate, and so there is some transmission of pressure though this is much less than if the septa were not there at all. But a succession of septa, each adding its effect to the next, does successfully contain pressure variations. Professor Gordon Newell was able to demonstrate this in the common earthworm, *Lumbricus*, by injecting saline fluid under pressure into the coelom of one segment and measuring the pressure change in another segment some distance away. There is no increase in the internal pressure of the distant segment until pressure in the first rises to the point at which the septa are ruptured, a pressure two or three times as great as that normally generated in an earthworm. In earthworms, the distance between adjacent septa is a good deal less than the wave-length of the peristaltic movements so that many septa are interposed between two adjacent regions of the body which are in opposite phases of the locomotory cycle. The existence of so many septa means, too, that one segment does not behave very differently from its immediate neighbours and the locomotory wave proceeds by a smooth progression along the body instead of in a discontinuous way.

The increased efficiency of the locomotory waves, with each region of the body acting as a self-contained unit, has immediate consequences for the life of the animal. Unsegmented coelomates such as echiuroids and sipunculids can construct burrows but none of them engages in prolonged or continuous burrowing and they are essentially sedentary animals. The segmented earthworm, on the other hand, is a vagrant burrower and is clearly adapted for repeated and prolonged activity.

The effect of the subdivision of the coelom upon the morphology of the worm is enormous. Much of the advantage of having an undivided coelom is lost and it now becomes necessary for the worms to provide reproductive organs and excretory organs in almost every coelomic compartment, an independent blood supply must be provided for the organs in each compartment and, since much of the musculature becomes repetitive, the nervous system also has a separate ganglion and separate nerves in relation to each division of the coelom. Almost the entire structure of the animal develops a repetitive pattern in relation to the septal compartments of the coelom. This is known as metameric segmentation and is one of several types of morphological repetition we meet in the animal kingdom, but it is only in the annelid worms that it bears a relation to peristaltic burrowing.

Hydrostatic isolation of different regions of the body appears also in the structural plan of several phyla which are distantly related to the vertebrates, though in them the reasons for this are quite different from those in annelids, and the animals have a very different structure. They are known collectively as oligomerous animals, signifying that they have only very few segments compared with annelids which have many and are polymerous. Perhaps the clearest example of the oligomerous body plan can be seen in

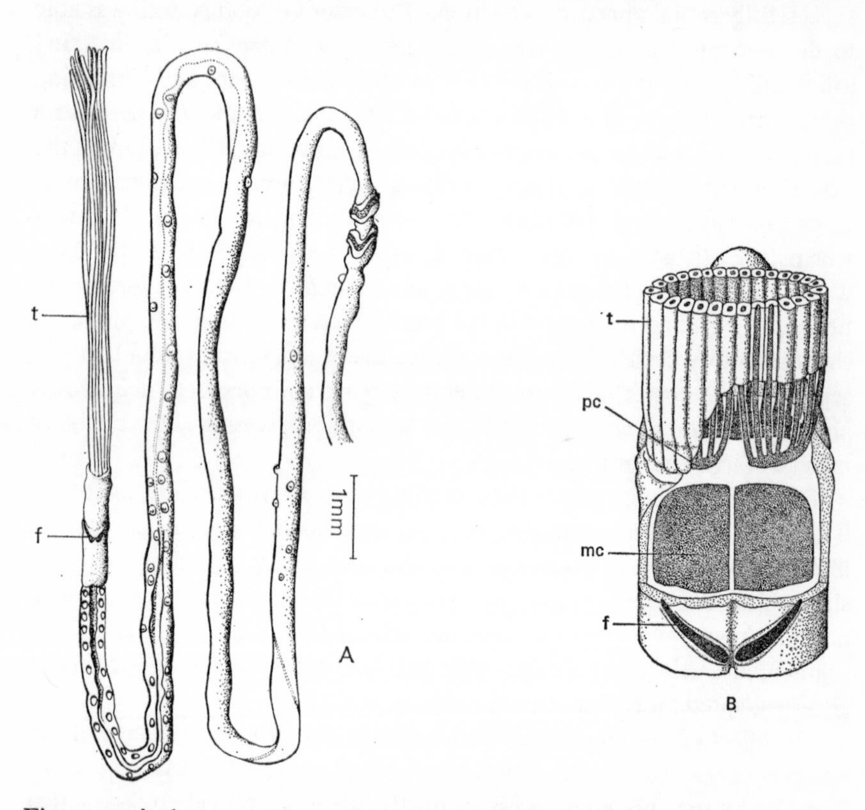

Fig. 5.14 A, the pogonophoran *Lamellisabella* (the posterior end of this very long worm is omitted); B, coelomic compartments of anterior end. f, frenulum; mc, mesocoel; pc, protocoel; t, tentacles

the Pogonophora, a phylum of very curious marine animals that have come into prominence in recent years now that they are known to be very common and something of their structure is understood.

Pogonophorans (Fig. 5.14) are very long and very slender and live in a secreted tube, the greater part of which is buried in the mud with only the top inch or so exposed. They were originally dredged only from the sea bed of the deeper parts of the ocean, but lately have been found in the relatively shallow water of the continental shelf including the approaches to the English Channel. What must be quite unique in non-parasitic metazoans is their total lack of a mouth or digestive system although they have a tentacular food-collecting organ, and it is supposed that the food they collect is digested outside the body and then absorbed by the tentacle cells. The pogonophoran body is divided into three regions, each with its own, isolated, hydrostatic system and, unlike annelids, each with a quite different function. The tentacular apparatus is borne on the most anterior region of the body and is presumably held erect while it is exposed at the mouth of the tube and is sieving food particles from the water. At the base of the tentacles there is a fluid-filled reservoir with branches running into each tentacle. Compression of the reservoir forces fluid into the tentacles and renders them stiff and turgid. The musculature of the reservoir relaxes and allows the tentacles to collapse while the animal is withdrawn into its tube and is digesting its catch. The second region of the body is short and bears a pair of chitinous bars which, when this segment is dilated, are pressed against the sides of the tube and wedge the animal while it is collecting food. This region of the body, like the first, contains a coelomic compartment. The third region of the body is very long and contains a general perivisceral coelom. This part of the body engages in peristaltic movements by which the worm creeps up and down its tube. It is, in fact, the equivalent of the whole body of an unsegmented coelomate.

The pterobranch hemichordates (Fig. 5.15) are another rather obscure group of animals and also have a body composed of three regions, each with its independent coelomic compartment. Like pogonophorans, pterobranchs live in a secreted tube and are tentacular food collectors. In the pterobranchs, however, the tentacles are borne on the middle of the three segments. The most anterior segment, or pre-oral lobe, is a very mobile prehensile organ which the animal appears to use to pull itself towards the opening of the tube or even, in some species, to crawl around on the outside of it. The third region again has a general perivisceral coelom but, whatever its past history, modern pterobranchs appear to make very little use of it.

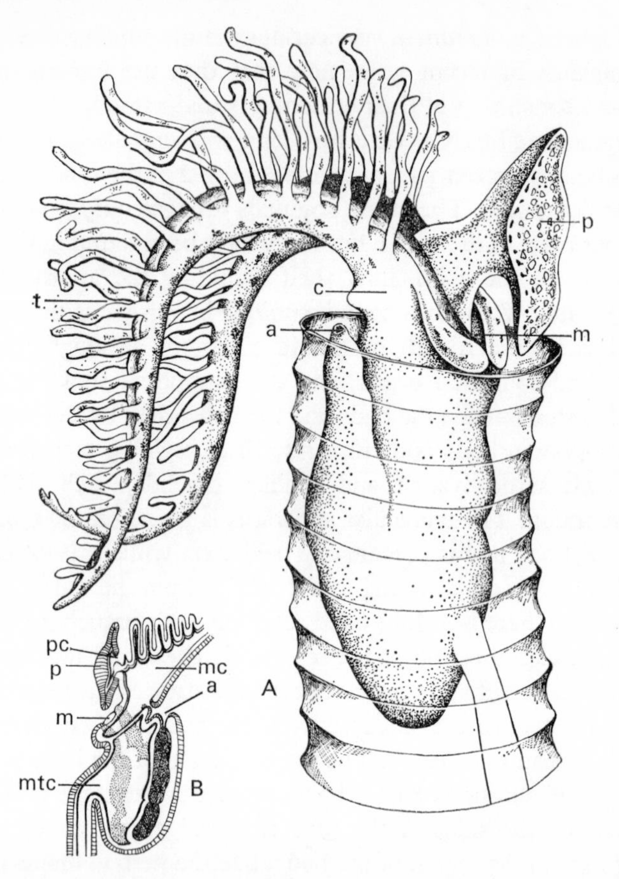

Fig. 5.15 A, the pterobranch *Rhabdopleura* in its tube; B, arrangement of the coelomic compartments. a, anus; c, collar; m, mouth; mc, mesocoel; mtc, metacoel; p, proboscis; pc, protocoel; t, tentacles

Phoronids (Fig. 5.16) are tubicolous, tentacular feeders, but so far as we can tell have only two body regions, one with a coelomic hydraulic apparatus for the tentacles, the other with the perivisceral coelom which is used, as in pogonophorans, in the production of peristaltic movements when the worms move up and down their tubes.

The Pogonophora, Hemichordata and Phoronida are not at all closely related to each other, despite a considerable similarity in their basic architecture. It is most likely that this relates to their common habit of living in tubes and feeding by collecting suspended matter from the surrounding water by a tentacular apparatus. Two independent hydrostatic systems, at least, appear to be necessary: one for the tentacles, the other for the major part of the body which generates peristaltic waves. In both the Pterobranchia and the Pogonophora, a third independent coelomic compartment is used, for a prehensile organ in the former and for the curious wedging apparatus in the latter.

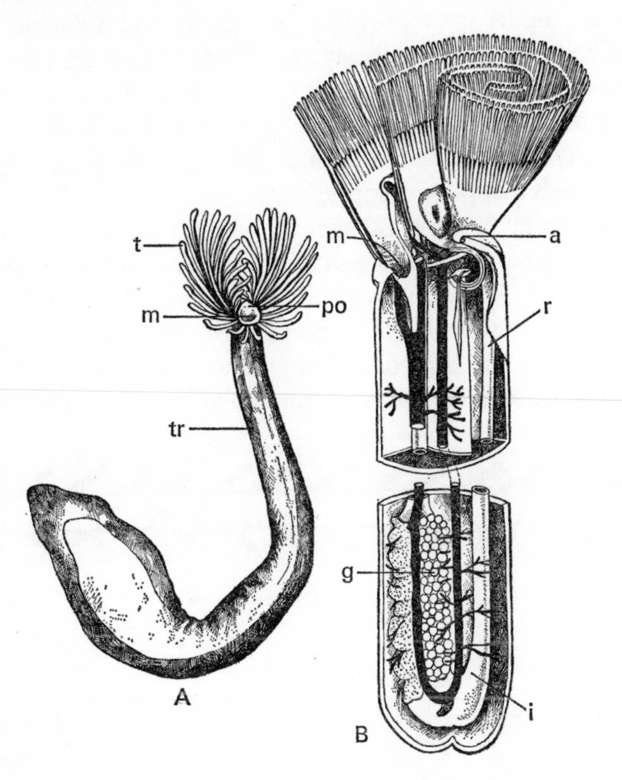

Fig. 5.16 A, *Phoronis*; B, internal structures. a, anus; g, gonads; i, intestine; m, mouth; po, pre-oral lobe; r, rectum; t, tentacles; tr, trunk. Blood vessels shown in black

The 'segments' of oligomerous animals are obviously not the same as the 'segments' of annelids. In the former, each region of the body has quite different functions and a totally different structure. In annelids each segment is essentially a replica of all the others and each in turn performs precisely the same tasks both in locomotion and, minor specializations apart, in other activities. But whatever the fundamental differences between the oligomerous and metamerous worms, both types of body architecture have provided the basis for a riot of modification and experimentation and have been enormously productive of new types of animals.

Appendages and the arthropods

The climax of evolution in the invertebrates is undoubtedly reached with the appearance of limbs. These have been developed in several groups of segmented animals, but the most successful have been those of arthropods in which the limbs have proved most plastic and valuable structures capable of being modified endlessly and devoted to a great variety of uses. We are most familiar with the arthropod limb as a jointed structure armed with hooks or claws, but it can probably be traced back to a simple lobe arising directly from the body wall and this is essentially the form taken in the other animals that possess appendages.

The Polychaeta (a class of annelid worms) show most clearly the repercussions upon the basic segmented body plan of the development of appendages which, in these worms, are simple lateral lobes arising from the body wall, one on either side of each segment. When the polychaete worm *Nereis* is crawling, the parapodia on opposite sides of the body are alternately swung backwards and so push the animal forwards (Fig. 5.17).

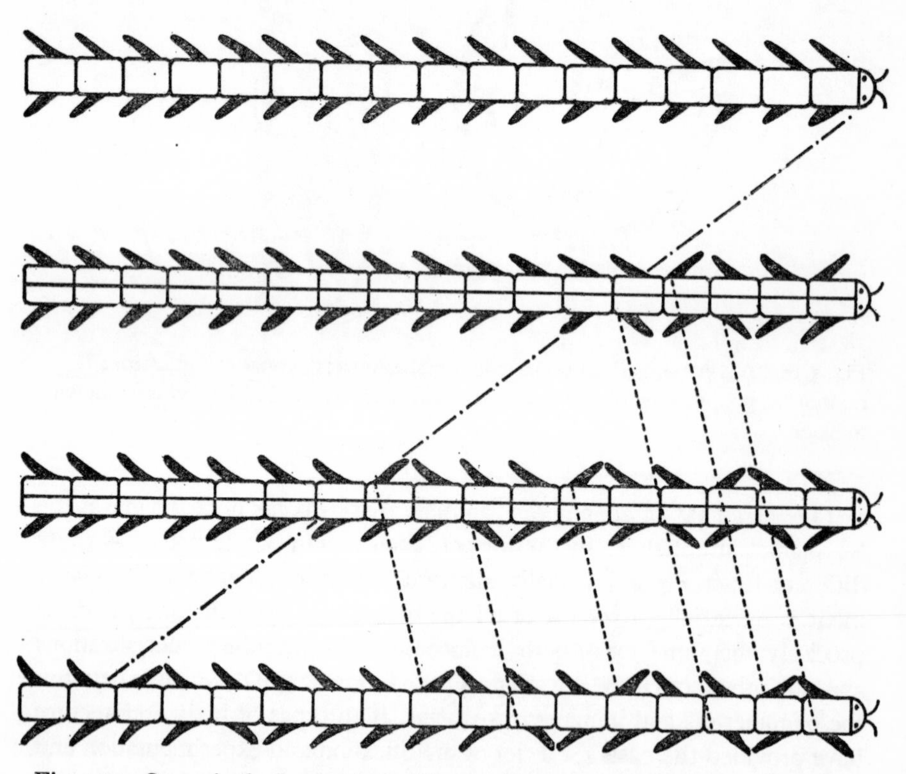

Fig. 5.17 Stages in the locomotion of *Nereis* with the aid of lateral appendages

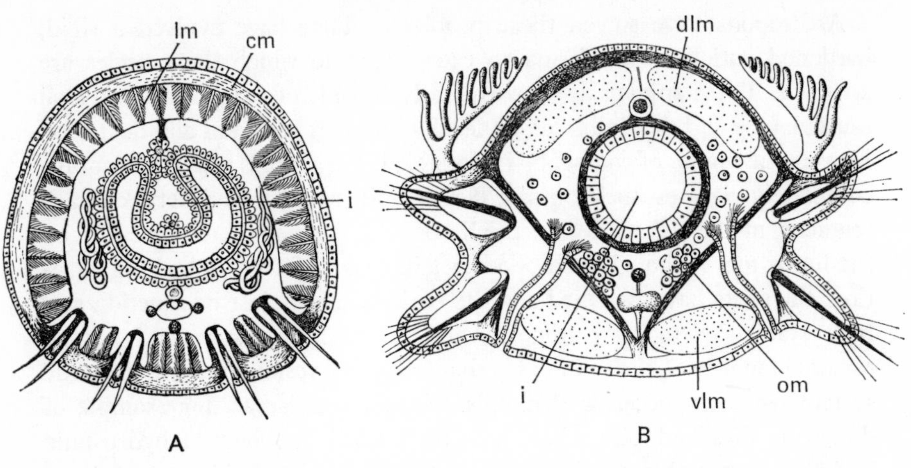

Fig. 5.18 Cross-sections of (A) an earthworm and (B) a polychaete. cm, circular muscle; dlm, dorsal longitudinal muscle; i, intestine; lm, longitudinal muscle; om, oblique muscle; vlm, ventral longitudinal muscle

Since there are generally several parapodia on each side of the body engaged in performing this power-stroke, the worm is subjected to a fairly uniform and continuous forward thrust. For the parapodia to be moved in this way, they must be hinged at their point of insertion into the body wall and must be moved by muscles arising outside the parapodium. Both the provision of fixed hinges and of specific skeletal muscles are unusual features in soft-bodied animals. These problems are resolved by bracing the hinges with oblique muscles and by inserting the parapodial muscles in the mid-ventral line where the body provides the greatest rigidity. It is a partial solution but an adequate one. With the parapodial muscles crossing the coelom and running into the parapodium, the circular and longitudinal muscle coats of the body wall are disrupted (Fig. 5.18). The circular muscles are greatly reduced and are confined to the intersegmental regions between the parapodia, while the longitudinal muscles are restricted to the dorsal and ventral sides of the body. With this arrangement of the body-wall muscles, peristaltic movements are impossible and, indeed, parapodial crawling is incompatible with peristaltic burrowing. The parapodium appears to have been evolved in worms living in an oozy surface mud which has too thin a consistency for peristaltic locomotion through it to be possible. It cannot be claimed that this has been a particularly successful experiment, for many polychaetes have abandoned parapodial locomotion, but the development of these appendages and their disruption of the body-wall musculature has left its traces throughout the Polychaeta.

Arthropods have solved these problems. They have evolved a rigid, hardened cuticle which forms an exoskeleton to which the muscles are attached. These animals are consequently capable of using a true skeletal musculature and the limbs can be used with great precision and dexterity. Movement would, of course, be impossible if the animals were encased in a completely rigid exoskeleton and flexible joints are formed between each segment, at the articulation of the limbs with the body, and generally on the limbs too. Although arthropods retain traces of a segmental organization, the septa have vanished and the coelom, having lost its function as a hydrostatic skeleton is confluent with the vascular system and has no separate identity. This is not to say that arthropods cannot use a hydrostatic system when the occasion demands. Many insect larvae, for example of dipterous flies and butterflies, are soft bodied and have a hydrostatic skeleton comparable to that of a worm; barnacles, spiders, and some centipedes all employ hydrostatic systems for specific tasks, but in every case, the fluid skeleton is provided by the haemocoel and subdivision of it by septa is never regained.

Having evolved what appears to be a perfect mechanical system, arthropods have gone from strength to strength. Of all the known animal species, more than 80 per cent are arthropods, and they have occupied every available ecological niche. The exoskeleton introduces certain new problems as, for example, that of growing while encased in a hardened cuticle. This problem has been solved by the animals moulting at regular intervals. We shall see in a later chapter that the existence of a rigid exoskeleton has occupied a central role in arthropod evolution, in both a positive and a restrictive way.

Regression and simplification

The history of the Metazoa is in some ways an unfolding of more and more complicated types of animal architecture with each new development permitting the animals to exploit more fully the potentialities of their fundamental structure, to colonise environments that were previously closed to them, or to use the environment in a new way. But this is not the whole story and among all major groups of animals there are some which show the reverse trend and have become secondarily simplified.

The Annelida provide within a single phylum the clearest examples of modifications of the fundamental body structure, both progressive and regressive, with changing habit. We have already seen that two types of annelid have very advanced structures: the oligochaetes which show, in its perfected form, metameric segmentation of a soft-bodied animal as an adaptation to vagrant burrowing in the soil, and some polychaetes, such as *Nereis*, while still metameric, have engaged in one of the first experiments in the use of appendages with the modification of the muscular architecture that this has involved. Polychaetes are a very varied group of worms and many are quite unlike *Nereis*. A few, like *Lumbrinereis* and the Capitellidae resemble earthworms and have diminutive parapodia and parapodial musculature. The reduction in the size of the muscles entering the parapodia allows a more extensive body-wall musculature to be developed and it forms almost complete muscle coats. This, together with the retention of a full set of intersegmental septa, permits these worms to burrow and behave mechanically in much the same way as earthworms. In many other polychaetes, regressive evolution has gone much further. The parapodia lose their locomotory function, the circular and longitudinal muscles of the body wall become complete, the septa are nearly all lost, and the worms live a sedentary life in semi-permanent burrows. In their essential structure and habits they are little different from unsegmented coelomates such as the echiurids and sipunculids. An even greater regression appears in a group of worms closely related to oligochaetes – the leeches (Fig. 5.19). They have abandoned burrowing altogether and live on the surface of stones and leaves, or (the parasitic leeches) attached to the body of their host while they are feeding. Leeches (Fig. 5.20) creep with the aid of a pair of suckers, one at the anterior end around the mouth, the other at the posterior end beneath the anus. The posterior sucker is attached to the substratum, the circular muscles of the body wall contract and the leech becomes very long and thin. The anterior sucker is then attached, the posterior sucker detached and the longitudinal muscles now contract to their greatest extent so that the posterior sucker is drawn forward. Generally the body is flexed during this movement so that the posterior sucker is attached to the substratum immediately behind the anterior one. In this way, each 'stride' is almost equal to the fully extended length of the worm. No differential pressures are set up in different regions of the body in this type of creeping, unlike peristaltic locomotion, and the whole of the longitudinal musculature and the whole of the circular musculature behave as single units. The mechanical function of the septa has disappeared and with it the septa also; they make only a transitory appearance during embryological development. Furthermore, since the leech creeps over the surface rather than burrows into it, only small forces need be generated. The damping effect of a tissue layer within the body is therefore immaterial and the *raison d'être* of the coelom has also vanished. It is reduced to a series of narrow channels which in blood-sucking leeches replace the blood system. Leeches live in a similar situation to turbellarians and their structure is not much different; the wheel has turned full circle in the elaboration of the architecture of worms.

Much of the evolutionary history of the oligomerous animals, too, is dominated by regressive changes, though in these the predominant factor has not been so much change in locomotory habit as the adoption of a completely sessile habit and the loss of locomotory devices altogether. The three phyla Phoronida, Brachiopoda and Bryozoa are closely related. Of these, the Phoronida are tubicolous and retain at least two independent

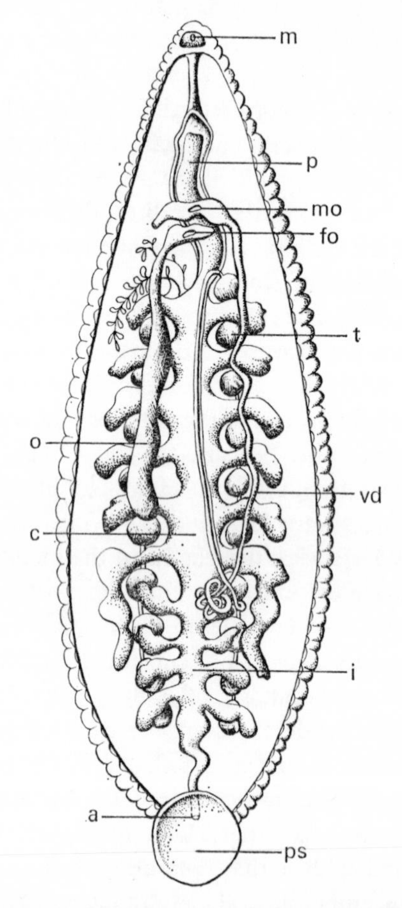

Fig. 5.19 Internal structure of a leech. a, anus; c, crop; fo, female genital opening; i, intestine; m, mouth surrounded by anterior sucker; mo, male genital opening; o, ovary; p, proboscis in sheath; ps, posterior sucker; t, testis (ten pairs); vd, vas deferens from testes

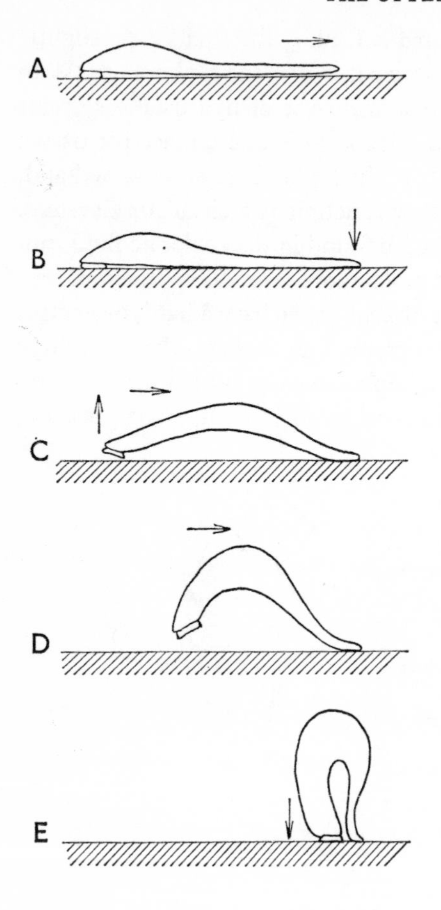

Fig. 5.20 Stages in the locomotory cycle of a leech. The posterior sucker is fixed to the ground and the anterior end is extended (A) and then fixed by the anterior sucker (B). The posterior sucker is then removed (C) and fixed to the ground immediately behind the anterior one (D–E). The cycle is then repeated

hydraulic systems, the other two phyla are both for the most part immobile and have suffered a corresponding modification of structure and reduction of the coelomic hydrostatic systems. The three phyla are united in the possession of a tentacular feeding organ arranged on a horseshoe base, the lophophore. A current of water is drawn through it while it is held erect and food particles in the water are sieved out. The ability of the phoronids to migrate up and down their tubes (the reason why it is necessary to have at least two independent coelomic compartments) is clearly protective; the worms must be exposed for feeding but can be safely withdrawn into the tube at other times. But this is not the only way protection from predators can be achieved. In brachiopods (Fig. 5.21) a hard, calcareous, bivalve shell similar to the bivalve shell of a mollusc, is secreted, concealing the body and the lophophore. The shell is generally attached to the ground by a stalk and the animal is quite exposed but protected by its shell. When the animal is feeding the shells gape slightly and a water current is drawn in and circulates over the lophophore. With such an adaptation the brachiopod makes no use of hydrostatic systems and much of the coelom is occluded with muscle and connective tissue. Not all though; the coelomic channels in the lophophore are retained and, as so often happens, they acquire a new function as a circulatory system. Cells containing a respiratory pigment are found in the coelomic fluid, not the blood, and the coelomic fluid is in much more vigorous circulation.

The Ectoprocta (Fig. 5.22), like the Brachiopoda, have a hard, protective casing, but instead of drawing water currents into the case, the main body of the animal, the polypide, and its lophophore can be protruded from and retracted into the case. This operation is achieved by hydrostatic pressure,

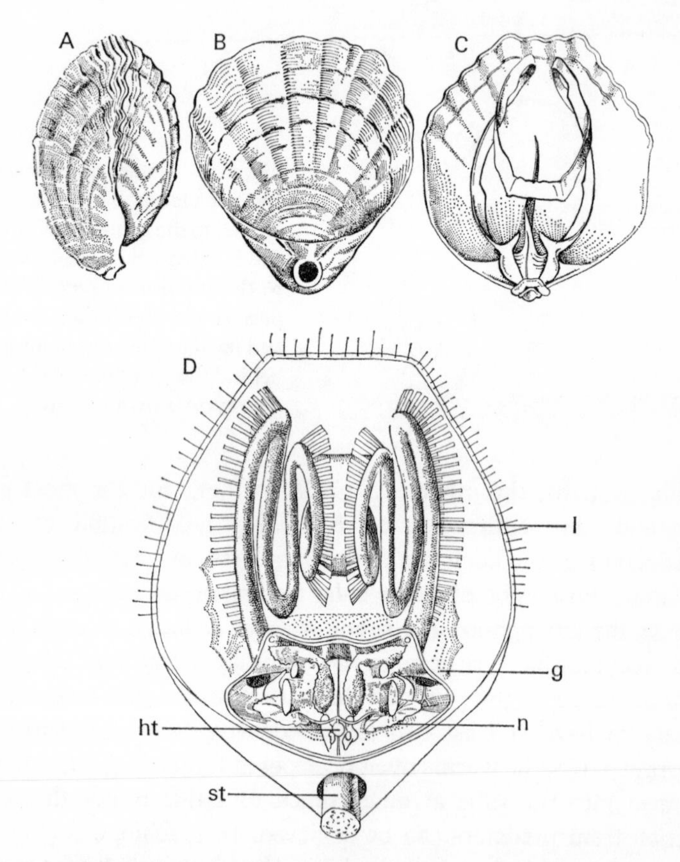

Fig. 5.21 Brachiopoda. A, side view; B, ventral view of a brachiopod; C, internal view of upper shell with skeleton of gill; D, internal structure. g, gonad; ht, heart; l, lophophore (gill); n, funnel of excretory organ; st, stalk

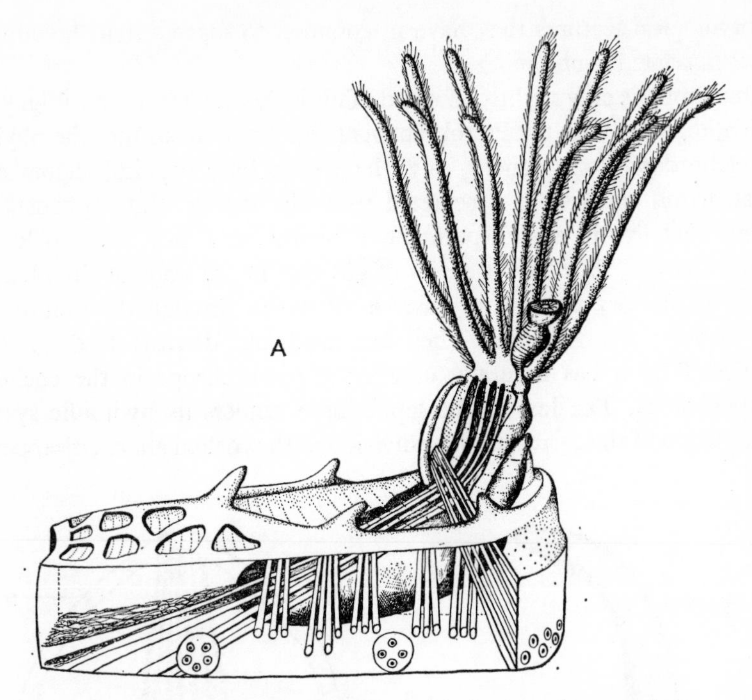

Fig. 5.22 Ectoprocta. A, a single individual with the tentacles extended

and withdrawal of the polypide into its box is by retractor muscles. Since eversion of the polypide and erection of the lophophore are parts of the same process, there is no occasion on which an independent lophophore hydraulic mechanism might be used and, as a result, these animals have a single, undivided coelomic compartment. The most intriguing problem in ectoprocts is how, if the body wall is a rigid case, the body-wall muscles can compress the coelomic compartment to evert the polypide. Ectoprocts are colonial and small and, in most, each abuts with its fellows on four sides, leaving only one surface exposed. In some which are not so closely fused together the body wall is sufficiently flexible for conventional body-wall muscles to produce sufficient distortion to cause polypide eversion, but in the others, various devices have been evolved to provide the greatest calcareous covering for the animal and at the same time leave a flexible diaphragm which can be drawn downwards by modified body-wall muscles to produce the necessary displacement of coelomic fluid.

Changes that have taken place in the remaining oligomerous animals also relate to the adoption of a sedentary life, but in them the structural modifications have been so radical that there is little apart from the few basic embryological features they have in common to suggest that the animals are even related to one another.

One group is only slightly changed. This is the Enteropneusta (Fig. 5.23) which, together with the Pterobranchia (Fig. 5.15), constitutes the phylum Hemichordata. Pterobranchs, as we have seen, have a typical oligomerous construction and live in a secreted tube-like shelter. Enteropneusts are worms that live in a semi-permanent burrow, and they have no lophophore. Instead, they have a series of gill slits in the walls of the pharynx and a respiratory current of water is drawn in through the mouth and out through the gill slits. They are mud and detritus feeders. This change of habit has naturally occasioned some change in the coelomic compartments. The loss of the lophophore renders its hydraulic system redundant and the corresponding division of the coelom almost disappears.

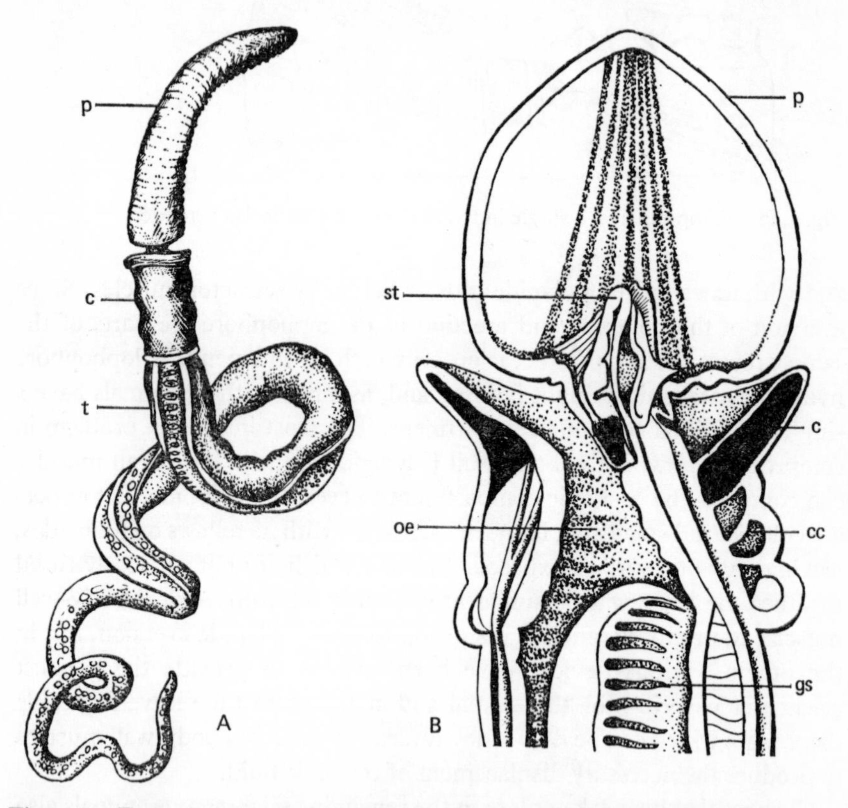

Fig. 5.23 Enteropneusta. A, *Balanoglossus*, whole animal; B, internal structures of head region. c, collar; cc, coelomic spaces in collar; gs, gill slits; oe, oesophagus; p, proboscis; st, stomochord; t, trunk

All that remains is a few ill-defined spaces in the collar from which, in pterobranchs, the lophophore arises. The pre-oral lobe remains a prehensile organ especially in those enteropneusts in which it is very long. It searches out the environment ahead of the animal, is constantly active and, since peristaltic locomotory waves are almost confined to it, acts as the chief locomotor organ. Its coelomic cavity is retained. The general perivisceral coelom is present but not very spacious in most enteropneusts and, as in the pterobranchs, cannot be regarded as of great significance.

The Echinodermata form a large and important phylum of such a peculiar structure that their precise relationships are an enigma. They are clearly related to the oligomerous worms since they develop three sets of coelomic compartments in precisely the same way as them during their embryonic development, but here all resemblance ends. Echinoderms have evidently had a long and complicated history. Presumably they were originally bilaterally symmetrical animals that adopted a sedentary life and as an adaptation to this became superficially radially symmetrical with a consequent disruption of their original bilateral symmetry. Now almost all have ceased to be sessile and some have even acquired a superficial bilateral symmetry once more. The result, as can be imagined, is near chaos and the animals can only be described as profoundly asymmetrical. With all these changes the coelomic compartments have long since lost whatever may have been their original function: those on the left side of the body fail to develop far and are for the most part insignificant, while those of the right side have acquired special functions.

The most remarkable feature of echinoderms is the water-vascular system. This is a system of coelomic canals which serve a circulatory and hydraulic function. It does not serve as a hydrostatic skeleton for the body-wall musculature, however. These are not soft-bodied animals and the body is protected by calcareous plates which are laid down in the thick connective tissue layer beneath the epidermis and severely restrict any change of shape. In the sea-urchins the plates are usually fused together to form a rigid test or shell completely investing the whole animal. Depending upon the degree of fusion of the plates and the immobility of the body wall, the body-wall muscles are reduced or absent. Blind ending branches from the water-vascular canals project through the body wall and serve either as food-collecting organs or as numerous little feet on which the animal creeps, as in starfishes. The main cavity inside the animal is also coelomic but it is quite separate from the water-vascular system and has no dynamic function except in the soft-bodied holothurians which have returned almost to a worm-like condition.

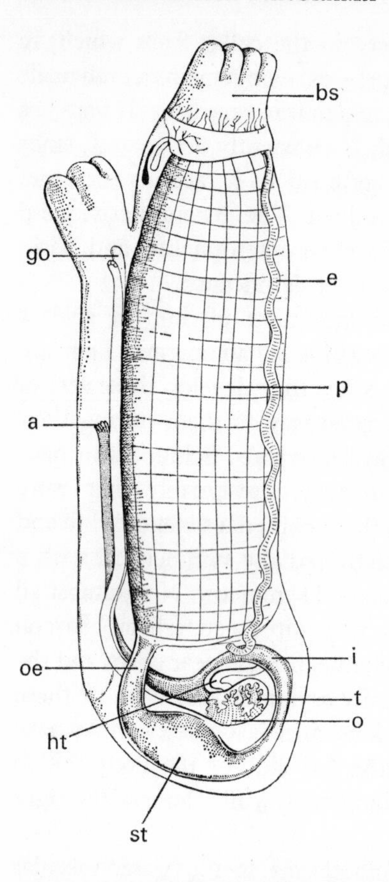

Fig. 5.24 Solitary ascidian.
a, anus; bs, buccal siphon;
e, endostyle; go, genital openings;
ht, heart; i, intestine; o, ovary; oe,
oesophagus; p, pharyngeal gill slits;
st, stomach; t, testis

The urochordates present an even greater enigma than echinoderms. They show a total adaptation to a sedentary life. Like enteropneusts, they have pharyngeal gill slits, but these are developed to an extraordinary degree. In the sea-squirts (Fig. 5.24), most of the animal is pharynx perforated by innumerable and complicated gill slits through which a current of water is passed and which filter out food particles. The viscera are a small mass at the base of the pharynx where the animal is attached to a rock. The salps are pelagic and in these the viscera are confined to the side of the body, the greater part of which forms a cylinder and contractions of the body-wall muscles drive a current of water through the cylinder so that the animal swims by jet propulsion.

In all these examples of a secondary loss of structure with a change of habit, the animals become simple in certain essential features of their morphology, but the reversal is never complete and they can rarely be regarded as simple animals. Vestiges of their former structure remain and are often pressed into service for new tasks. Understanding how the peculiarities of their structure have been derived presents a considerable problem which in many cases is still not completely solved, but at least we can be reasonably certain that we can make sense of the great variety of animal morphology and see it as variations on only a few themes.

Chapter 6

Structural gadgetry and functional adaptation

WE HAVE already seen that there are only a few basic structural plans on which animals are built and in each phylum it is possible to discern an underlying unity of structure that is not shared with members of other phyla. But basic though these structures may be, they are modified in a myriad of ways and their potentialities are exploited to the full with the result that some animals are so bizarre that it is difficult to tell to which phylum they belong and some are so odd that they scarcely look like animals at all. In addition to the uniqueness of their basic structural plan, many groups of animals have invented a special organ or structure and this, too, in its coming and going remains a characteristic feature of the group whatever form it may take. In all this variety it is safe to say that there are always good functional reasons why an animal is modified in the way it is, though in a good many instances this is an act of faith rather than a demonstrable fact simply because we do not know enough about the animal's biology to be able to understand its structure.

Molluscs

The molluscs illustrate as well as any group of animals the variations that can be played on a basic structural plan. Molluscs may well have originated from animals something like turbellarians, and their divergence as a distinct and separate phylum was undoubtedly associated with the adoption of a vegetarian diet. One of the characteristic and unique features of the Mollusca is the possession of a radula, a special rasp-like structure used for scraping algal growths from rocks; in all probability it was one of the first purely molluscan features to have been evolved. It consists of a ribbon armed with rows of chitinous teeth like the teeth of a file, and it can be rolled forwards out of the mouth and scraped across the substratum. With later, changing feeding habits, the radula is lost in some molluscs, but it is a structure no other animals have invented.

There are other, more far-reaching changes of structure that result from the change from a carnivorous to a herbivorous diet. Carnivores have a highly nutritious diet, herbivores do not and must therefore consume large quantities of food for which they need a large digestive system. Molluscs have a long intestine and other organs associated with digestion piled up on their backs and with such heavy loading of the ventral locomotory surface, the crawling forms move slowly. Not surprisingly, such succulent animals are protected by a hard calcareous shell and most of the features we particularly associate with molluscs are a consequence of the evolution of a shell. Such a bulky animal requires far more oxygen than can diffuse across the small part of the body wall outside the protective shell, consequently it has gills to increase the rate of oxygen uptake and a blood circulatory system to distribute the oxygen around the body. The delicate gills are protected by a fold of the body wall, the mantle, and for some inexplicable reason into this same mantle cavity the genital and excretory ducts and the rectum open. Here we have the archetypal mollusc with a foot on which it creeps, a head, and a visceral mass protected by a shell. Five basic plans have been evolved from this starting point and these are represented today in the five existing molluscan classes.

Primitive gastropods preserve much of the architecture of the supposed original mollusc, but the visceral mass has been rotated on the foot so that the mantle cavity is now at the front instead of the back. This rotation of the visceral mass can actually be observed taking place in some gastropod larvae. While having the mantle cavity at the front certainly provides a space into which the head can be tucked and withdrawn completely into the shell, we do not really understand the point of this morphological change and much of the subsequent evolutionary history of the Gastropoda seems to be concerned with undoing the effects of torsion. The chief consequence of torsion is that the nervous system is twisted into a figure of eight. One feature of the evolution of the nervous system in practically all animals is the concentration of the ganglia distributed around the body into one or two important nervous masses; this is very difficult when the nervous system is twisted. This is countered in two ways; either by rotating the visceral mass back to its original position, thus unravelling the nervous system which may then be concentrated into a large brain, or by simply concentrating the ganglia despite the loop in the nervous system and re-organizing the innervation of the various organs around the body. The former solution has been adopted in the beautiful shell-less nudibranchs; most of them have a posterior anus though the original gills and the mantle cavity have been obliterated in the process of detorsion. The sea-hare

Aplysia shows a half-way position with the rectum opening to the side and the nervous system partly unravelled. The pulmonate snails and slugs (including the common garden snail *Helix*) do not unwind; the mantle cavity and anus remain in their original anterior position and the nervous system of course remains twisted, though as the ganglia have fused together to form a nervous mass encircling the gullet, this is no longer obvious.

Sanitation in the mantle cavity is another problem for gastropods. Both faeces and excreta are expelled through the mantle cavity at the risk of fouling the gills. In some primitive gastropods the rectum is shortened and the anus is at the inner end of the mantle cavity, the mantle and shell are slit or perforated immediately above the anus, as in the keyhole limpets, so that the faeces and excreta are voided without passing down the length of the mantle cavity or approaching the gills. Most gastropods have adopted a more radical solution to this problem. One of the original pair of gills is lost so that it is possible to arrange an orderly circulation of water through the mantle – in at one side, through the gill, and out at the other side of the mantle. The anus opens into the current of water that has already passed the gill and is in the process of leaving the mantle. Often the rectum is extended to the edge of the mantle now that there is no danger of faeces contaminating the gills. Usually one of the kidneys is suppressed and the single kidney duct opens along with the rectum into the exhalant stream of water.

Far from losing one of the gills, bivalves show an enormous development of them and of the mantle chamber, for in these molluscs the gills are generally involved in food collection as well as in respiration. Water passing through the gills carries a quantity of suspended particles, these are trapped in the narrow passages in the gills and tend to clog them. Most molluscs go to some trouble to collect these trapped particles together and transport them by ciliary currents to the edge of the mantle, but these particles contain a substantial proportion of organic materials, the remains of dead plants and animals, as well as micro-organisms, and constitute a source of food. A few gastropods and the majority of bivalves exploit this food source by diverting the cleansing currents into the mouth so that particles removed from the gills are consumed instead of being rejected at the edge of the mantle. With this total commitment to a new method of feeding, bivalves have no use for a radula and it has disappeared, but a reliance upon particles suspended in the water as a food source has more conspicuous consequences for the morphology of these molluscs. Since they cannot fail to collect food so long as they breathe, they do not need to move in search of food and bivalves are sedentary animals. The foot lacks the flattened creeping surface that it has in gastropods and, instead, is diminutive or is used as a digging organ in those bivalves that burrow. There is also a great reduction of the head with its sense organs. Bivalves have not undergone the rotation of the visceral mass on the foot that is characteristic of gastropods and the chief developments in this class of molluscs relate to the increasing development of the gills, perfection of the particle-collecting and sorting mechanisms and, as in gastropods, modifications that result eventually in the complete separation of incoming and outgoing water currents.

The third major group of molluscs, the cephalopods, include the nautilus, squids, cuttlefishes and octopuses. Once again there has been a radical modification of the basic molluscan plan. The head and visceral mass are well developed but the foot is unrecognisable and has been transformed into a ring of tentacles around the head. At one stage in the past, cephalopods were extremely numerous, as the number of fossil ammonites and belemnites testifies, and they had an external shell. In modern cephalopods the shell is generally reduced or even lost. The nautilus has a large, coiled, external shell with many chambers, although the body of the animal is confined to the last of these. As the animal grows it shifts forwards in the shell and periodically lays down an additional calcareous diaphragm behind it, so creating one more chamber. In *Spirula*, the shell is still chambered but, as in other cuttlefish and squids, it is completely internal, and in most of them it is reduced in size and acquires a new function. The chambered shells of *Nautilus* and *Spirula* are gas-filled and act as buoyancy organs, but in the cuttlefish *Sepia* the shell is completely internal, the gas-filled chambers are much reduced and it acts as an internal skeleton. The cuttlebone of *Sepia* is often washed ashore on our beaches and is commonly used to cheer up the otherwise rather dull life of the domestic canary. The shell is even more reduced in the squids like *Loligo* and it is not calcified but is simply a horny quill. Octopuses lack a shell altogether, although tiny remnants of it serve as points of attachment for the retractor muscles of the head.

The cephalopod mantle is on the ventral side of the animal and contains one, or, in *Nautilus*, two pairs of gills and in other respects is quite conventional. Water is drawn in at the sides of the mantle opening and out through the centre of the opening which is drawn out into a tube or siphon. The muscles of the mantle are very powerful and a rapid contraction of them squirts water violently through the siphon. The siphon is quite flexible and can be directed in any direction so that, depending upon which way it is pointed when the mantle muscles contract, the animal is shot in the opposite direction by jet propulsion. This is often used as a defence mechanism, particularly when it is coupled with the injection of a cloud of pigment or 'ink' into the water that is expelled through the siphon, so that the animal is jetted violently backwards leaving behind a smokescreen. Squids have enormously developed giant nerve fibres that conduct nerve impulses to the muscles of the mantle very rapidly and allow the animals to react with great speed in an emergency.

Another remarkable feature of cephalopods is their large brain – the largest and most complicated of any invertebrates – and their large, image-forming eyes. Cephalopods are hunters and carnivores, they hunt by sight, grasping their prey with their tentacles. One other feature of cephalopods associated with their greatly developed brain is their advanced powers of learning which will be discussed in a later chapter.

There are three peaks of evolution among animals. Two of them are the higher insects and the mammals, the third is the Cephalopods. The remaining molluscs are a far cry from this pinnacle; they include the limpet-like chitons, solenogasters which superficially resemble worms, the tusk shells, and *Neopilina*, a living fossil discovered in the great depths of the ocean as recently as 1952. All of these are regarded as quite distinct derivatives from the primitive molluscan ancestor. Perhaps the most extraordinary is *Neopilina* for this mollusc, which looks something like a limpet, has a large coelom and some signs of segmentation that are quite unknown in other molluscs. Chitons also show vague signs of segmentation in some structures. They are rather like elongated limpets though they have never undergone torsion and the anus is in its proper, posterior position. They have a shell composed of eight plates in a longitudinal series and if dislodged from the rocks (they cannot cling to rocks nearly as effectively as a limpet) many of them roll up in a ball like a hedgehog or woodlouse. Naturally, the muscles of the shell plates are arranged in eight divisions and this implies some repetition of other organs in the body. The mantle cavity encircles the foot around which the mantle is draped like a skirt.

Anatomically, *Neopilina* bears some resemblance to chitons, save that the shell is in one piece. The anus is posterior and the mantle cavity encircles the foot as in chitons, but the five pairs of gills are exactly comparable to those of other molluscs, unlike chiton gills which appear to be secondary structures and often developed in a very irregular fashion. The muscles that retract the foot in *Neopilina* are arranged in eight pairs and the nervous system bears a strict relationship to the muscles. Gills, kidneys, reproductive organs, muscles and nervous system are all repetitive and might indicate that molluscs have evolved from completely segmented animals such as annelids. If this is true, it makes nonsense of most of our present views of molluscan evolution, but it must be remembered that detailed knowledge of the internal structure of *Neopilina* is based on an examination of very few specimens, and we know nothing of its embryological development and little of its habits of life. We know very well from other animals that it is essential to understand how and why an animal is constructed in relation to its physical environment before jumping to conclusions about its position in any evolutionary picture, and in our ignorance about *Neopilina*, it is early to discard our concept of the archetypal mollusc. At the same time, the recent discovery of living monoplacophorans – the group to which *Neopilina* belongs – which we had supposed were safely extinct by the end of the Devonian period some 300 million years ago, is a potential time-bomb for malacologists and we have certainly not heard the last of it.

Although the evolutionary origins of the two remaining molluscan groups are uncertain, their structural adaptations do not have far-reaching implications as they do in *Neopilina*. The Scaphopoda are burrowing molluscs with a single tubular shell, usually shaped like an elephant's tusk (hence the popular name 'tusk-shells') and open at both ends. The animal lives head downwards in the sand or mud of the sea bed generally in deep water. It feeds on microscopic organisms like foraminiferans which it captures with long thread-like tentacles developed on the head. As in bivalves to which they may be distantly related, scaphopods have a diminutive head which is little more than an extensible tube with a mouth at the tip. The foot, which is used as a burrowing organ is naturally well developed and the mantle cavity extends the whole length of the animal but there are no gills. Water is drawn slowly into the mantle cavity by the action of cilia. Periodically the muscles of the mantle contract forcing the water out again through the lower opening by which it entered. Since this end of the animal is buried, the violent exhalation of water disturbs the substratum where the animal is feeding and must help to bring a further selection of micro-organisms within range of its tentacles.

The final group of molluscs, the Aplacophora, look rather like short worms and are hardly recognisable as molluscs. They have no shell and the only reminder of the ability of molluscs to secrete calcareous structures is the presence of small spines of a calcareous material embedded in the body wall. In some, the foot is a shallow ridge inside a groove on the ventral surface, which is all that remains of the mantle cavity and in *Chaetoderma* both these vestiges have disappeared. Most have no gills and in the few that possess them, they are new structures and unlike the gills of other molluscs. These are rather rare animals; there are only about two hundred species in the group and these are not often encountered. Some burrow in sand in the sea bed where they feed on worms and other small invertebrates, others coil around corals and hydroid coelenterates.

Arthropods

There are more species of molluscs living in the world than of any other phylum with a single exception – the Arthropoda. The most notable feature of molluscs is the extreme plasticity of their body architecture. The half-dozen basic architectural plans can, with some stretch of the imagination, be derived from a single hypothetical archaic mollusc, but it must be remembered that there is no evidence that such an ancestor ever existed and, to a considerable degree, this supposed archetype is little more than the lowest common denominator of modern molluscs. The situation in arthropods is rather different and their amazing diversity and success is all the more remarkable when set alongside that of molluscs for the fact that they are all built on the same plan. There is even a strong suspicion that this plan has been acquired in more than one way from different pre-arthropodan animals, and, if true, this is further testimony to the enormous value of the arthropodan type of body architecture.

Arthropods are metamerically segmented animals with a pair of appendages on each segment and a hardened cuticle which, as we have already observed, permits them to make extensive use of true skeletal muscles. Because of the use of a rigid instead of a hydrostatic skeleton, movements of the limbs can be quite independent of any changes going on in the segment to which they belong and, depending upon the complexity of their muscle systems, and the various directions of hinging of the joints, can have great freedom of movement. In its simplest form, each appendage is a flattened, bilobed paddle used in swimming. We find this situation in the little brine shrimp *Artemia*, and the swimmerets of the abdominal segments of the common prawns and shrimps are also of this type. But this kind of appendage is not useful for creeping over the ground, and running consistently through arthropodan history as far back as the trilobites near the start of the fossil record there is a development of a more complicated bilobed appendage with the inner branch forming a jointed leg. The outer branch may be modified in a variety of ways: it seems to have been a gill in trilobites, but in modern mysid shrimps and the euphausid krill it is twirled around as a propeller and used in swimming, while it is lost altogether or severely reduced in many higher arthropods. The form of the jointed limb can be changed endlessly and the appendages of different segments are often modified in different ways. In the lobsters, the limbs of the last segment are broad and flattened and contribute to a tail fan, the remainder of the abdominal segments carry flat, bilobed swimmerets, then in the thoracic part of the body there are four pairs of walking legs and a fifth pair at the front of the series modified as giant claws. In front of the claws other appendages are modified as antennae, jaws and food-manipulating structures.

The more primitive arthropods have many segments, but throughout the phylum there is a tendency for the number of appendages to be reduced and also for adjacent segments to become fused together. In all arthropods three segments have shifted forwards in front of the mouth and these, together with a non-segmental structure corresponding to the annelid prostomium and a variable number of segments behind the mouth are all fused together to form a head. The appendages of these head segments are variously modified as sense organs or jaws though some make only a transitory appearance during embryonic development and are not represented in the adult animal at all. The appendages of the first segment and, in insects and myriapods, of the third segment also, are of this type. Crustaceans, insects and myriapods show comparable developments: the appendages of the second segment are antennae (or in crustaceans, those of the third segment also), while those of the fourth, fifth and sixth segments are jaws of various types. Trilobites had antennae on the second segment, but all succeeding segments bore bilobed limbs, though the more anterior of these near the mouth must have served as jaws in some way. Arachnids (spiders and allied animals) have a different structure from other arthropods. In place of antennae on the second segment they have a pair of prehensile organs, the chelicerae, and on the third segment there are a pair of appendages, the pedipalps, which take various forms in different arachnids but which also are primitively prehensile. The succeeding segments bear legs and, unlike all other living arthropods, no segmental appendages are specifically set aside as jaws. Instead, as in trilobites, the basal part of a number of limbs near the mouth is modified so that they are able to crush food material. This peculiarity of arachnid structure is, of course, correlated with the fact that the great majority take their food in a partially digested, liquid form.

The most important jaws are the mandibles, the appendages of the fourth segment. They show subtle variations of structure which are of some significance in deciding the relationships between different arthropods.

Crustacean mandibles have a small jointed appendage attached to their outer edge, those of insects and myriapods do not. This mandibular appendage, insignificant at first sight, is in fact the remains of a jointed limb and it indicates that, as in the arachnids, only the basal part of this limb functions as a jaw, the rest is very much reduced and may almost disappear altogether. In the other arthropods the whole appendage is transformed into a mandible; it is usually still jointed in myriapods but is fused into a single piece in insects. This fundamental difference in so basic a structure as the mandible alone is sufficient to suggest that there must be a very ancient and profound cleavage within the Arthropoda and the several branches of the phylum have undergone separate, if sometimes parallel developments. One example of these parallel developments is the way the jaws work. They may execute a rolling movement in the same way as the limb bases move during normal locomotion or they may make transverse biting movements. The type of biting movement does not depend on the fundamental structure of the mandibles and evidently the same kinds of modification have been taking place in jaws that originated in different ways.

In most groups of arthropods the walking legs show parallel development for there is a widespread tendency for them to become reduced in number. An animal like a millipede with a very large number of legs can generate a considerable traction. Millipedes are burrowers in loose soil and crevices and are obliged to develop strong locomotory forces, but animals moving on the surface have no such needs; speed is more important than power. Fast movements are very difficult for an animal with many legs because it cannot take long strides without tripping over itself. With fewer, more widely spaced legs, each leg can take a longer stride (the legs in fact can be longer) and faster movements are possible. This has happened in the chilopods which otherwise superficially resemble millipedes. In the fastest moving chilopods, like *Scutigera* which hunts insects, there are only fourteen pairs of legs and these are of different lengths so that each steps outside the one in front. Similar tendencies can be seen in the Crustacea. The more primitive swimming crustaceans have numerous appendages, but walking species have far fewer. Isopods, like woodlice and the shore slater *Ligia*, have only seven pairs of legs. Lobsters, crayfishes and most crabs have reduced the number to four pairs, and spider crabs and prawns use only three pairs of legs for walking. Like most of the higher crustaceans, most arachnids use only three or four pairs of legs for walking and insects have only three pairs. In all these walking arthropods the legs are long, so giving an increased stride, and they project at the sides of the body rather than being directly underneath it. In all of them, too, the successive legs are of different lengths so they do not foul each other.

Despite the fact that vertebrates manage with only four legs, or sometimes walk on only two, arthropods seem to have come to the conclusion that for fast running on land, six legs are the irreducible minimum. One obvious advantage of this system is that it is possible for two legs to be removed from the ground at once, and still leave the body firmly supported on four others. When a quadruped walks, as soon as one foot is removed from the ground, the animal's centre of gravity must be shifted so that it falls within the tripod formed by the other three legs. Constant adjustment of posture is essential while the animal is walking and even more when it is running and two legs (or in a gallop, all four) are off the ground simultaneously. The degree of flexibility of the body needed for this has not been developed by arthropods for reasons that we shall see in a moment; even so, it is still within their capabilities to walk after the loss of one or two legs, but like a three-legged mammal which can limp along, their efficiency is reduced.

The fact that running over the surface is best done with three pairs of long legs has important consequences for arthropod structure. It frees other appendages and the segments to which they belong for other purposes. Particularly when only three or four pairs of legs are used in locomotion, it is essential to provide a firm base for them and the muscles that move them, and the leg-bearing segments are often more or less fused together to form a thorax. Sometimes the thoracic segments are also fused on to the head segments so that the whole anterior part of the body is in a single, solid piece, as it is in crabs and lobsters. The lack of joints between these segments besides providing rigidity also allows a protective shield to be developed – the carapace – which encases at least the head and thorax and may even provide protection for the whole body. Since the segments behind those bearing legs are of secondary importance they are often highly modified or even reduced almost to vestige. In a prawn, which swims as well as walks, the 'abdomen' is large and the abdominal segments bear fully formed, flattened appendages which are used as swimmerets, but in crabs which are designed as walking animals, the abdomen is reduced to a small triangular flap tucked under the rest of the body. In scorpions the abdominal segments are retained and form a tail which carries a sting and poison gland, but most other arachnids show a more or less severe reduction of the tail in precisely the same way as in higher crustaceans. Indeed, in the harvest men, or opilionids, all the segments of the body are fused together to make a globular body and all external signs of segmentation are lost.

Polymorphism and the coelenterates

Coelenterates have exploited the potentialities of their structure in yet another way. In the molluscs several different types of body plan have been evolved from some early molluscan ancestor and in the existing classes of Mollusca we can see how the changes have been rung on each of these types of architecture in turn. In arthropods there is essentially only one basic segmental plan which has lent itself to almost endless modification, although underlying the diversity of these animals the fundamental segmental organization can almost always be detected. Coelenterates have three types of organization: a globular form in the phylum Ctenophora and two others in the Cnidaria. These are the medusa and the polyp, which we shall explain later, but the point of interest in the cnidarian coelenterates is that both structural forms run throughout the phylum instead of representing types of organization that have been developed in one class or another as we have seen in the molluscs. Often, indeed, a single coelenterate species exists in both forms, either successively at different stages in its life history, or simultaneously, when both forms occur side by side in the same colony. When a species exists in more than one form it is said to be polymorphic, and polymorphism is a surprisingly widespread phenomenon found in most groups of animals. Usually the differences between polymorphs of a species are rather trivial and nowhere else in the animal kingdom has it been exploited as it has in the cnidarian coelenterates.

The coelenterate polyp is more or less cylindrical, with a mouth surrounded by tentacles at one end and the base either attached to the substratum as in *Hydra* or a sea-anemone, or, in the colonial hydroids, extending into a stalk by which the polyp is connected with other members of the colony. The medusa is a free-living, swimming form. It takes the form of a bell or umbrella with the mouth at the end of a process, the manubrium, hanging from the centre of the concave surface like the clapper of a bell. The tentacles of the medusa fringe the rim of the bell and the animal swims by pulsations which expel water through the mouth of the bell. In both the medusa and the polyp the animal's mouth leads into a single central cavity. This may be undivided, as in hydroid polyps, or partially subdivided by a series of radial septa or mesenteries producing a number of compartments opening off the central cavity, as in anthozoan polyps. In the more primitive medusae there may be four gastric pouches opening off the central 'stomach', but more often there are radial canals running from it to the edge of the bell where they join a circular canal in the rim. Fully developed, medusae look very different from polyps, but stripped of their specializations both can be seen to be easily derived from a hollow ball with walls composed of two cell-layers separated by the mesogloea and a single opening to the exterior, the mouth. Such an organism is found in an early stage of the development of most coelenterates from the egg, and a very young sea-anemone that has not yet become attached to the substratum is able to stretch its mouth so wide when feeding that the polyp becomes almost a flat disc looking remarkably like a medusa.

Medusoid and polypoid forms have a different significance and play different roles in the various groups of coelenterates. In the Anthozoa, which include sea-anemones and reef-building corals, the medusoid has been completely lost, but both forms exist in the other classes. In the scyphozoan jelly-fishes, the medusa is dominant, is usually large, and has undergone great elaboration, but almost all scyphozoans pass through a diminutive polypoid phase which may live for some months multiplying vegetatively before budding off a number of small medusae. This process of strobilation, a form of asexual reproduction, will be considered again in a later chapter. In one group of scyphozoans, the Stauromedusae, the medusa remains attached to the substratum by a stalk, sitting mouth upward and resembling very closely polyps in their habit. The relationship between polyp and medusa is much more complicated in the Hydrozoa. In some, such as *Obelia*, there is a regular alternation of polypoid and medusoid phases during the life-cycle, much as in the Scyphozoa. The polyp buds off medusae which have gonads and produce eggs and these develop into polyps, but in other hydrozoans, one phase may be emphasized at the expense of the other. In many polyps there is a tendency to suppress the medusa. In *Campanularia* a medusa is formed but it never becomes detached and in *Tubularia* and *Hydractinia* what may be regarded as an attached medusa has become a very reduced structure indeed, while in *Hydra* there is no vestige of the medusoid generation and the polyp itself is capable of producing eggs and spermatozoa. Medusae can also be produced by asexual budding of other medusae, from the sides of the manubrium or from the bell margin. The situation is complicated and disorderly, however. Closely related hydroids may reproduce in quite different ways and a single species may have alternative methods of reproduction: the medusa of *Rathkea* has gonads and produces eggs which develop normally, but it also is noted for the habit of budding medusae from the manubrium. Further complication is introduced by the fact that although many hydroid polyps and medusae have been described and named, we do not always know which polyp belongs to which medusa and we are left with a double system of nomenclature, one for medusae and another for polyps. Thus the little fresh-water jelly-fish *Craspedacusta* is the medusa of *Microhydra*. As investigations proceed more medusa will become identified with their polyps and perhaps in due course the system of naming these animals will become more rational.

In the colonial Hydrozoa, polymorphism is not restricted to the appearance of alternative medusoid and polypoid forms, for the polyp may be specialized in a variety of ways to perform different functions in the colony. The commonest type of polyp and, of course, the only one in solitary hydroids such as *Corymorpha*, the giant deep-sea *Branchiocerianthus* and, in fresh-water, *Hydra*, is a feeding polyp with tentacles, mouth and gastric cavity. Many colonies include a second type of polyp without mouth or tentacles, which is specialized for the formation of medusae by budding. These are known as gonozooids and in many colonial species there is a third type of polyp modified for defence and food capture. It is without mouth or digestive cavity and generally also lacks tentacles. The precise form of these dactylozooids or tentaculozooids varies enormously. They are generally elongated and very extensible and are always provided with batteries of nematocysts, the characteristic stinging cells of coelenterates. Sometimes they have short, stubby tentacles, but more often they are without them and are so long that they look like tentacles themselves. In *Plumularia* these polyps are much smaller and cluster around the feeding polyps. They have club-shaped ends beset with nematocysts or adhesive cells, or both, and very similar polyps are found also in *Clathrozoon* colonies.

Specialization of polyps to perform different functions in the colony is possible because all its members are in organic continuity with one another. The stalks and stems from which the polyps arise are all hollow and the cavity in them communicates with the gastric cavity of the feeding polyps. Food collected by them is available to the whole colony. This feature of the organization of coelenterate colonies underlies polymorphism in the Siphonophora in which it has developed to an even greater degree than in other hydrozoans. Siphonophores are floating colonies containing both polypoid and medusoid forms. They are supported in the water by one or more pulsating bells or by a gas-filled float, and the remainder of the colony hangs beneath, either in a long strand or a compact whorl. The swimming bells take a great variety of forms in different species and are all modified medusae, lacking mouth, manubrium or tentacles. The gas-filled float, where it exists, was once thought to be another modification of the medusoid form, but, in fact, is derived from an apical invagination in the developing larva and therefore has a totally different nature. It contains a gas gland which secretes the gas, generally resembling air in its composition. Polyps are represented by the gastrozooid which is the feeding individual with a single, long tentacle arising near the base of the polyp instead of in the more usual position near the mouth. Other polyps, also with a single, long, basal tentacle are modified as defensive dactylozooids without a mouth but with batteries of stinging nematocysts. A third group of polyps is modified as gonozooids, budding medusae as in many other hydroids. The medusae, with their gonads, often remain attached to the gonozooid and so represent yet another member of the colony. They have no manubrium or mouth, and even if shed from the parent gonozooid exist simply to release the eggs or spermatozoa and have only a short life in the plankton, never feeding.

Siphonophore colonies vary quite widely in the arrangement of individuals in them. In the Calycophora, including such genera as *Muggiaea*, there are one or more swimming bells from which a long stem grows bearing clusters of individuals at intervals along it. Each cluster includes a gastrozooid and a few sexual medusoid individuals. Sometimes the latter also serve as additional swimming bells. The Physophorida have a small float generally in addition to a number of swimming bells arising from the stem growing downwards from it. Some, such as *Agalma* and *Forskalia*, have numerous bells, but in others the number of swimming bells is reduced and the float enlarged, and in *Anthophysa* and some other members of the Physonectae there are no bells at all. As in the Calycophora, the flotation and swimming region is followed by a series of clusters of individuals including gastrozooids, gonozooids and dactylozooids. Those with only a single large float, like the Portuguese man-of-war, have only a very short stem so that all the individuals are closely clustered beneath the float. They include a few extremely long trailing dactylozooids which encounter prey swimming some distance beneath the animal, smaller dactylozooids with tentacles, bunches of gastrozooids without tentacles, and numbers of gonozooids. Finally, in the Chondrophorae, we find even greater modification of the coelenterate colony. This group of siphonophores includes *Velella* which occurs in great numbers in the open oceans and is sometimes blown ashore in large numbers. Like the Portuguese man-of-war, it has a float and in *Velella* it is drawn up to form a sail by which the animal is blown along. *Velella* has only a single gastrozooid in the centre of the underside of the disc. This is surrounded by circles of gonozooids which, unusually, have a mouth, and the edge of the disc is encircled by dactylozooids.

Echinoderms

The echinoderms present a quite different picture from the groups of animals we have considered hitherto in this chapter. Whereas many phyla can be viewed as showing the development of more and more complicated structure by the elaboration of a simple ancestral type of organization, with echinoderms we are painfully aware that much of their evolutionary history, and perhaps the most significant part of it, was completed before the start of the fossil record and cannot now be reconstructed. To some extent this is true of most major groups of animals, but as a rule it is possible to deduce what must have gone before with some confidence; this is hardly possible with echinoderms. As we saw in the last chapter, they are basically asymmetrical animals with several unique features. Their asymmetry can be explained if it is supposed that they were originally bilaterally symmetrical animals that became attached to the substratum and, at a later stage in their evolution, resumed a free-living existence. Most of their history seems to consist of adjusting and changing a body architecture that had already been profoundly modified for a sessile existence to meet the requirements of a great range of other modes of life. This is perhaps the most surprising thing of all about echinoderms, for our experience with other animals is that the adoption of the sessile habitat is an evolutionary dead end which leads to such acute specialization and simplification of structure that further evolutionary advance is impossible.

There are several reasons for supposing that echinoderms originated from bilaterally symmetrical animals that settled on the substratum and became fixed. Many fossil echinoderms lived in this way and among existing forms, the crinoids, or sea-lilies, spend part if not all of their lives attached to the substratum by a stalk and the larvae of many asteroid starfishes adhere to the substratum by a sucker during their metamorphosis. Both crinoid and asteroid larvae when they attach themselves, do so at their anterior end and several of the peculiarities of echinoderm structure can be explained only if it is supposed that the original free-moving ancestors of echinoderms did so as well. At first sight it is a fantastic notion that an animal should settle on its head for this means that the mouth is close to the ground and must be shifted to the opposite end of the body to feed in this new position, with a consequent disruption of the whole body proportions and internal structure. There are several precedents for this, however, and very good reasons why this should be so. Once a sessile animal has become cemented to the substratum, there is no possibility of it moving again during its lifetime and every reason why it should select an appropriate place to settle with some care. The larvae of all rather immobile animals, and not simply those that become permanently attached to the substratum, spend some time exploring possible sites and even postpone settlement and metamorphosis until they have found a situation that meets their requirements. This testing of the substratum calls for the use of sense organs which are clustered at the anterior end of the animal close to its brain. Once an acceptable site has been found it is important for the animal to become attached to it as quickly as possible before it is swept along by water currents. Since the anterior end of the body is in contact with the substratum while it is being tested, this is where the adhesive organs or cement glands are most commonly found. The animal is now where it wants to be but has its mouth close to the ground. During subsequent growth the position of the mouth is readjusted, often with a considerable change in the structure of the animal. This is the situation in acorn barnacles and ectoprocts, and something of the sort appears to have happened in the Echinodermata. It is not necessary to suppose that the entire phylum evolved from an organism corresponding to a present-day larva, although this was once a widely held view. The same considerations apply to adult animals as much as to larvae and if it is true that echinoderms evolved from a pterobranch ancestor, there is an additional reason for expecting them to have become attached to the substratum at the anterior end, for in the pterobranchs, the chief locomotor organ is the prehensile pre-oral lobe and this is the most likely structure by which such an animal would have become attached.

Like their oligomerous ancestors, echinoderms have three pairs of coelomic compartments but these have become reorganized, presumably during the adjustment to the gut that accompanied the adoption of a sessile existence. The coeloms of the right-hand side of the body are very much reduced and those of the left-hand side are highly modified. Most importantly, they form a system of canals encircling the mouth and radiating out along the arms. These are the water-vascular system and the perihaemal system. The water-vascular system is a characteristic feature of all echinoderms and nothing like it is found in any other animals. With water circulating in this system of canals, the blood system becomes redundant and in most echinoderms it is severely reduced. One other characteristic feature of echinoderms is the development of an internal calcareous skeleton of overlapping or abutting plates in the connective tissue beneath the epidermis. Often the skeletal plates are more or less completely fused together. This must originally have been a protective device for the sessile echinoderms, but while it has been retained in most unattached forms, it has called for special adaptations to overcome the problems of feeding, breathing and moving in these animals encased in an almost rigid armour.

Although we cannot be certain about some of the extinct echinoderms, these problems are generally solved in the living ones by the use of podia. These are thin-walled, finger-like processes extending from the water-vascular canals through the epidermis to the exterior. They therefore carry extensions of the coelom through the body wall and since the coelomic epithelium lining them is ciliated, water is circulated in and out of them from the water-vascular system. Simple diverticula of this sort can serve as gills without much modification and possibly their original function was respiratory, but in modern echinoderms they have become multi-purpose organs. The walls of the podium contain muscle fibres so that it can be retracted and, if parts of the musculature contract independently, the podium can be bent in any direction. The podium is extended by hydrostatic pressure and coelomic fluid is forced into it by contraction of the internal bulb, the ampulla. In crinoids and brittle stars it is criss-crossed by muscle fibres which can isolate a part of the main canal and contract a short section of it, so inflating the corresponding podia. In sea-urchins, starfishes and sea-cucumbers the podia are generally used in locomotion and are capable of more powerful movements. Often they terminate in a sucker and by their directionally co-ordinated stepping movements transport the animal over the substratum. They can be dilated more powerfully than in the crinoids and brittle stars in which their chief role is feeding, and there are special anatomical developments to bring this about. The side branches of the radial water-vascular canal do not lead directly into the podia, but into a bladder or ampulla which communicates with the podium. The ampulla is muscular and when contracted forces water into the podium and distends it; there is a one-way valve to prevent fluid being driven back into the main water-vascular canal. This is by no means the limit of the functions of podia. Those around the mouth are almost always sensory although in a few sea-urchins and sea-cucumbers they are used in feeding and often podia in different parts of the body perform different tasks and are modified accordingly.

The major water-vascular canals are associated with what, in most echinoderms, is the main portion of the nervous system. In a starfish, for example, on the under surface of the arms there is a double row of podia on either side of an exposed nerve cord. These radial nerves join a circum-oral nerve ring corresponding with the circum-oral water-vascular canal. Lying between the radial nerve and the radial water-vascular canal there is a second, quite separate coelomic canal known as the hyponeural sinus and this is closely associated with a radial strand of 'haemal' tissue. In sea-urchins, sea-cucumbers and brittle stars, the radial nerve is protected by calcareous plates or has sunk beneath the epidermis to become internal, but in all echinoderms there is a regular association of nerve strand, water-vascular canal, coelomic sinus and blood vessel constituting what is known as the ambulacral complex. Such a consistent and characteristic association of structures cannot be fortuitous and at first sight it is puzzling that they should be linked together in this way. The answer is to be found in the Crinoidea and very probably in many fossil echinoderms.

Crinoids are stalked and generally sessile, though a few, like *Antedon* in European waters, break free of their stalk and are unattached throughout their adult lives. The viscera are confined to a small central capsule and five pairs of movable arms radiate from the upper surface of this. The ambulacrum is not primarily nervous but is a food groove flanked by simple podia which trap food with the help of mucus and stuff it into the food groove in which it is transported by cilia down the arm, across the small capsular body, into the mouth situated in the centre of its upper surface. In other echinoderms the podia have acquired other functions and different feeding methods have been devised, with the result that in the four other existing classes of echinoderms, all of which are free-living, the food groove has lost its original purpose. The nervous elements in it have become of increasing importance and in starfishes and sea-urchins constitute the most significant part of the nervous system. However, since at an early stage in echinoderm history podia and ambulacral grooves were functionally linked as parts of a food-collecting and transporting system, this association has survived in animals where there is much less obvious need for them to be linked together.

The oddities of echinoderm structure have imposed severe restrictions on the kinds of animal that have been developed in the phylum, although, as we have already observed, it is surprising enough that they should have been as successful and varied as they have been with a past history of a completely sessile habit. The modern echinoderms all show at least strong traces of a radial symmetry to remind us of their past. In most of them there are five ambulacra, though in modern crinoids the arms bifurcate, in a few ophiuroids the arms branch repeatedly (in the basket stars), and some asteroid starfish have large numbers of arms. The five modern classes of echinoderms are constructed in five different ways though, unlike the situation in the Mollusca, there is an unmistakable family likeness between them.

The greatest variation is found in the four unattached classes: the

Asteroidea (starfishes), Ophiuroidea (brittle-stars), Echinoidea (sea-urchins, heart-urchins and sand dollars) and Holothuroidea (sea-cucumbers). In asteroids the skeletal plates in the body wall are not fused together so that in some degree the arms are flexible. The entire body is hollow so that the gonads and digestive diverticula extend into the arms. Ophiuroids, also with arms, bear a much closer superficial resemblance to crinoids for the viscera are confined to the small central disc and the arms are solid. An interesting development in these brittle stars is the appearance of large internal calcareous pieces in the arms, which articulate upon one another like vertebrae and have strong muscles inserted upon them so that the long flexible arms can be used as locomotory organs in place of the podia which are used primarily in feeding and respiration. Echinoids and holothurians have a quite different appearance from other echinoderms and are also much more varied. They have no arms and the ambulacra run in the body wall. Sea-urchins are globular with the skeletal plates in the body wall more or less completely fused together. They are completely dependent upon the podia, or, in some cases, the movable spines covering the body, for movement. Many of them have become burrowing animals and this has been accompanied by a tendency for them to become flattened rather than globular. The heart-urchins which live in a permanent burrow in sand or mud are somewhat compressed, but the greatest development in this direction is in the sand dollars which are the size and shape of a flat, round biscuit. They feed at the surface of the sand, but migrate up and down, burying themselves when not feeding. Animals with a globular shape would clearly be ill-adapted for such behaviour, but flattened, they burrow edge-on and can disappear into the sand surprisingly quickly.

Holothurians show the greatest variety of all the echinoderm classes and show the greatest emancipation from the consequences of their sessile past. The skeletal plates are reduced to small spines and spicules and the body wall is leathery. They have thus regained a flexibility which other echinoderms lack and in general appearance and sometimes in habit they are like fat worms. The podia can be supplemented or sometimes replaced by the body-wall musculature for locomotion and the animals can crawl on the surface of rocks, burrow into sand or, in one order, the Elasipoda, are pelagic. The least modified forms have five ambulacra with five double rows of podia arranged symmetrically and running the length of the body, but some creeping forms move only on one side of the body which becomes flattened like the sole of a gastropod's foot and only the three rows of podia that can be accommodated on it are retained. In *Holothuria*, the podia are confined to this ventral creeping surface and they are distributed over the whole of it. In the synaptid holothurians which are long, worm-like burrowers, the podia have vanished altogether and the animal moves by peristaltic contractions of the body-wall muscles in exactly the same way as a coelomate worm. Another unusual development in many holothurians is the evolution of respiratory trees. These consist of a series of long tubes extending throughout the body between the viscera and arising from the rectum. Water is alternately drawn into them and expelled, and the apparatus serves as a water lung. Most burrowing holothurians must therefore live in U-shaped tubes so that the anus can draw in fresh water, but the synaptids have no respiratory trees, relying instead upon oxygen diffusion across the tentacles around the mouth and across the body wall, and they are not restricted to U-shaped burrows.

Radiative evolution

The evolutionary history of the animal kingdom is essentially an account of the development of new modifications of structure and physiology that permit the animals to exploit new environmental circumstances. Even at the level of major groups of animals – even at the level of groups of phyla – it seems likely that the appearance of new types of body architecture was related to the adoption of new ways of life in a very general way, as, for example, the adoption of a burrowing habit, or creeping over a hard substratum, or living in a tube. Once the major structural types had become established, their subsequent evolution has consisted of exploiting the potentialities of the structure they have inherited. Some types of structure have proved immensely successful because of their great plasticity, as in the Mollusca. Other types, like that of the Arthropoda, have proved to be full of inherent potentialities without any radical departure from the basic structural plan. All types of structure have their ultimate limitations, however, and this is nowhere more clearly seen than in the coelenterates and echinoderms where one is surprised at the success and variety that has been achieved from what, at first sight, can only be regarded as very unpromising material. The old view that evolution is inevitably progressive must be revised. Some modifications of animals, successful in the short term, must be regarded as mistakes in the long run. Much of the history of gastropods, for example, appears to be concerned with undoing the effects of the torsion of the visceral mass upon the foot, and modern echinoderms are notable for their return to a mobile life after becoming adapted to a sedentary existence.

These changes are long-term trends, and it is easy to regard evolution as a steady onward progression. In fact it now seems likely that evolution is much better regarded as radiative. From a single starting point animals branch out in a number of directions like the spokes of a wheel, to exploit various features of their environment. These changes, of course, demand appropriate modifications of the animals to fit them for their new life. So far we have considered radiative evolution only as it affects major structural features and major groups of animals. This process continues in finer and finer detail down to the level of individual species and much of this will become apparent in the following chapters.

Chapter 7

The provenance of the invertebrates

INVERTEBRATES are found almost everywhere that life of any sort can be supported: crustaceans and worms in temporary pools of ice-melt in Antarctica or on glaciers, spiders among the permanent snows of high mountains, shrimps in brine pools, and nematode worms living as parasites in virtually all plants and animals, as well as living free in the soil, the sea or in fresh water. In all these environments, whether they are commonplace or very peculiar, the animals are to some extent specialists, and the odder the environment the more specialist the animals in it must be in order to make a living. This is a reflexion of one of the most important principles of zoology, the principle of adaptation. Animals are constructed to do particular jobs and this is true even to their most detailed features. Furthermore, these adaptations affect all aspects of the animal, not merely its structure: how it breathes, collects and digests its food, breeds, and so on. The most usual way of viewing adaptation is from the standpoint of the animal, that is considering how an animal is suited to its environment and way of life, but equally we may view it from the standpoint of the environment and see what kinds of adaptations must be made by any animal that lives in it.

Major Divisions of the Environment

It now seems certain that the earliest life developed in the sea and that the first animals lived there. The most fundamental living processes in all animals can still only take place under conditions similar to those found in the sea and life away from the sea presents immediate difficulties, for somehow the animal must provide its cells and tissues with conditions suitable for the vital processes to go on even though the external environment is quite different.

The first problem is water. Sea water is a 3–3·5 per cent salt solution and the blood and body fluids of almost all animals have something approaching the same salt concentration. Even the hardest fresh water has only a small fraction of this salt concentration and animals living in it are in constant danger of a fatal dilution of their body fluids by water diffusing into the body and salts diffusing out. On land, animals have the converse problem of preventing the concentration of their body fluids by water loss through evaporation. To some extent undesirable water loss or gain can be mitigated by rendering the body surface impermeable, but no animal can seal itself off completely from its environment, at least, not permanently. The animal must breathe, and since the oxygen molecule is larger than a water molecule, the respiratory surfaces where gases diffuse in and out of the body are also areas where uncontrolled water diffusion takes place.

The inevitable and continuous loss or uptake of water by terrestrial and fresh-water animals entails a continuous regulation of the concentration of the body fluids and this is done chiefly by the excretory organs. Terrestrial animals conserve water by excreting a very concentrated urine, and fresh-water animals are able to bale out much of the water that diffuses into them by excreting copious quantities of very dilute urine.

Both methods of controlling the concentration of salt in the body fluids introduce complications. Fresh-water animals, although they can get rid of excess water via the kidneys, are still faced with the problem of salt conservation, for no animal can excrete pure water and there is a continuous, if small, salt loss in the urine. Salt loss from this and other sources may be made good from food (though this merely passes the problem of salt accumulation to the food organism), but a more significant method is probably the uptake of salt from the water by special salt-concentrating cells, generally in the gills. At least, this has been demonstrated in some brackish-water crustaceans as well as in mosquito larvae and several fishes and is likely to occur in many brackish-water and fresh-water invertebrates.

Animals that have successfully adapted to terrestrial conditions are able to produce concentrated urine and, often, very dry faeces, so reducing water loss from these sources, but concentration of the urine is dangerous and entails further adaptations. Most aquatic animals excrete a high proportion of ammonia in the urine; it is a very soluble and convenient excretory product, but it is also very toxic. If the urine is concentrated there is a danger that the concentration of ammonia in the body will reach such a level as to poison the animal. This is avoided in animals that suffer a water shortage by converting some, at least, of the ammonia to less poisonous substances such as urea (in most terrestrial animals), uric acid (in the garden snail *Helix*), or a related substance, guanine (in spiders).

However, these are all a good deal less soluble than ammonia so that although by excreting them the animal avoids the danger of poisoning itself, it is now faced with the danger that the urea or uric acid will crystallize in the tiny kidney ducts and block them. As a safeguard, concentration of the urine does not take place in these ducts, but is postponed to the last minute. Very often the hind part of the intestine, the rectum, is responsible for removing water from the faeces, and often it receives the kidney ducts also and removes water from the urine along with the faeces at a site where there is no danger of clogging narrow ducts. This appears to be the ultimate adaptation and is found in all the more successful terrestrial invertebrates: insects, spiders and some tropical earthworms.

To a much greater extent than their fresh-water relatives, terrestrial animals are able, or perhaps obliged, to avoid the problems of uncontrolled water movements. This is done in a variety of ways. Relative impermeability of the body surface is almost universal, by the development of a thickened cuticle, as in land arthropods, by enclosing the greater part of the body in a protective shell, as in snails, or by enveloping the body in mucus which if it does not prevent transpiration at least reduces it to a low rate, as in many worms or slugs. The very much greater availability of oxygen on land than in water helps. Air contains about 20 per cent oxygen whereas air-saturated water no more than 0·7 per cent, and exposed, thin-walled gills, which are the principal sites of uncontrolled water diffusion in aquatic animals but are necessary for respiration, are lacking in terrestrial animals.

Animals that do not have these advanced anatomical and physiological specializations may still live on land providing they confine their activities to habitats where the humidity is high. Thus the few land planarians and nemerteans, many slugs, and the onychophoran *Peripatus* live in forest litter. The more aquatic woodlice, *Asellus* and *Porcellio*, remain under stones or bark where the humidity is always high, and earthworms that are not particularly well adapted to terrestrial conditions are confined to moist soil. This is the case in the common earthworm of temperate regions, *Lumbricus terrestris*, which actually excretes quite dilute urine and must burrow deeper in summer as the surface soil dries out. A related species, *Allolobophora icterica*, of the south of France, aestivates, lying coiled up in a mucus-lined cavity in the soil, dormant, during the hot summer and resuming activities in the autumn after autumn rains have made the soil habitable for it. Not all earthworms are so much at the mercy of their environment, however, and the tropical earthworm *Pheretima* has a number of anatomical and physiological features that enable it to survive and remain active even under desert conditions.

The extremely low salt concentration in fresh water and desiccating conditions on land are not the only hazards of these environments; both are much less stable and more variable than the sea. Air and soil temperatures may vary seasonally from −30°C to +30°C and in deserts daily temperature fluctuations of 30°C are quite common. Water temperatures are more equable and rarely fluctuate beyond the range 0°–20°C, generally much less, and show only slight daily fluctuations. Small, shallow bodies of fresh water are particularly variable, they may freeze completely in winter and dry out in summer. They experience wide fluctuation in their chemical composition and, depending upon their flora, wide variations in their oxygen and carbon dioxide concentrations.

This instability and variability of non-marine environments makes them difficult for animals to colonize, especially if at some seasons of the year the character of the environment is radically changed. To be sure, none of these problems presents insuperable difficulties, but each calls for special adaptations. The extreme adaptation is an ability to produce dormant forms which enable the animal to survive periods when active life is insupportable in the environment. There are many instances of such dormancy, especially in small bodies of water. Fresh-water sponges form gemmules, resistant cysts containing a few cells from which the entire sponge can be reconstituted, during periods when ponds dry up. Rotifers overwinter as dormant eggs. Many snails seal the opening of the shell and become dormant in cold or dry weather.

The young stages of all animals are particularly vulnerable and the breeding biology of animals living away from the sea is almost always modified to protect the young from the rigours of the environment. Apart from the difficulty the very young stages have in regulating their water and salt balance until the necessary organs have developed, an additional hazard in fresh water is that rivers and streams and even the great majority of lakes flow eventually to the sea. To animals that are adapted to life in fresh water, sea water is lethal and the one-way flow of fresh water results in a steady transfer of floating animals from their proper environment towards the sea. Slow or weak swimmers are therefore confined to still waters, or must be attached to vegetation, or live in the substratum. The hazard is particularly great for larvae and a swimming or floating larval stage, so common in marine organisms, is almost completely abolished in their fresh-water relatives. On land the problems are different. Spermatozoa and eggs cannot simply be released into the surrounding medium, as they often are in marine animals. Special arrangements must be made for fertilization of the eggs. Furthermore, the developing embryo cannot absorb salts or even water from its environment so that all the raw materials for development must be provided in the egg, and to make matters worse, the young are unable to fend for themselves until they reach a fairly advanced stage of development because of the peculiarity of their environment. They must begin their independent life as miniature adults. The adaptations that answer these problems include some form of copulation or exchange of spermatozoa, fertilization of the eggs in the female genital tract or in a cocoon, and the eggs usually contain large quantities of yolk or by some other means are provided with food reserves adequate for a prolonged period of development. These adaptations have a profound effect on the anatomy of the reproductive system. Copulatory organs, special glands to secrete egg shells or cocoons, structures to receive the spermatozoa, and further, the development of more or less complicated breeding behaviour and courtship are all common features of land animals and even to some degree of animals in fresh waters.

Estuaries

The region where a river enters the sea and where there is a transition from fresh water to sea water is a specialized, difficult environment which few animals have mastered. Most fresh-water animals can withstand only slight exposure to salt water and although marine animals can generally withstand some dilution of the water in which they live, all have a fairly limited tolerance in this respect. There is, of course, a salinity gradient from the sea beyond the mouth of the estuary to fresh water above the head of the estuary so that marine animals penetrate some distance into the estuary from one end and fresh-water animals from the other. However, only a small number of species are able to survive and flourish in the middle reaches of the estuary and these are all physiologically specialized. The precise nature of the salinity gradient varies from one estuary to another depending upon the flow of fresh water, the shape of the estuary and the depth of the main channel. In narrow estuaries with a deep channel, a tongue of dense sea water extends some way up the channel into the less dense river water flowing out over it. Marine animals may then extend much further up the estuary (providing they live in the bottom of the channel) than they do in broad shallow estuaries where there is more complete mixing of salt and fresh water and the salinity is the same at all depths.

Adaptation to these brackish conditions might not be so difficult were it not for the fact that the position of this gradient moves up and down the estuary twice daily with the tide and seasonally, depending upon the flow of the river, so that animals living at a fixed point in the estuary are exposed to wildly fluctuating conditions. The tidal influence in estuaries affects animals swimming in the water much less than those on the substratum. As the tide rises, sea water enters the estuary and dams up the river, so shifting the gradient upstream but carrying with it the swimming animals which remain in the same body of water more or less continuously and so experience a constant, if reduced, salinity. They therefore have fairly constant conditions to adapt to and these adaptations are often quite precise. Three closely related species of the amphipod *Gammarus* inhabit different levels in estuaries and, indeed, in some rivers three sub-species of *G. zaddachi*, differing from one another in only the most minute way, follow one another in sequence over the few miles of an estuary, one gradually replacing the other in more saline water.

On the bottom, animals have to contend not only with fluctuating salinity but also with mud. Rivers in their lower reaches carry large amounts of fine silt which is precipitated to the bottom as the river water meets sea water. This heavy and continuous shower of sediment excludes many animals which would be smothered or have their feeding and respiratory organs clogged, and the fact that the only environment available, apart from the water itself, is fine, soft mud, is a further limitation on the variety of animals that can survive these conditions. In fact very few animals can tolerate the unfavourable conditions of the middle reaches of estuaries. Those that can, however, have little competition, abundant food, and so exist in enormous numbers.

Mid-estuarine mud banks on the Atlantic coast of Europe are dominated by a polychaete *Nereis diversicolor* and contain also a small oligochaete *Clitellio arenaria*, the bivalves *Macoma balthica* and *Scrobicularia plana*, large numbers of an amphipod *Corophium volutator*, and a gastropod *Hydrobia ulvae*. Often they contain no other animals. *Hydrobia* creeps on the surface of the mud, the others burrow into it and all six animals feed on the surface deposit on which there grows a film of bacteria and algae. Estuaries are generally extremely productive and there is a superabundance of food (even if the diet is unexciting); for this reason we have this rare example of several species living together and all exploiting the same food supply.

Downstream, the more resistant marine forms gradually appear: the shore crab *Carcinus*, a few more polychaetes, and if there are rocks, the acorn barnacle *Balanus improvisus*, mussels and limpets. Small members of these may even penetrate as far as the middle estuary region. Upstream, there is a gradual increase in the number and variety of fresh-water animals, first insects, particularly the larvae of dipterous flies, and then planarians, oligochaetes, gastropods and other members of the river fauna.

The sea-shore

The narrow strip of land that is periodically covered and uncovered by the tide at first sight appears to be a transitional zone between the land and the sea, just as estuaries are transitional between fresh water and the sea. In fact, the sea-shore is much more a rather odd marine environment and, apart from a few insects, arachnids and centipedes, almost all the animals living on the shore are essentially marine animals. Neither terrestrial nor fresh-water animals can withstand exposure to sea water but a surprisingly large number of marine animals is unharmed by temporary exposure to air. The beach at low tide is a scene of almost complete lifelessness, with the animals either concealed or inactive: they clamp themselves to rocks, like limpets, tightly close their shells, like barnacles and mussels, hide under stones or weed, like shore crabs, or withdraw into the sand like bivalves or worms. As soon as the beach is covered by the advancing tide, it springs to life and the animals emerge from their hiding places and resume their activities.

The sea-shore is an extraordinary microcosm. Within the space of a few yards, from extreme low-water mark where the beach is rarely exposed to the air, to the highest level reached only by spray and the splash of waves on very high tides, there is as great an ecological change as exists between the foot and the peak of a mountain. Each level on the beach experiences a different tidal regime. Many of the animals found on the lowest part of the shore are purely marine species exposed by an exceptionally low tide; they generally survive the experience but are not particularly adapted to shore life. The regular inhabitants of the upper or middle parts of the beach are specialists. Many are adapted only to a particular degree of exposure to the air and to one set of tidal conditions. Sometimes several closely related species may live on the same beach, each with its own specializations, and be arranged in a sequence from low to high water mark. Thus, three, or sometimes four species of the periwinkle *Littorina* are to be found zoned in this way on most rocky shores of Britain. The lowest part of the beach is colonized by the large edible winkle *Littorina littorea* living on bare rocks, and the brightly coloured *L. obtusata* on the weeds. *L. saxatilis* is confined to the upper part of the beach and overlaps with the tiny species *L. neritoides* which, if it is present, extends well above high water mark where it may be found in crevices in the cliffs. Both the last two species can withstand desiccation and breathe air for long periods, and in *L. neritoides* the gill chamber is even modified to form a simple lung. *L. saxatilis* is also adapted to its quasi-terrestrial life in another way by being viviparous and so not migrating to the lower part of the beach for breeding. Many other animals show a similar restriction to a particular section of the beach, and even a casual glance at most rocky shores will show that different tidal levels are populated by different animals.

Having said this, it must be added that the more closely any beach is examined the more complicated and confusing the zonation appears. This is because the precise distribution of a species depends upon many factors besides the length of time it is exposed by the falling tide. On shaded beaches, or even on shaded parts of a beach, animals can survive exposure much better than on sunny beaches and so extend further up the beach, and for the same reason the intertidal distribution of animals is different in regions where the lowest tides occur in the early morning and evening from those where the lowest tides are in the middle of the day. Animals also differ in their tolerance to wave action with the result that the fauna on one side of a boulder may be quite different from that on the other. Excessive wave action excludes seaweeds and the animals associated with them; the absence of weeds allows massive numbers of acorn barnacles to settle and colonize the rocks. In short, the average rocky beach is one of the most complicated and varied environments that exists and has a correspondingly large and varied fauna. Not surprisingly, it is an extremely difficult environment to analyse and understand.

The precise nature of a beach depends in the first place upon the rate of water movements, the geology of the area and the shape of the coastline. Soft rocks, such as chalk and sandstone, are eroded by wave action, and the boulders and stones gradually ground to powder on the beach and carried away in the sea. Hard limestones and granite are eroded much more slowly. Sediment in the water settles out as water currents become slower, and the slower the current, the finer the particles that are deposited. Hence sheltered bays and inlets have sandy or muddy beaches, and headlands or narrow channels through which fast currents run have no sediment at all but have beaches of bare rock and large boulders. Each type of beach presents different conditions and has its own special fauna.

Sandy beaches at first sight appear lifeless. The surface layer of sand is churned up and disturbed as waves run up the beach and, apart from a few shrimps living in the surf, few animals are able to live in this unstable part of the environment. Almost all the animals live beneath the surface, some burrowing and feeding there, others emerging at the surface to feed when the beach is covered at high tide. One essential feature of animals of sandy beaches, chiefly polychaetes and bivalves, is an ability to burrow. Worms burrow by peristalsis or some variant of it, the molluscs burrow with the aid of their inflatable foot which is particularly extensible in these species. A few animals, like the polychaete *Nephtys*, have no settled home, but the majority are sedentary and live in more or less permanent burrows or tubes, and often the surface of a sandy beach is peppered with holes of various size and dotted with worm tubes projecting above the surface testifying to the population lying beneath the surface, waiting for the returning tide.

In the more sheltered inlets, sands grade imperceptibly into muds as finer particles of sediment settle out of the water. New species replace the sand-dwellers, but conditions of life are not greatly different. In the quiet waters that are necessary for mud to be deposited there is virtually no wave action and little disturbance of the surface layer and the deposit includes dead plant and animal material in far greater quantity than is deposited on sandy beaches where water currents are faster. As a result, muddy beaches are densely populated by animals that can exploit this food source; the rich cockle beds are well known, but there are also other bivalves living in the mud and large numbers of worms in great variety as well as crabs and gastropods creeping on its surface.

Beaches exposed to severe wave action are rocky barren places. All but the largest boulders, and sometimes even those, are swept away, seaweeds are torn from the rocks and few animals can survive the constant pounding of the sea. But beaches with even a slight degree of protection from wave action are varied and richly populated. Seaweeds grow in profusion and provide shelter for many animals, and between and underneath the stones there are often patches of sand and mud which harbour animals normally associated with much more sheltered beaches. The animals living on or among the rocks are faced with a variety of environmental hazards while they are uncovered by the tide, some of which we have already mentioned: exposure to the sun which heats and dries them, heavy rain showers may dilute a pool of water in which they have taken refuge, and there is the mechanical action of the waves as the tide returns to dislodge and damage them. Only a few mobile animals remain exposed on the rocks while the tide is out. These are chiefly gastropods, winkles, limpets and dog whelks, all protected by a heavy shell and with a sucker-like foot by which they adhere to the rock. Most animals that can move do so and seek the shelter of weeds or, better, creep into crevices and under stones and rocks, and here, too, are to be found most of the delicate and unprotected attached animals, sponges, ascidians and hydroids. Almost the only attached animals living on the exposed rock surfaces are acorn barnacles and mussels, both of which are able to close their shells tightly and so seal themselves from the outside world during the inter-tidal period.

The oceans

Although the sea-shore is populated by marine animals, many with close relatives below the shore line, it gives us only a glimpse of the richness of life in the sea. A further diversity of environments exists in the oceans themselves, each with its own fauna showing appropriate adaptations to fit it for life there.

Since the majority of marine animals spend at least part of their lives in the plankton, often as eggs or larvae, it is perhaps appropriate to start there. In the surface waters of the sea wherever there is an adequate supply of inorganic nutrients, principally phosphates and nitrates, there is a rich microscopic plant life photosynthesizing, multiplying and dying. Small animals swimming and floating in these waters graze on the plants and larger animals feed on the smaller. All these animals must remain afloat and their many ways of doing so account for much of the diversity of planktonic organisms. Many, of course, swim, often simply by beating appendages, but even more frequently by cilia. Most larvae are ciliated but as they enlarge and become denser, the cilia become less effective for maintaining them in the rich surface waters and sometimes bizarre forms are developed by the elaboration of the ciliary tracts, as in molluscs after the formation of the shell and most echinoderm larvae as the calcareous skeleton begins to make its appearance. Gas-filled floats are used by some colonial siphonophore coelenterates such as the Portuguese man-of-war. Other siphonophores and medusoid coelenterates, including the scyphozoan jelly-fish swim by means of a pulsating bell. Most pelagic salps swim by jet propulsion with the contraction of the muscles encircling the barrel-shaped body squirting water through and out of the animal. Remaining in the surface waters may also be aided by the development of projections and extensions of the body which act as a parachute and reduce the rate of sinking.

Since the plants depend on sunlight in order to photosynthesize, vigorous plant growth is confined to the surface waters of the sea, extending at most to a depth of one hundred metres. This is the region of most intense production of the primary sources of food material on which all life in the ocean ultimately depends. The intense light at the surface of the sea is injurious to many animals, however, and often they migrate from the deeper waters to the surface at dusk, feed during the night and then return to the dimly lit waters where they spend the day. Diurnal migrations of this sort are particularly marked in pelagic crustaceans. Since many of them begin their return migration to deeper water while it is still dark, it is evident that their movements are not simply a response to the varying light intensity, indeed, in the laboratory they continue to migrate up and down at the appropriate times even in constant darkness or constant illumination. This suggests that the animals must have some sort of internal clock which regulates their movements, but beyond that we are still very ignorant of how these migrations are controlled.

Although plants can multiply and grow only in the illuminated region of the sea, this is not the only place where they are available as food. As the plants die, their remains sink towards the sea bed and since temperatures are low, bacterial decay is slow. This plant material supports a fauna living at intermediate depths and on the sea bed where all the animals feed on detritus or are predators. In intermediate waters, well below the planktonic community at the surface, there are often dense shoals of prawns, so dense, in fact, that they are recorded on ships' echo sounders as what used to be regarded as a mysterious 'deep reflecting layer' before their presence was known.

The sea bed extends from the shore line, sloping gently across the continental shelf to a depth of some two hundred metres, and then dropping sharply to the abyssal plain at a depth of about four thousand metres, though this 'plain' has peaks which may break the surface as oceanic islands, and canyons more than ten thousand metres deep. The shallow continental shelf has a varied character, particularly near the coast, with rock, gravel, sand and muddy substrata, and abundant food so that it supports a rich and varied fauna. In the deep still abyss there is only the finest sediment, an ooze of foraminiferan shells, fine mud or, in the greater part of the Pacific, a red clay which contains virtually no organic matter and is an almost lifeless desert. On the continental shelf conditions are not much different from those we can observe on many beaches, except that the animals are permanently covered by the sea and therefore need none of the special adaptations of intertidal animals. Conditions in the depths of the ocean are very different. Food is generally scarce and the water cold. Animals are much less abundant than in shallow water and their growth rate is slow. Many have special adaptations for living on or in the oozy mud and because the waters are completely still, delicate and fragile animals can live there. Hydroids, glass sponges and crinoids have very long stalks by which the animals anchor themselves in the mud for there are no solid objects to which they can be attached, and many of the crustaceans and pycnogonids have exceedingly long legs which presumably distribute the weight of the animal over a wide area and prevent it sinking into the mud.

Fresh waters

Many of the environmental conditions in the sea are found also in fresh waters and, apart from the fact that all fresh-water animals must regulate the salt concentration in their body fluids, their adaptations are not very much different from those of their marine relatives living in comparable situations. There are basically two types of fresh-water environment: rivers and streams with running water, and ponds and lakes where, although there may be wind-generated currents, there is no steady one-way flow of any significance. As a result of this, lakes may support a self-contained, self-supporting community in a way that is impossible in any part of a river or stream.

Rivers pass through a number of distinct stages from source to sea. Starting as mountain or hill streams, they are fast-flowing, turbulent, cold and contain no sediment in suspension and, as a rule, little dissolved salts. These torrential streams erode their banks and the stream bed so that as they descend from their source, they carry an increasing load of suspended material. Gravel and sand is deposited again as the river becomes less precipitous and the current decreases, and in succeeding stretches of the river finer particles fall to the bottom. Lowland rivers are broad, shallow, slow and muddy; they are also warmer than mountain streams and contain dissolved salts leached from the land through which the river has run. Finally, as the river reaches the sea, its waters are dammed up and travel very slowly indeed, mud and silt are deposited and the beginnings of the estuarine mud banks are formed.

Unlike the sea or in lakes, the greater part of river vegetation is rooted, growing from the banks or in shallow slow moving water, from the river bed. Shallow lakes may have an extensive rooted vegetation, but these tend to silt up and, in fact, are generally in the process of evolving into bogs and marshes. In deeper lakes rooted plants are confined to the edges and planktonic plants play a major role, as they do in the sea. Unless there is a great dearth of phytoplankton, lakes and ponds have a rich crop of planktonic animals in the spring and summer. These include many rotifers, *Daphnia* and other cladocerans, and copepods such as *Cyclops*. A few other invertebrates may appear in the plankton in small numbers, but the diversity of fresh-water plankton is much less than that in the sea and far fewer animals have planktonic larvae. Rivers are not completely devoid of plankton but it is confined to the middle reaches. There is none in the fast-flowing headwaters, but in most rivers of reasonable length diatoms, protozoans and rotifers are common in the slower moving lowland parts of the river, disappearing again near the river mouth. Because of the one-way flow of rivers, river plankton is continuously carried downstream and has no means of moving in the reverse direction. In other words, no region of a river has a permanent, self-sustaining plankton population; if the planktonic organisms multiply, the benefit is felt downstream, if they die they must be replaced by organisms flowing into that stretch of the river from further upstream. Ultimately, the river plankton must come from lakes and quiet backwaters and represents an irrevocable loss to the lake community. Despite this continuous drain on the lake plankton, lakes may still have a balanced community. The plankton is very seasonal and most of the organisms in it multiply at a very high rate and so there is great over-production. There is a correspondingly high mortality and many fishes, aquatic insects and other pelagic feeders live on the plankton. A small additional loss of plankton in the outflow of the lake therefore makes very little difference to its total economy. Furthermore, as the planktonic organisms breed during their passage downstream, their numbers observed in the river may be very much greater than the number leaving the lake and so give an exaggerated impression of the loss to the lake community.

Not all lakes have a rich crop of plankton. Some are extremely productive of life and are termed eutrophic lakes, others are almost sterile and are oligotrophic. Whether or not a lake is productive depends upon the distribution in it of the inorganic salts which are essential for the growth of the planktonic plants, and this depends, in turn, very largely upon the shape, size and geography of the lake. In summer, the surface water is heated by the sun, and, becoming less dense as a result, floats on the denser, cooler water below. As this process continues, the surface waters become warmer and warmer, and less and less dense until a lake of reasonable depth is divided quite abruptly into two separate bodies of water, an upper warmer layer and deep, cold layer. The slight wind currents set up in summer may stir up the surface water, but are insufficient to cause much mixing of the upper and lower water layers. The planktonic plants, upon which life in the lake ultimately depends, are confined to the warm surface water where there is sufficient light for photosynthesis. As they grow and multiply, they use up the phosphates, nitrates and other inorganic salts in the water, but when they die, or the animals that have consumed them die, their bodies sink to the bottom with the result that the nutrient salts accumulate in the deeper, cold water instead of being returned to the planktonic community at the surface. By the end of the summer, the surface waters have become depleted of salts and, naturally, the productivity of the lake declines. In autumn, the surface waters cool down, become denser again and the discontinuity between the upper and lower bodies of water disappears. Autumn storms may now churn up the whole lake, mix all the water and redistribute the salts so that a fresh crop of phytoplankton is possible next season. Sometimes, if there is sufficient daylight, there may even be a brief upsurge of the plankton before winter begins. This is the situation in a eutrophic lake where there is at least annual mixing of the water and consequently a rich annual crop of animals. Particularly in deep, narrow lakes, however, disturbances caused by storms may never affect the bottom with the result that mixing of the water is always incomplete and salts trapped in the bottom layers of water are never returned to the surface where photosynthesis is possible. Lakes of this type have almost no plankton and since the contribution of rooted vegetation to the total economy of the lake is small, there is very little primary production and few animals.

Rivers and lakes provide a multiplicity of environments for animals living in them although they are nothing like so varied in this respect as the sea. There are other obvious differences. Fresh waters are dominated by insects which are to be found playing an important if not overwhelming role in almost every situation on land or in fresh water but are almost unknown in the sea. The fresh-water invertebrate fauna is also much less varied, for many marine groups have no fresh-water representatives. There are no fresh-water cephalopods, echinoderms, protochordates or barnacles, and almost no sponges, bivalves or coelenterates. Those invertebrates which have established themselves in fresh water, however, have generally done so very successfully. There is only one family of fresh-water sponges, the Spongillidae, but these small, encrusting animals are widespread and common; there are few coelenterates in fresh water, but *Hydra* is one of the best known fresh-water invertebrates.

The low concentration of salts in fresh waters prevents animals from developing hard calcareous shells. The swan-mussel, *Anodonta*, has a horny rather than calcareous shell; the crayfish, *Astacus*, living in streams and brooks deposits calcium carbonate in its exoskeleton but to a much smaller extent than its marine relatives. These physiological differences apart, fresh-water animals live much the same lives as their marine relatives. *Anodonta* in the muddy bottom of lakes and slow rivers is a conventional filter-feeding bivalve, *Astacus* is not much different in its habits from a marine lobster. The fauna of comparable situations in the sea and in fresh water show similar adaptations even though the animals composing it are quite unrelated. Thus in fast-flowing streams there are fresh-water limpets, *Ancylus*, living on boulders and, although pulmonates, they have a conical streamlined shell and a foot modified as a sucker as in the much more primitive marine limpets. Many animals are concealed beneath stones in fast water currents. In fresh-water streams these include turbellarians, leeches and gammarid amphipods. Their marine counterparts might be turbellarians and amphipods, but the leeches would be replaced by polychaetes, gastropods and small crabs.

The land

The land provides as varied a range of environments as the sea, but so few groups of invertebrates have successfully adapted themselves to a terrestrial existence that for the majority, their representatives on land are confined to a handful of situations where conditions are humid and relatively equable. Even more than in fresh water the invertebrate fauna is dominated by insects and, apart from them, only spiders, myriapods, and a few other arachnids, nematodes, pulmonate gastropods and, to a lesser extent, earthworms have successfully mastered the problems of a terrestrial existence. As we have already observed, what makes the land a difficult environment to colonize is the shortage of water and the high rate of water evaporation, and the enormous and rapid fluctuations in temperature and other physical conditions.

For animals that are able to burrow, the soil provides some protection from the fluctuations of temperature and humidity besides providing an abundant food supply in the form of rotting vegetation. Measurements made in the United States have shown that when the ground is covered with snow, the soil temperature at a depth of six feet does not fall below 4°C although air temperatures may reach −25°C. Nearer the surface, of course, soil temperatures follow air temperatures more closely, but always with some time lag that tends to even out the more rapid and erratic fluctuations above ground. Dry soils in desert conditions and water-logged or compacted soils with poor aeration are colonized by few animals, but elsewhere the soil fauna is numerous. In favourable conditions there may be hundreds of thousands of earthworms and thousands of millions of soil nematodes to the acre, as well as many millipedes, tardigrades, isopods and other arthropods. Not all of these animals are permanent burrowers.

Many, including centipedes, some millipedes, woodlice and slugs, shelter in crevices, burrows and under stones to emerge and forage at the surface when conditions there are tolerable.

One of the most equable of terrestrial environments is in caves. In the inner recesses there may be no more than 1–2 °C difference between winter and summer temperatures and often the humidity is permanently very high. However, these conditions are found only where daylight does not penetrate and there are no green plants. Like the depths of the ocean, deep caves ultimately depend on organic material brought in from outside to sustain life, but there is very little of it in caves and although they provide a suitable physical environment for a variety of animals, they have a meagre fauna. A few isopods, millipedes and mites feed on such organic debris as exists, while net-spinning spiders catch flies and millipedes and also eat each other. Cave pools contain a few oligochaetes and amphipods, but with the relative humidity near 100 per cent, such animals are almost as much at home in air as in water.

The leaf litter on the floor of temperate and tropical forests supports the richest fauna of any terrestrial environment. Humidity is high, physical conditions, particularly within the tropics, are almost constant, and there is an abundance of food. Oligochaetes, slugs, isopods, millipedes, and a variety of arachnids are all common, and it is here that we find the least adapted of terrestrial animals: terrestrial planarians and nemerteans, and the onychophoran *Peripatus*. Fallen trees, once they have begun to rot, provide a very similar environment, and a similar collection of animals can be found under the loose bark or in the more rotten parts of these as in leaf litter.

Dry soils and drying conditions on land or vegetation are the province of air breathing arthropods and gastropods. Land gastropods have no gills and their mantle is converted to a lung. The greater part of a snail's body is protected against desiccation by its shell and in arduous conditions it can withdraw completely inside and seal off the opening with mucus. Most snails and slugs are restricted to humid environments but some remain active in full sunlight in quite arid conditions on sand dunes or in dry grasslands wherever there is vegetation. Evergreen forests are one of the few terrestrial environments where there is ample vegetation but is hardly colonized by snails. Extremely arid land near the sea in the tropics of the Pacific region are also populated by the robber crab *Birgus*. Although this is an air breather it is not completely emancipated from the sea, but returns to it for breeding. The only other invertebrates, apart from insects, that are successful in completely dry conditions are arachnids, spiders and scorpions, and some centipedes. Spiders and perhaps also mites are among the most versatile of land invertebrates and appear in almost as great a range of environments as those most successful animals, insects. Young spiders drifting on their long silken threads have even been collected in the aerial plankton more than one thousand feet above the ground and many are undoubtedly transported great distances.

The ecological niche and animal communities

Most animals show a high degree of adaptation to their environment as, indeed, they must if they are to survive. In order to live on a wave-swept boulder on the shore, the limpet must have a foot arranged as a sucker and a stream-lined shell to prevent it being detached and swept away and also a shell shaped in such a way that it can be pressed closely to the rock during the low tide period to prevent desiccation. It feeds by rasping encrusting algae growing on the rocks and for this purpose must have a specially designed rasping organ, the radula, to resist the constant wear it receives. By pursuing the whole biology of the limpet, it is possible to show that the details of its structural peculiarities, its physiology, behaviour, breeding habits are related to the particular environment in which it lives and no other animal has exactly the same adaptations or makes its living in exactly the same way.

The precise situation in which an animal lives is known as an ecological niche; for the Atlantic limpet, *Patella*, this is on rocks between tide marks, browsing on encrusting algae. On the Pacific coast of North America, where there are no *Patella*, the same niche is occupied by a different and more primitive limpet, *Acmaea*, and on tropical shores by a pulmonate *Siphonaria*. Generally, because of our ignorance, it is impossible to define very exactly the ecological situation of an animal and niches are often described in rather general terms, either in relation to the physical features of the environment or in relation to food. A habitat niche might therefore be a subtidal marine muddy substratum, wave-swept intertidal rocks, a tundra moss cushion, or the root mat of a forest. A food niche might be that of pelagic grazer of phytoplankton, large terrestrial carnivore, or mud eater. Both of these concepts are useful for they enable the ecologist to make valuable comparisons of the fauna of similar environments in different parts of the world or in different situations.

The acorn barnacle, *Balanus*, and the mussel, *Mytilus*, both filter food organisms from the water, both live attached to intertidal rocks at the same tidal level, both live closely packed together and dominate the situations in which they live. From an ecological point of view they are comparable animals apparently equally capable of occupying the same habitat niche and neighbouring beaches or even neighbouring parts of a beach may be dominated by *Mytilus* or *Balanus*.

Although *Balanus* and *Mytilus* apparently have comparable ecological requirements, they are not identical. *Mytilus*, for example, is better able to withstand sand-scouring than acorn barnacles so that rocks flanking sandy beaches are more likely to be populated by *Mytilus*, particularly at the lower tidal levels close to the sand substratum. Barnacles are also less able to compete with fucoid seaweeds for space; the constant movement of fronds over the rock face interferes with the young, settling barnacles and often an isolated *Fucus* plant in the middle of a dense patch of barnacles is surrounded by bare rock. It is clear, then, that a habitat niche describes the ecological conditions under which it is possible for an organism to live. Whether or not the organism is found in that niche on a particular beach may depend on a number of chance complicating factors as is evidently the case with *Balanus* and *Mytilus*, or be a reflection of the fact that the habitat niche has not been defined precisely enough and that one locality differs from another in a subtle way that is important to the animal but has not been appreciated by the ecologist.

Food niches, too, may be defined in a broad or a narrow way. Animals that filter fine particles from the water occupy a 'filter-feeding niche', but on closer examination it often appears that species living side by side exploit this food source in quite different ways. The sabellid polychaetes *Sabella* and *Branchiomma* may sometimes be found on the same beach, both have a crown of ciliated tentacles which collect food particles from the water, but in *Sabella* the tentacles are held stiffly upwards in a cone, a current of water is drawn through them and particles are filtered from the water current; the tentacles of *Branchiomma* are extended more or less flat on the substratum, it relies much more on particles that settle on it, and less on water currents and filtration. Similarly, with bivalve molluscs, the great majority of which are filter feeders, but some, like the cockle, *Cardium*, have short, wide, stiff siphons and draw in water from above the surface of the substratum, while others, like *Tellina*, have long, narrow siphons and collect particles that have already settled to the bottom.

Broadly defined, habitat niches and food niches satisfy the requirements of a number of different animals and it is very unusual to find an area populated by only a single species, even though one may be so numerous as to dominate all others. An area of intertidal rock apparently covered with closely packed barnacles, on close examination proves to harbour also limpets hemmed in by the barnacles and with barnacles attached to their shells, small periwinkles *Littorina saxatilis*, and on southern coasts *L. neritoides* also, nestle between the barnacles or in empty barnacle cases, and the predatory gastropod *Nucella* is generally present, feeding on the barnacles. In addition, the empty barnacle shells are occupied by tiny gastropods *Otina* and *Cingula* and large numbers of a small bivalve *Lasaea*. The spaces between the barnacles are populated by small crustaceans, nematodes, mites, insect larvae and the primitive wingless insect *Anurida maritima*. There is in this single habitat a number of species which are almost always found living together with a dense crop of barnacles.

This is not to say that these animals are found only with barnacles. *Littorina* is numerous on barnacle-free beaches, *Nucella* preys also on *Mytilus*, and the smaller animals can be found in any suitable crevice where small quantities of detritus accumulate. Nor is it essential that there should be a single dominant species; in all reasonably defined environments there are recognizable communities of species even though none of them is particularly numerous. This is true of the crevice fauna of barnacle-strewn beaches. The inner recesses harbour the polychaetes *Amphitrite gracilis* and *Cirratulus cirratus*, the crustacean *Tanais*, the small gastropods *Cingula*, *Onoba* and *Leucophytea*, the bivalve *Lasaea*, and a myriapod, a pseudoscorpion and a beetle which have invaded this marine environment from the land.

Whatever its composition, the community is a collection of animals occupying the same habitual niche and within that niche each species is adapted in its own way to exploit the available food resources. One result of this is that few species living together in the community have precisely the same requirements and competition between them is reduced to a minimum. Indeed it is possible that in many communities the constituent members are sufficiently different from one another that there is virtually no competition.

Chapter 8

Food and feeding

ALL THE activities of animals: moving, excreting, secreting, even growing, involve the use of energy and this energy comes ultimately from the animals' food. Green plants are able to use the energy of sunlight to build complicated organic molecules from simple salts as a result of photosynthesis. Animals cannot do this and directly or indirectly depend upon plants for their energy and building materials. Herbivorous animals feed directly upon plants, the herbivores are eaten by carnivores, and these, in their turn, may be eaten by further carnivorous animals. This series of relationships between eater and eaten constitutes a food chain, and the organisms in a simple food chain form what is known as a pyramid of numbers, with the organisms at the base of the pyramid, the plants, being most numerous and the final carnivores at the top of the pyramid existing in the smallest numbers.

Usually the natural situation is a good deal more complicated than this. Several carnivores may feed on a single food organism, and each carnivore takes a variety of food. Furthermore, the eggs, young, faeces and dead bodies of the ultimate carnivores contribute to the diet of the lower carnivores. Under these circumstances, the interrelationships between food and feeders is so involved that the term 'food web' describes the situation much more accurately than 'food chain'. Unravelling the interrelationships in a food web is very difficult and so far has been attempted in only a few simple situations. The whole system may be viewed as a flow of energy generated by the plants, the primary producers, passing through a variety of animals which make use of this energy until, finally, all the complicated molecules constructed by the plants are broken down and the chemical energy contained in them is dissipated.

By no means all animals feed on living cells. Dead plants and animals may be consumed directly by scavengers, but the remains that escape the scavengers are attacked by bacteria, fungi and other micro-organisms and are broken down to small fragments. These organic particles, together with a variety of microscopic organisms constitute 'detritus' which is a very important source of food to many animals, particularly in the sea.

There is an almost endless variety of food, and animals are modified in as many ways to exploit these food sources. It must be remembered that any source of large organic molecules can provide nourishment for some animal, providing it can break them down to simpler molecules and so gain the energy locked up in them. This accounts for some bizarre feeding methods. Materials which cannot be used by one animal and are voided by it, undigested, in the faeces, may be usable by another animal. The value of regarding the relationship between the organism and its food as one of energy flow is that it removes all prejudgement about what is and is not available in the environment as a food resource.

Basic sources of food

So far as animals are concerned, there are several basic food sources even though not all of them lie at the base of the pyramid of numbers.

In aquatic environments, both in the sea and in fresh waters, a most important source of food is the phytoplankton composed of very small green plants, diatoms and flagellates, which float in the upper layers of the water where they receive sufficient light for photosynthesis. Very often they are the most important producers in these environments. Phytoplanktonic organisms are directly grazed upon by members of the zooplankton, including many small crustaceans like the copepod *Calanus* in the sea, and *Daphnia* in fresh water. Particularly in the sea, almost all groups of animals exploit this food source at some time in their life history, more especially as larvae and young forms. Large green plants, which are also primary producers, are confined to shallow coastal waters in the sea though here they are of major importance and are eaten by many animals. In fresh water and even more on land large green plants are numerous and are by far the most significant primary producers.

In assessing the value of plants as a food source which supports the entire animal population in an area, it is usual to consider the net production rather than the standing crop. The standing crop of plant life represents the total resources of the environment at any time, but the plant and animal community is a continuing phenomenon and it is important to take into account the rate at which the environmental resources turn over. This is measured by the net annual production which represents the total amount of solar energy bound in the form of organic molecules as a result of the photosynthetic activity of plants (making due allowance for the energy used up by the plants themselves) and which is available in one way or another to animals during the year. It is comparable to measuring the prosperity of a country by its gross national product rather than by its total capital resources.

Detritus is a vague term which embraces a great variety of materials from freshly fallen leaf litter and the remains of dead animals to colloidal organic molecules in water. Herbivorous animals are responsible for the production of an enormous quantity of detritus for as much as 90 per cent of the organic material they eat passes out of the gut with the faeces. Even the more efficient carnivores utilize only 50–70 per cent of the food they consume, the remainder being lost in the faeces.

Dead plant and animal material is broken down by bacteria and fungi to small particles which are added to the general mass of detritus, but in addition, the bacteria themselves form a significant food source for some animals. Bacteria are common in all environments and are particularly numerous on interface surfaces: around particles suspended in the water, on the surface of algae, and especially on soft, muddy substrata where the particles of silt are very fine and so, in total, present an enormous surface. Some animals, even quite large ones, can be sustained on a diet of bacteria alone. The molluscs *Hydrobia ulvae* and *Macoma baltica* are both common in estuarine muds where they feed on a mixture of organic detritus, bacteria and silt. The faeces have a high organic carbon content and low organic nitrogen; the molluscs can make no further use of these discarded materials. The faeces are attacked by bacteria which feed on the organic carbon and the organic nitrogen content of the faeces increases. If such faeces are fed to the molluscs, the bacteria are digested and used as food while the remaining unchanged organic carbon compounds remain undigested. In the same way, it proves very difficult to starve the polychaete worm *Nereis diversicolor*. If deprived of animal and vegetable food, it secretes quantities of mucus on which bacteria feed and multiply. The worm then consumes the mucus and the bacteria on it and can sustain itself almost indefinitely in this way.

About 10 per cent of the net primary production in the sea goes into solution in the water. To what extent this is lost as a food source is still uncertain. By using radioactive isotopes it has been possible to show that a wide range of marine invertebrates are able to absorb amino acids from very dilute external solutions and it has been estimated that the worm *Clymenella* living in rather polluted mud with a high organic content may gain as much as 70 per cent of its nutritional intake by the absorption of amino acids in solution in the surrounding water.

The primary consumer

Strictly speaking, primary consumers are those organisms that feed on the primary producers, the plants. In practice it is often difficult to maintain a firm distinction between primary consumers and other feeders. Herbivores may feed on large plants or phytoplankton and the plants may be living or dead. Dead plant material is, of course, a major constituent of detritus but most detritus feeders and, indeed, most animals feeding on small particulate material make no distinction between food of plant and animal origin.

To deal effectively with plants, particularly the larger ones, animals must have some means of breaking the tough cellulose of the plant cell wall to release the cell sap and protoplasm inside. The enzyme cellulase, which digests cellulose, is surprisingly rare in animals when it is considered how many are herbivores, though this enzyme is found in members of most phyla. The great majority of herbivores have a symbiotic relationship with micro-organisms that live in the gut and manufacture cellulase.

Except for animals living on microscopic plants, herbivores need some kind of jaw and tooth apparatus for piercing, sucking, rasping or scraping up their food, as well as for biting and chewing. Such an apparatus is often adaptable to a variety of foods and it makes little difference whether plant or animal material is taken. Some of the most elaborate 'mouth-parts' of this sort are found in crustaceans and many of them are omnivorous. Arthropods, with their segmented structure, have the advantage that if a number of segments are fused together in the mouth region the appendages of each of them may be used to chew and manipulate the food. The isopods *Idotea* and *Ligia* live on the shore and feed largely on seaweeds. Four pairs of segmental appendages are used as mouth-parts: the mandibles, maxillulae, maxillae and maxillipeds. The mandibles do the biting while the maxillulae and maxillae help abrade the food. The mandibles also have hard protuberances comparable with molar teeth, which grind up the food, and bristles on the mouth-parts push the bits of algae bitten off by the jaws inwards so that they come within range of the grinding apparatus. The maxillipeds form the outermost part of this jaw apparatus and prevent food falling out of the mouth as well as brushing up such fragments that do escape.

Some isopods, the woodlice, have successfully invaded the land where they live in damp places under bark and stones, and in the soil. They are all primarily vegetarian and are especially important as consumers of leaf litter in woodlands. Millipedes share this role with them, though they have less elaborate and generally weaker mouth-parts with only a pair of mandibles and a pair of maxillae. As a result, they tend to confine their attention to the softer, partly rotten leaf debris. *Glomeris* is common in meadowland where it chews up rotting litter, though we are not certain how much it relies on the dead vegetation and how much on the bacteria and fungi that are breaking it down. Other millipedes, particularly the common *Julus* and *Cylindrojulus* feed on freshly fallen and year-old leaves which have not rotted appreciably. They are found in mixed woodlands, but given a choice, *Julus* prefers oak leaves and *Cylindrojulus* pine needles.

In the shallow sea algae encrust almost all hard surfaces and provide a rich meadow for grazing animals. Here the vegetation must be scraped up rather than bitten off and a different sort of jaw apparatus is needed. In the sea-urchins like *Echinus* this takes the form of a curious and complicated structure known as 'Aristotle's lantern' because it looks something like a lantern and was first discovered and described by Aristotle himself. The apparatus consists of five chisel-shaped teeth which project slightly through the mouth. They are moved by an extremely complicated system of muscles and levers which constitute the 'lantern'. They scrape up encrusting algae but can also be used for tearing up soft food such as a dead fish, or for biting parts of larger seaweeds.

The moving belt radula of gastropod molluscs and chitons appears to be an ideal adaptation to rasping algae from rocks. The radula is a flexible ribbon studded with teeth arranged on it in regular transverse rows. It lies in the floor of the mouth extending from the radula sac where the ribbon and teeth are constantly produced to replace worn-out parts of the radula, and over a protrusible tongue, the odontophore. The teeth of the radula normally lie flat, but as the odontophore is protruded and applied to the substratum the teeth are erected when the radula is bent over the end of the odontophore. The odontophore is rocked back and forth so that the teeth are alternately erected and scraped against the substratum, and collapsed backwards, drawing food into the mouth. This system can be modified in many ways, depending upon the nature of the food material. The pulmonate snails and slugs living on land and in fresh water have abundant soft plants to feed on and in them the radula has very many small teeth with broad cusps. The limpet, *Patella*, rasping algae encrusting rocks has fewer but stouter teeth; it also has an enormously long radula because the teeth are worn away very quickly. The ormer, *Haliotis*, feeds in a similar way but only the teeth in the middle of the rows do the rasping, the teeth on either side are much smaller and finer, and may act as a sieve to prevent large fragments of food entering the gullet. The sea-hare, *Aplysia*, eats large pieces of seaweed and the radula works in conjunction with the hardened sides of the mouth which bite off fragments drawn in by the sharp radula teeth. A further refinement of the feeding technique is found in some saccoglossan gastropods in which the radula has become a tool for slitting open individual plant cells, enabling the sap to be sucked out by the muscular pharynx. The radula has a single tooth row and as the leading tooth is erected over the odontophore it slices through the cell walls of the plant. *Limapontia* and *Actaeonia* feed on *Cladophora* in this way. They hold the algae by the edges of the mouth while the radula is at work and gradually move along the algal filament leaving behind a row of empty cells.

One of the fruits of recent wrecks of oil tankers has been the realization of how efficient the grazing animals of shallow waters are. After the wreck of a tanker on the coast of Lower California a few years ago, the sea-urchins were either killed or moved out of the area and for several years there was an enormous increase in the amount of the giant kelp *Macrocystis* in the area, previously held in check by the grazing urchins. The excessive use of emulsifiers on English south-western beaches after the wreck of the *Torrey Canyon* killed most of the limpets and within a few months what had previously been clean rocky beaches became bright green with the growth of the slimy alga *Enteromorpha* which is normally restricted to muddy estuaries where few limpets live. The experience is very similar to that on land where we discovered how much verdure was consumed by rabbits only after they had been nearly exterminated by myxomatosis.

Feeding on small particles

Animals feeding on small particulate food have the same problems and comparable adaptations to one another whether the food material be live micro-organisms or dead plant and animal fragments. In the sea there is a constant rain of this material towards the bottom and the animals exploiting this food source may be divided into two classes: those that filter particles from the water and those that eat it after it has settled and become part of the bottom deposit.

Efficient filter-feeding demands a filter, some means of drawing a water current through it, and a transport system to transfer the trapped particles to the mouth. The suspended material often contains a proportion of inedible, inorganic particles and the efficiency of the filtering system can be further improved if it incorporates some method of sorting potential food from unwanted material.

Since a great many aquatic animals in any case produce a water current over their gills for respiratory purposes, it is quite common to find the gills themselves modified as food collecting organs. Bivalve molluscs have carried filter feeding to a high degree of efficiency by just these means. The mantle cavity is divided into a lower inhalant chamber and an upper exhalant chamber by the greatly extended gills. The cilia on the gills produce the water current and are specialized in different parts of the gill filament to do a variety of jobs, the lateral cilia draw water between the gill filaments, the long latero-frontal cilia filter out the particles carried in the water stream and transfer them to the frontal cilia, which transport the food towards the edge of the mantle. Bivalve gills are generally so big that they are folded into a W-shape to accommodate them in the shell. The apex of each fold, towards which the food particles are carried, is modified as a food groove and the cilia in this transport the food, generally wrapped up in mucus by this stage, forwards to the mouth. Sometimes the food grooves are protected so that only small particles enter them, large particles also tend to fall off the gills on to the mantle where tracts of cilia carry them to the edge of the mantle and so to the exterior. This is the first stage of sorting the food and increasing the nutritional content of material that is eventually taken into the mouth. Two lobes which cover the mouth, the palps, are also generally important sorting organs. They have a complicated series of ridges and ciliary tracts on their inner surface and as particles are carried over them they are sorted on the basis of size and density.

A few gastropods have evolved ciliary feeding mechanisms which closely parallel those of bivalves. The slipper limpet, *Crepidula*, is a filter-feeder. Its gill filaments have no latero-frontal cilia and particles are filtered by a sheet of mucus covering the gill through which the water current passes. Some bivalves also lack latero-frontal cilia and probably filter water through a mucus sheet in the same way.

It is perhaps not very surprising that some gastropods should have evolved food-collecting devices similar to bivalves; they have the same basic structure. What is more surprising is that the polychaete worm *Sabella*, the peacock worm, has a filtering device almost identical as bivalves. *Sabella* lives in a tube and at its anterior end has a conical crown of filaments which are stiffly erected when the animal is feeding. Each of these branchial filaments bears a row of pinnules on its sides. These are ciliated and produce the water current through the branchial crown and filter out the food particles. The trapped particles are carried to the base of the pinnules into a food groove running down the filament towards the mouth. There is even a sorting apparatus at the base of the filament. The sides of each filament are drawn out into two folds which are pressed against one another; the food particles must pass between these folds to reach the mouth. The largest particles cannot enter the sorting apparatus and are discarded. Medium-sized particles enter between the folds but not to the base of the groove to which only the smallest particles have access. This very fine material is carried to the mouth and eaten; the medium-sized particles pass into a ventral sac where they are used to provide tube-building material.

The tentacular crown of the lophophorates – the phoronids and ectoprocts – and also of the Entoprocta, is very similar to the *Sabella* crown and is used in food collecting in much the same way. Only a few have been studied in detail and it is obvious that in minor ways the feeding devices differ in different animals even though the principle is the same in them all. In the ectoproct *Flustrella*, for example, there are no cilia transporting material to the mouth, instead the particles that accumulate in the crown are drawn in by the sucking action of the oesophagus and the beating of the oesophageal cilia.

Flustrella is a completely non-selective feeder; whatever is trapped in the tentacular crown is eaten, and the same is true of the tunicates in which there has been an incredible elaboration and development of the filtering equipment. In an earlier chapter we saw that the pharynx of sea-squirts such as *Ciona* or *Ascidia* is enormously enlarged and constitutes by far the largest part of the body. The gills in the wall of the pharynx first appear as the usual crescent-shaped slits, but each slit becomes very long and convoluted as it develops and eventually the greatly enlarged pharynx wall is fully occupied by these slits with only thin bars of tissue separating them. Cilia associated with the gill slits draw a current of water into the pharynx through the mouth and out through the gill slits. Running along the ventral side of the pharynx there is a ciliated gutter, known as the endostyle, which is richly supplied with mucus cells. A sheet of mucus secreted by these cells is carried by cilia up either side of the pharynx to a groove running along its dorsal surface. All the water passing through the gills is filtered through this moving belt of mucus which is rolled up in the dorsal groove and carried to the mouth. The mucus sheet is a very effective filter and removes all particles larger than 1 μ (0·001 mm) entering the pharynx. Tunicates are completely non-selective feeders and the only way they can control their food intake is by ceasing to secrete mucus if the incoming water becomes excessively turbid.

Almost all ciliary feeders use mucus in some capacity, if not to filter the food, as in tunicates, then to consolidate and bind together particles that have been trapped by the cilia and are being transported to the mouth. The chief function of the cilia is often simply to produce a water current and this can be done by mechanical means. The polychaete *Chaetopterus* lives in a parchment-like 'U'-tube on the sea bed. It is a curious looking worm and the dorsal parts of the parapodia on the 12th segment are greatly enlarged and elongated like a pair of arms so that when extended they can contact the walls of the tube. They secrete mucus which is drawn out into a sort of bag and held across the tube, and all water passing through the tube must go through the mucous bag. A water current is produced by three enormous muscular fans formed on the underside of the 14th, 15th and 16th segments. When the mucous bag becomes clogged with particles it has filtered from the water, it is rolled up by the arms and passed forward along a ciliated groove to the mouth. Animals do not need to be as highly modified as *Chaetopterus* to feed in this way. The Californian echiuroid worm *Urechis* lives in a U-shaped burrow in the mud and although shaped like a simple cylindrical bag feeds in precisely the same way. It creeps forward to the mouth of its burrow and secretes mucus around the edges of the opening. It then retreats, leaving behind a conical mucous sheet extending from the front of the animal to the mouth of the burrow. Water is pumped through the net by peristaltic movements of the body and when the net becomes charged with particles, it is rolled up and eaten. *Nereis diversicolor* which is a conventional errant polychaete with no special modifications in that direction has been observed to feed occasionally in precisely the same way.

Crustaceans have no cilia and seem unable to produce quantities of mucus. They have small hairs or setae on the body and limbs and these are sometimes used for filtering particles from the water. The acorn barnacle, *Balanus*, feeds in this way. The six pairs of thoracic limbs are modified as combs, the last two or three pairs are short, but the others are long enough to project from the shell and comb the water. The larger appendages of *B. perforatus* can strain particles only 35 μ (0·035 mm) in diameter, but the animal can also collect much finer particles that are filtered by the shorter limbs. These are closely beset with very fine setae spread across the entrance to the mantle cavity and with spaces between them no more than 1–2 μ wide. Water currents are produced by movements of the larger cirri and also by the pumping action of the operculum. In fact, by pumping with the operculum and using the filtration apparatus of the shorter appendages, it is possible for these barnacles to feed without extending the longer appendages from the shell at all.

Many small crustaceans are planktonic and feed on the small phytoplankton. While they certainly use setae for filtration, they also rely heavily on a totally different method of accumulating and trapping food particles.

The fairy shrimp *Chirocephalus* swims, feeds and breathes simultaneously by the beating of its numerous, close-set appendages which are flattened and leaf-like. The limbs beat in a metachronal rhythm, with each limb beating slightly before the one in front of it. As a limb is moved forwards the space between it and its neighbour is enlarged and water is sucked into this space. When the limb swings backwards this space is closed and the water is driven out again, some outwards and some inwards towards the mid-line of the animal. The water going outwards is filtered through setae on the base of the limbs. The water passing inwards meets water coming from the limb on the opposite side of the body and is deflected forwards. It is deflected forwards because the limbs behind are already producing an inward spurt of water, but those in front, beating a little later, are not yet doing so. Consequently all food particles trapped between the limbs are eventually driven inwards and forwards towards the mouth. Setae on the maxillules form a further sieve which prevents particles escaping at the anterior end of the animal, so that there is a continuous accumulation of particles at the mouth. The water flea, *Daphnia*, feeds in essentially the same way except that it has far fewer trunk limbs.

Copepods like *Calanus finmarchicus*, which is the chief food of the herring, swim chiefly by means of the antennae and five pairs of thoracic appendages. None of these appendages is particularly concerned with feeding. Movements of the antennae, mandibular palps, maxillae and maxillules set up a large vortex in the water on either side of the body. A pair of smaller swirls of water moving in the opposite direction is set up by the vibration of the maxillules and maxillipeds. The effect of these swirls of water is to concentrate food particles in the centre of the vortex, as anyone can verify by stirring up some sugar in a glass of water. Thus it is that particles in the water are brought together near the mouth and are almost stationary. The maxillipeds drive this food material forwards where it is filtered out by closely set setae on the maxillae. They are then scraped off by the maxillules and transferred to the mouth.

All the animals we have considered so far trap, by one means or another, the food particles suspended in a large volume of water and to do this they must produce water currents. The food particles are constantly settling on to the sea bed wherever movements of water are slow enough to permit this and some animals can get sufficient food by passively waiting for it to fall on them. Some ophiuroids, for example, simply sit with their arms raised in the water with strands or sheets of mucus hanging from them. This mucus with whatever particles may have stuck to it is then transferred to the mouth by the podia. The sea-cucumber, *Cucumaria elongata*, lives in a burrow in muddy substrata and, when feeding, it extends a crown of ten tentacles on the substratum waiting for particles to settle on it. There are eight long tentacles and two smaller ones. Periodically a long tentacle is bent over and stuck into the mouth, and particles on it are scraped off as the tentacle is taken out of the mouth again. The smaller tentacles do not trap many food particles themselves, instead they wipe the larger ones occasionally and transfer particles to the mouth. All the tentacles are sticky with a mucus secreted in the pharynx and the whole operation is like a small child with a sticky finger and a bowl of sugar. Another echinoderm, the crinoid *Antedon*, is a slightly more active feeder than this. It sits with its ten arms spread out horizontally waiting for settling particles. When a particle touches a podium or tube-foot (a misnomer in this animal since they are not used for locomotion) it bends over and flicks the particle towards a food groove which runs down each arm, converging on the mouth. The particles are enveloped in mucus and are transported by cilia in the groove.

Bulk feeders and deposit feeders

In the stiller parts of the sea, in lakes and slow moving rivers, there is a continuous rain of dead plant and animal material from the upper layers of the water, which eventually becomes incorporated into the substratum. Even on land much the same thing happens when fallen leaves and dead vegetation begin to rot and become part of the soil. This rich source of food is exploited by many animals in all these environments. They may live on the surface of the substratum as an 'epifauna' or seek the shelter of it by burrowing as an 'infauna', and they may swallow the deposit indiscriminately or select edible material from the substrate and ignore the inorganic particles of sand and clay.

Perhaps the most famous bulk feeder is the earthworm, if only because it attracted the attention of Charles Darwin who devoted a whole volume to it. At one time it was thought that earthworms burrow by eating their way through the soil. We now know that it is not essential for them to take in food in order to burrow, but burrowing plays an important part in feeding. During the elongation of the anterior part of a burrowing worm, grooves on the under side of the prostomium lead the soil to the mouth. The prostomium can then be raised and the food sucked into the mouth by the muscular pharynx. The soil contains a great deal of organic material: dead plants and animals, bacteria, fungi, protozoans, and small animals, and it features in the diet of all earthworms. It is the sole source of food for

Allolobophora longa and *A. nocturna*, but *Lumbricus terrestris* is a more selective feeder and depends very largely on dead leaves lying on the surface of the soil. It emerges from its burrow at night to collect fallen leaves, dragging them into its burrow point first. Pieces of the leaf are then broken off and swallowed. A study of earthworms in an apple orchard showed that of the seven species of earthworms living there, only *L. terrestris* played an important part in leaf burial and that this species was responsible for burying 90 per cent of the annual fall of leaves – some two thousand pounds per acre.

The lugworm, *Arenicola*, is almost the marine counterpart of the earthworm. It lives in a U-shaped burrow on sandy and muddy beaches, or rather, an L-shaped burrow for generally an arm of the U is filled with sand. The animal lies at the base of the U, with its head towards the filled shaft from which it eats the sand. It pumps a current of water through the burrow by peristaltic movements, in at the open tail end of the burrow, over the body and out through the head shaft, percolating through the sand as it does so. Particles suspended in the water are filtered out and enrich the sand the animal eats. Furthermore, the worm occasionally clears out the sand from the head shaft which is then filled up again by material trickling down from the surface deposit. The surface layer of sand contains by far the richest supply of detritus and it is this, enriched by the filtration process, that *Arenicola* eats. With such a diet it is not surprising to learn that the faeces of *Arenicola* often contain more organic nitrogen than there is in the surrounding sand.

Most animals living in the substratum are in fact feeding only on its surface layer, though the ways in which they collect it vary enormously. The small delicate bivalve *Tellina* living in sandy beaches has two long siphons developed from the mantle. Water is drawn in through one and expelled through the other. The inhalant siphon is particularly extensible and is used as a vacuum cleaner sucking in particles from the surface of the substratum around the animal, while the exhalant siphon is held aloft to prevent contamination of the ingoing water. The inhalant current is very narrow so that it draws in only loose deposits on the substratum and these are very largely of organic material.

The little bivalve *Nucula* and its relatives collect the surface deposit in a different way. The palps, which in most bivalves are merely a sorting organ for particles collected by the gills, are enormously developed. They have two lobes, one of which is extended backwards and drawn out as a proboscis which picks up material from the sea bed. The material is passed to the inner lobes of the palps which retain their function as a sorting apparatus. The gills are reduced in size but they still have the ciliary filtering and sorting devices found in other bivalves and contribute filtered material to the diet.

The terebellid polychaetes *Amphitrite johnstoni* and *Terebella lapidaria* collect the bottom deposit in another way. These burrowing, tube-building worms have enormously extensible tentacles furnished with cilia, mucous glands and a very complicated musculature. The hollow tentacles are capable of a variety of movements: they can be flattened to adhere to the substratum, or rolled up to produce a ciliated gutter running along them. The tentacles are extended by ciliary creeping just as in planarian movement and when they are stretched out they are fixed to the substratum by mucus. Small particles encountered by the tentacles are carried to the mouth in the ciliated gutter, larger particles are assisted on their way by muscular contractions of the sides of this groove, and the largest particles are brought to the mouth by a complete retraction of the tentacles. None of this food material is placed in the mouth directly by the tentacles but is wiped on to the lips where the particles are roughly sorted according to size.

To some extent these terebellids are selective feeders, picking up individual particles, and this process is taken a stage further in burrowing heart-urchins such as *Spatangus* and *Echinocardium*. They have a variety of spines and tube-feet modified to perform such tasks as constructing the burrow and for respiration and feeding. Particles are removed from the area in front of and below the mouth, partly by spines which incorporate material into the sides of the burrow and also by the podia around the mouth. These tube-feet terminate in a sticky disc to which the particles adhere. The podium is then put into the mouth through a grid of spines which scrape the particles of food off as the podium is withdrawn. *Echinocardium* can also feed on the surface layer of the substratum where there are the richest deposits of detritus. The podia of the anterior part of the body are extended out of the burrow and grasp particles by the finger-like processes with which the podia end. Handfuls of material are brought down into the burrow and the podia may return repeatedly to a patch of algae or other organic material until it is all consumed. The particles of food collected in this way are bound together with mucus and passed along a tract of small flattened spines to the oral region where they are picked up by other podia and put into the mouth. This method of transporting mucus-bound food along a tract of small movable spines is intriguing for the process is almost identical with the transport of food along ciliary tracts save that it is greatly increased in scale.

Carnivores

Plant material is much less nutritious than animal tissues and consequently herbivores must eat a great bulk of food; fortunately plants are numerous and widespread. Carnivores eat nutritionally concentrated food in the form of animal material, they occupy a higher position in the pyramid of numbers and are less numerous than their herbivorous cousins, and they are known as secondary consumers. At first sight there seems a clear cut division between herbivores and carnivores, but this is not so. A great number of animals have a mixed diet and eat whatever comes to hand. Most animals that filter particles from the water or collect detritus make no distinction between food of plant and animal origin and might as well be regarded as carnivores as vegetarians. Even among animals that feed exclusively on animal tissues conventional ideas about carnivores must be discarded. It is usual to think of them as predators that actively hunt down their prey. This is true of some, but by no means of all of them. Some carnivores are grazers, feeding on colonial or gregarious sedentary animals, others simply wait and capture animals that accidentally enter a trap or snare.

Grazing on animals

That it should be possible for carnivores to graze on other animals, like a cow in a field, is at first sight surprising, but a good many animals, such as sponges, hydroids and ectoproct bryozoans, reproduce vegetatively and others, although breeding sexually, are highly gregarious, like mussels and acorn barnacles. All of these produce a counterpart to grass meadows on land or beds of seaweed on the shore for grazing carnivores and, particularly if the food animal is soft, very little modification is required to turn a herbivorous grazer into a carnivorous one. Many nudibranch gastropods fall into this category. Often they are specialists so far as their diet is concerned (dorids, or sea-lemons, feed on sponges; and aeolids, or sea-slugs, feed on hydroids) but they use the radula for drawing food into the mouth and the hardened lips for biting it off in exactly the same way as the sea-hare, *Aplysia*, which feeds on seaweeds.

Gastropods which seem to be primitively herbivorous grazers have very commonly turned to carnivorous grazing, though a number of them have adopted special means of countering the protective devices of their prey. Several gastropods feed on bivalves or acorn barnacles despite the fact that both are encased in calcareous shells. The dog whelk, *Nucella*, is common on rocky shores where there are barnacles. The mouth of this gastropod is at the tip of a proboscis which is used to force open the operculum of the barnacle. The radula rasps out the contents of the prey. *Nucella* is also able to bore through the shell of mussels and these form an important alternative food for it. The oyster drill, *Salpinx*, is another gastropod that bores through the bivalve shells. Its radula has hard, sharp teeth and in boring through the shell is aided by secretions that probably contain enzymes which weaken the shell by breaking down its protein matrix. Once a hole has been made, the proboscis is thrust into the soft tissues which the radula reduces to small fragments that are sucked up by the pumping action of the oesophagus. A number of gastropods feed on bivalves but are unable to drill through their shells. Instead, they either pull apart the two valves of the shell, as *Murex* does with its normal food the oyster *Crassostrea*, or wait until the bivalve gapes before attacking it, as the whelk *Buccinum* does. Once the valves of the prey are a little open, the whelk inserts the tip of its own shell between the valves, preventing them closing and then inserts its proboscis into the bivalve's tissues and feeds.

This is very similar to the method of feeding of the common starfish, *Asterias*. It takes up a characteristic hump-backed position over such molluscs as *Mytilus*, the common mussel, and with a large number of its tube-feet attached to the valves, exerts a steady pressure until the valves gape. The starfish then protrudes its stomach through its mouth and between the shells of the bivalve where the tissues are digested and absorbed by the stomach while it is outside the starfish's body. Starfish are voracious predators of bivalves and are very efficient feeders. A gap of as little as 0·1 mm between the valves of the prey is sufficient to allow the stomach of the starfish to be extruded into the bivalve.

Predators

Actively hunting and capturing mobile animals is a complicated business. A good deal of energy is used in capturing and killing live prey and if the hunter is to get a reasonable return for the energy it expends, it must eat animals that are large in relation to its own size. The specializations of predators, in fact, can be seen as means of minimizing the energy cost of capturing prey. Often sense organs concerned with prey capture are highly developed. Many predators rely on sheer strength to overpower their prey, but often they have evolved poisons which immobilize the prey quickly. Some ingenious methods of relatively effortless prey capture have been evolved, as in the spider *Scytodes* which shoots a sticky toxic mucus at its prey, immobilizing it and sticking it to the ground so that feeding can begin.

Having gone to so much trouble to capture its prey, it is obviously important that the predator should not lose it again and many have a capacious gut and swallow their food whole. Generally such a meal is followed by a period of inactivity while digestion is proceeding. This is true of coelenterates, most of which are carnivores. They do not have specialized sense organs for detecting their prey, nor do they actively hunt. Instead they simply seize animals that blunder into their outspread tentacles. Coelenterates, though not the closely related ctenophores, have batteries of explosive stinging cells on their tentacles and these cells, called nematocysts, contain a coiled thread which is violently expelled when the cell is stimulated. Most coelenterates possess several different kinds of nematocyst, the most important in feeding being a pointed, barbed thread which injects an immobilizing toxin into the prey and coiled threads which become entangled with the prey and hold it until the toxin has taken effect. Discharge of the nematocysts follows mechanical stimulation, their threshold is reduced by the presence of certain chemicals in the water and once the prey has been grappled, the nematocysts become very sensitive and the more the prey struggles, the more nematocysts are discharged at it. The tentacles are then wrapped around the prey, the mouth opens and the food is engulfed.

Anemones and many worms that swallow their prey whole are soft-bodied and distensible, so that they can take prey at least as large as themselves. A few animals that are rigidly encased still manage to consume big animals whole, but they require special adaptations to do so. The gastropod *Conus* emerges through a narrow slit-like opening in the shell and it is surprising that the animal can be retracted into it, to say nothing of its capturing and eating whole fish as large as itself. The radula of this gastropod is highly modified with only few, very large teeth. These are hollow and are used to inject one of the most venomous of animal toxins into a fish that comes within range. The fish is detected by smell and the proboscis on which the radula is borne tracks the prey until it is close enough for a strike to be made. The proboscis is then retracted dragging the prey into the mouth. It is clearly impossible for so large an object to pass into the stomach which is, of course, inside the shell, and, instead, the prey is digested and reduced to fluid in the mouth and the juices then passed to the stomach.

Many of the more advanced arthropods, like *Conus*, are unable to swallow large food because their bodies cannot be distended, but with their complex of mouth-parts they are well able to manipulate, dismember and masticate their food. The larger crustaceans require no different apparatus for dealing with animal food than seaweeds, and many of them eat both. On land, centipedes, scorpions and solfugids all eat live animals and masticate their food. Centipedes have three pairs of mouth-parts and the appendages of the first segment of the body are modified to form a pair of large poison claws with which the prey is captured and killed; the powerful mandibles then bite off pieces which are masticated and swallowed. Centipedes live largely on insects, but some of the larger ones can capture small vertebrates. *Scutigera* is capable of killing small geckos.

Scorpions have a pair of large pincers, the pedipalps, which are used to seize the prey and since most scorpions live on insects, the pedipalps are usually adequate to the task of subduing the prey. A few scorpions, such as *Scorpio maurus* and *Androctanus australis*, are much more prone to use the sting in their tail, not only for defence but also when dealing with a struggling animal and only the quietest prey is devoured alive. The food is picked to pieces by the pedipalps and mouth-parts, and the soft parts and juices are sucked into the mouth by the muscular pharynx.

Solfugids, like scorpions, are arachnids. They are not venomous but have large chelicerae, or pincers, and rely on their greater pugnacity and strength to overcome their prey. They will attack any insect, including those with very hard cases such as beetles, spiders, scorpions and even small vertebrates. Food is detected visually or by the groping pedipalps and first pair of legs. The prey is seized by the huge jaws and is then reduced to a soft pulp by the jaws and the chelicerae before being taken into the mouth.

The fluid diet

Several carnivores reduce their food to a semi-fluid broth before ingesting it and this has the advantage that a large meal may be consumed and digested very quickly once the preliminary stages are over. Carried to its logical conclusion, this method of feeding can remove the need for a masticating apparatus or indeed any mouth-parts beyond what is needed to penetrate the body of the prey.

The Onychophora feed largely on animals living in rotten wood, and particularly on woodlice. When *Peripatopsis* encounters a woodlouse, it holds it to its mouth by a powerful sucking action and the saw-like teeth slice pieces from the woodlouse's integument. These pieces are swallowed, but digestive enzymes are also poured into the wound and the juices alternately sucked up and regurgitated until the tissue of the prey is reduced to a fluid, which is swallowed, leaving the woodlouse an empty shell.

Spiders are all carnivorous and all have a similar way of killing and eating their food. The prey is bitten by the pincer-like chelicerae which inject a powerful toxin, paralysing it. The salivary glands secrete digestive enzymes and these are also injected, breaking down the protein in the prey, most of which is liquified as a result and then sucked up. The gut has numerous blind-ending side branches and a large liquid meal can be stored in them for complete digestion later. Apart from their method of feeding, the greatest interest of spiders lies in their methods of capturing their prey. Hunting spiders are conventional predators with excellent sight. Moving prey can be recognized at some distance and is either chased or, as in *Salticus*, leaped upon. But all spiders spin a silken dragline, used in much the same way as a climber uses a rope, and this is turned to advantage in capturing prey. *Segestria* lives in a burrow and stretches a few taut threads across the entrance; these warn the spider of any animals wandering into the burrow. *Atypus* and its relatives spin purse webs like a silken finger in which they shelter. Small insects crawling over the web are attacked through it, pulled through and devoured. Aerial traps to capture flying insects are used by some spiders. Sometimes only a few sticky threads are suspended from twigs, but the orb web spinners make the most elaborate and beautiful of all webs and they also exemplify, par excellence, the trapping rather than the hunting habit. The garden spider *Araneus* makes a large symmetrical web and either sits in the centre or off at the side holding a signal thread. As soon as an insect is snared, the vibrations in the web attract the spider which then runs down the web to feed or at least immobilize the prey for later attention. If too large an animal is trapped and the vibrations in the web too violent, the spider generally cuts the animal free. The most important sense for these spiders is a tactile one, but they can also recognize their prey by chemical means and distasteful insects are carried to the edge of the web and dropped over the side.

Other animals that have adopted a fluid diet are the blood-sucking leeches. They feed on birds and mammals, attaching themselves by a sucker that encircles the mouth and using their sharp chitinous teeth to penetrate the skin of the prey. The salivary glands secrete a histamine into the wound, and this causes the blood capillaries to dilate and ensures a full flow of blood which is sucked into the gut. An anticoagulant is added to the blood (and sometimes injected into the wound); this prevents the blood coagulating. Blood-sucking leeches like *Hirudo* feed at very infrequent intervals, perhaps only once or twice a year, and they take an enormous quantity of blood. *Hirudo* takes two to five times its own weight in a single meal. The oesophagus is enormously developed to contain this amount of blood and digestion proceeds very slowly. It is an extraordinary fact that these leeches have no digestive enzymes of their own. They rely entirely upon bacteria which are always present in the gut and the leech absorbs the products of digestion that are released as the bacteria attack the blood cells one by one.

Nematodes, like leeches, include both free-living and parasitic species, though the parasites live within their hosts. All nematodes take in their food by a powerful sucking action of the pharynx and while small plants and animals may be swallowed whole, the majority of nematodes have a fluid diet, a fact that makes the transition to a parasitic life particularly easy. Those free-living species that prey on plants and animals attach themselves by their lips, pierce the wall of the food organism with teeth or a spear-shaped stylus, and suck fluid from the prey. This is the common method of feeding in all predators that take a fluid diet and the resemblance to the habits of spiders or onychophorans is enhanced in a nematode such as *Aphelenchoides* which feeds on other nematodes, because it injects a secretion into its prey which paralyses it and initiates digestion of its tissues. Parasitic nematodes are morphologically very like their free-living relatives save that they have no teeth or spear. In vertebrates they are common intestinal parasites, many attaching themselves to the wall of the intestine where they suck blood and the soft tissues.

The digestive system

The close parallels that exist between quite different animals that have similar feeding habits extend beyond their external features to the structure and functioning of the digestive system.

One example of this kind of convergence will already have become clear. Most carnivores consume their food quickly and digestion starts immediately. Often the salivary glands secrete proteolytic enzymes into the mouth so that preliminary breakdown of the food begins as soon as it is ingested. Indeed, a number of animals inject the salivary enzymes into their prey and digestion actually begins before the food is taken into the mouth. This is true of such disparate carnivores as spiders, cephalopods and whelks.

Filter-feeders show some remarkable parallels in the way they deal with the mucus strand of food particles as it enters the stomach. In ectoprocts, lamellibranchs and the few filter-feeding gastropods, indigestible particles that will ultimately form the faeces do not leave the stomach immediately but are bound together with mucus to form a little rod which is rotated by the intestinal cilia. It projects back into the stomach and engages the incoming mucus string which it winds in like a capstan as well as stirring up the stomach contents. This mechanism has become quite elaborate in lamellibranchs: the rotating rod is constituted of solidified mucus and has become separate from the faecal matter. This mucus rod, or crystalline style, has also acquired an additional function in molluscs. Since the gills collect food particles from the water currents that pass through them, the animal cannot help but pass food into its stomach so long as it respires, that is continuously. Enzyme cells cannot secrete continuously and the problem that this poses is solved by the periodically secreted enzymes being incorporated into the crystalline style. As the style rotates in the stomach it is ground against a hard shield on the stomach wall and the tip of it disintegrates releasing the enzyme. In this way a continuous supply of enzyme is provided in the stomach to deal with the continuous movement of food into it.

Many animals collect particulate matter automatically and continuously, and all of them have the problem of arranging for continuous digestion; most of them have nothing comparable to a crystalline style. The very familiar situation in mammals is misleading and the digestive process of invertebrates are rather different. In mammals there is a fairly orderly progression of food through the digestive tract where a succession of enzymes is added, reducing the food to small molecules which are then absorbed. Digestion in other animals is much less orderly, and all stages of it go on together in the stomach. Sometimes the food is reduced to soluble molecules and absorbed in the same way as in mammals, but in many invertebrates particles are engulfed by cells lining the gut and digested inside them. The extent to which digestion is intracellular varies widely and almost always there must be some preliminary extracellular digestion in bulk in the stomach to reduce the food to fragments small enough to be taken up by the cells. In filter-feeders taking very small particles extracellular digestion may be minimal, and intracellular digestion overwhelmingly predominate. Providing there is a sufficient number of engulfing cells for a proportion of them always to be unoccupied, the problem of providing a continuous supply of digestive enzymes hardly arises and animals relying largely on intracellular digestion are well able to deal with food that is collected continuously.

Bulk digestion by extracellular enzymes is economical of space. For intracellular digestion an enormous surface must be provided to accommodate a sufficient number of cells, each of which is occupied for a time digesting a few particles and is unable to accept others during that period.

In both crustaceans and molluscs there is a very great development of digestive glands. These contain many tiny branched ducts leading to small chambers lined with secretory cells and cells engaged in intracellular digestion. Turbellarians and anemones also rely heavily on intracellular digestion and while there are no digestive diverticula, in the former the gut becomes highly branched and in the latter the edges of the mesenteries which accommodate the absorptive cells have a clover-leaf cross-section enormously increasing their surface area.

As with other activities, collecting food and dealing with it presents the animal with a number of problems. There are surprisingly few ways in which food can be collected, and, depending upon the nature of the food and the method of getting it, the animal is presented with a further set of problems with even fewer alternative solutions to them. The more detailed the process, the less freedom of action an animal has, and it is this that accounts for the remarkable parallels that have appeared between quite unrelated animals engaged in similar activities.

Chapter 9

The movement and locomotion of invertebrates

ALTHOUGH vertebrate striated muscle is the best understood example of the way in which chemical changes result in movement, this highly refined system is the result of evolution from primitive beginnings. The workings of these apparently simpler systems have proved more difficult to unravel than that of the concentrated and ordered interlocking filament system of the vertebrate muscle fibre, although there is now evidence that the molecular basis of all types of muscle contraction is essentially the same, that is, the progressive interdigitation of filaments of at least two types of protein. Adenosine triphosphate (ATP) is involved probably in the phosphorylation of one or other of the molecules and this is followed by spatial changes and loss of the phosphate grouping. In this way ATP is broken down to adenosine diphosphate (ADP) and the energy released is available to bring about the movement of the different filaments past one another.

It has also been shown that ATP can provide the energy required for flagellar movement since flagella which have been extracted with glycerol and from which most of the chemical components other than the 9+2 fibrils have been removed, can be activated by the addition of ATP in physiological concentrations. Even so, there is no generally accepted explanation of the way in which the motion of cilia and flagella is brought about and how much it is due to shortening of the fibrils or how much to their relative displacement in a way similar to that in muscle. Similarly, ATP can bring about changes of form in amoebae but again there is no clear picture of the mechanism of cytoplasmic contractility at the molecular level, although a generalized cytoplasmic contractility is a property of protozoan and sponge cells and cytoplasmic streaming is widely known in plants. Possibly all of these phenomena have as their basis molecular configurational changes, made under the influence of ATP or some such energy-releasing substance, analogous to the spatial changes taking place during the reactions between allosteric enzymes and their substrates. Thus liberation of stored chemical energy as mechanical movement may be essentially the highly developed specialization of a molecular process which is a common property of cells.

Size and Movement

However, before discussing in more detail the mechanism by which muscular force brings about locomotion, it is useful to examine some aspects of the relation between the size of an animal and the dynamics of movement and locomotion.

When an object moves through a fluid both inertia and viscosity must be taken into account. The ratio of inertial to viscous forces is denoted as Reynolds number (R) which is given by the expression

$$R = \frac{LV\rho}{\mu}$$ where L is a characteristic length of the system, moving with a steady velocity V in a medium of density ρ and viscosity μ.

When the moving object is undergoing vibratory motion an expression (Rv), similar to Reynolds number, can be used to calculate the ratio of inertial to viscous forces.

$$Rv = \frac{nL^2\rho}{\mu}$$ where n is the frequency of vibration in cycles per second.

If a cilium has a diameter of 0·2 microns (the important dimension in propulsion through a fluid) and beats at 20 cycles per second then Rv becomes numerically equal to

$$\frac{20 \times (2 \times 10^{-5})^2 \times 1}{10^{-2}}$$ (where $\mu = 0{\cdot}01$ poise).

That is, the ratio of inertial to viscous forces is of the order of 1:1,000,000. This implies that, in the movement of microscopic organisms, inertial forces are negligible and viscous forces are dominant.

On the other hand when the value of L is of the order of a few millimetres instead of microns the ratio of inertial to viscous forces rises to a value at which inertial forces are dominant.

The Platyhelminthes includes species in the size range covering the transition from viscous to inertial resistance to movement, and the effects of this on the design features of the locomotory mechanism can be seen in this group of animals. The smallest are comparable in size with protozoans and many small invertebrate larvae and, therefore, come into the category of organisms to which viscosity is all important. We shall not be greatly concerned here with the movement of these very small metazoa or with microscopic organisms but it is of importance to note the difference in nature of the physical properties of the medium and moving object which affect the locomotion of organisms of different sizes.

The small acoel and rhabdocoel turbellarians move by ciliary action but in the larger turbellarians, especially the polyclads, muscular movement is mainly responsible for locomotion although cilia can still provide the motive power for slow rates of progression over the substratum in some cases.

It may be thought contradictory to this that cilia are extensively employed in the animal kingdom for moving large volumes of water needed to supply filter feeders with sufficient food. This occurs in sponges, annelids, bivalve molluscs and tunicates especially, but crops up sporadically in other groups such as coelenterates and echinoderms. Generally no work is performed against gravity and the velocity of the water current is low and hence its kinetic energy is small: little work is therefore done in moving much water. The construction of ciliary fields required for current production and filtering necessitates a high surface/volume ratio whereas the surface/volume ratio of animals decreases with increasing linear dimensions. While dependence on cilia for locomotion becomes impossible above a certain size because there is just not enough surface available to accommodate the cilia that would be necessary to provide the power required, the sedentary life of most ciliary feeders permits the development of the appropriate body form.

Application of force to the medium

In this chapter we are chiefly concerned, not with the molecular basis of contractility or with the movement of microscopic organisms but with the application of the force generated by muscular contraction to bring about movement and locomotion. While the process of locomotion necessarily involves the movement of parts, movement does not necessarily involve locomotion, which may be defined as a change of location of the whole organism.

The forces provided by the contraction of muscles can be applied to the substratum on which an animal is resting or the fluid by which it is surrounded in one of two ways, with or without the use of appendages acting as levers. Many of the various worms such as nematodes, oligochaetes, leeches and trematodes have no appendages capable of catching in the substratum and acting as the fulcrum of a lever, whereas the possession of such appendages is a diagnostic feature of the arthropods. In other groups such as echinoderms and some polychaetes, turgid appendages are present which also function as levers.

Muscular force applied without the use of levers

This method of the application of a force generated by muscular contraction depends either upon friction between the moving element and the solid substratum or the inertia of fluid against which the parts of the body are successively pressed.

Friction between the animal and the substratum is important in the locomotion of turbellarian flatworms and gastropod molluscs which employ waves of contraction of their longitudinal muscles passing backwards down the body. The mechanism of movement has been more thoroughly studied in gastropods than in turbellarians because the molluscs are larger and their locomotion is rarely complicated by any dependence on cilia. Essentially their locomotion consists of raising a small part of the foot from the substratum, moving it forwards and putting it down again. This can be done by waves of contraction passing backwards over the foot or by waves passing forwards. In the case of the forwards passage of the wave of contraction, the region of contraction is lifted off the substratum and the part of the foot behind it is pulled forwards. Forward motion of the animal can also be brought about by waves passing backwards down the animal but in this case it looks as if the region in contact with the substratum is that which is undergoing contraction. Waves of muscular contraction which travel in the direction of movement of the animal are termed direct and are seen in most pulmonate gastropods and also in *Haliotis*, *Trochus*, *Pomatias* and *Littorina*, while retrograde waves passing backwards down the animal occur in *Fissurella*, *Aplysia*, *Tethys*, *Nerita* and *Turbo*.

Certainly variations in pressure occur between the sole of the foot and a glass plate over which a snail is creeping for they can be easily demonstrated by causing the snail to creep over a hole in the plate, which is connected to a sensitive recording manometer. Pressure variation of a few millimetres of water can be seen to occur as the locomotory waves pass the connection to the manometer.

It should be noted that even where waves of contraction appear to take place in one plane, as in the flat sole of the gastropod foot, the muscles which have contracted are dependent for their extension on the hydrostatic pressure in the fluid contents of the body wall. This method of locomotion can be looked on as a special case of the use of a hydrostatic skeleton for muscular antagonism in that the elongation of relaxing sections of the foot musculature results from the internal hydrostatic pressure set up in the haemolymph. The diagrams (Fig. 9.1) should help to make clear how both direct and retrograde waves can bring about forward progression.

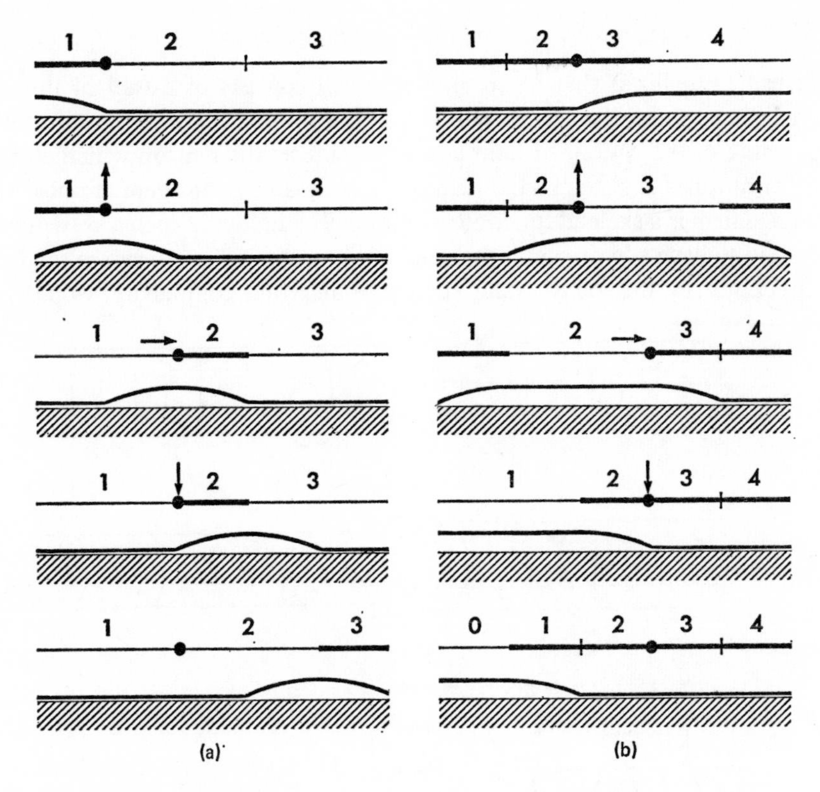

Fig. 9.1 Diagram to illustrate the passage of a wave of muscular contraction along the foot of a gastropod. In (a) the wave is moving in the direction in which the animal is travelling (from left to right). Successive muscle fibres are numbered 1, 2 & 3 and the one contracting is shown as a thick line. The junction of muscle fibres 1 & 2 is marked by a point. Note that the region at which contraction occurs is detached from the ground and re-attached when the contraction wave has passed. Forward locomotion (to the right) is denoted in the diagram by movement to the right of the point between fibres 1 & 2. In (b) the wave is moving in a direction (right to left) opposite to that in which the animal is travelling (left to right). The muscle fibres are numbered 1, 2, 3 & 4 as in (a) but their contraction occurs in the order 4, 3, 2, 1. The junction of muscle fibres 2 & 3 is marked by a point. Note that the region at which contraction occurs is, in general, attached to the ground and that detachment occurs at the point of relaxation. Forward movement (to the right) is denoted as in (a) by the movement to the right of the point between fibres 2 & 3

Muscular antagonism by means of the hydrostatic skeleton

It will be realized that in an animal which consists of a wall of muscle enclosing a fluid, the contraction of any one muscle affects the hydrostatic pressure of the contained fluid and hence affects the tension which all the other muscles have to exert. The use of a body fluid as an agent of muscular antagonism is a skeletal function, and animals employing such a system can be said to have a hydrostatic skeleton. If the muscles were to run in random directions in the body wall, enormous nervous complexity would be

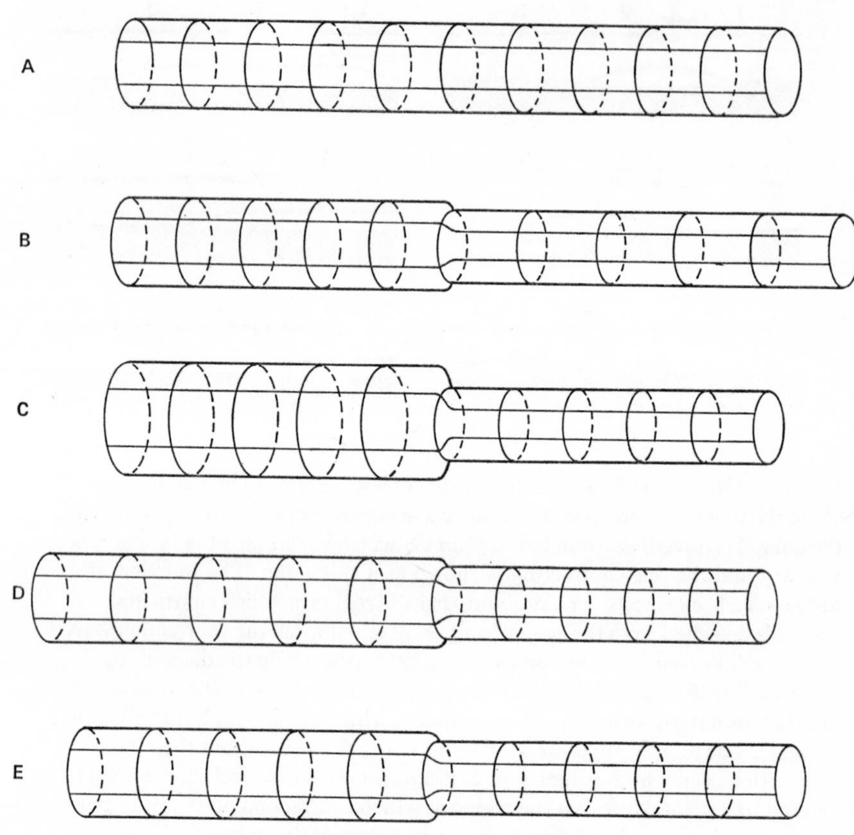

Fig. 9.2 Diagram to illustrate four possible results of the contraction of circular muscles at one end of a cylindrical animal. In A the muscles are all relaxed. In B the circular muscles of the right-hand end have contracted and this end has elongated. The left-hand end has remained unaltered. In C the length of the right-hand end has remained the same but the diameter of the left-hand end has increased. In D the length of the right-hand end has also remained the same. The length of the left-hand end has increased but not its diameter. In E the length of both ends has increased but their diameters have remained the same as in B and D

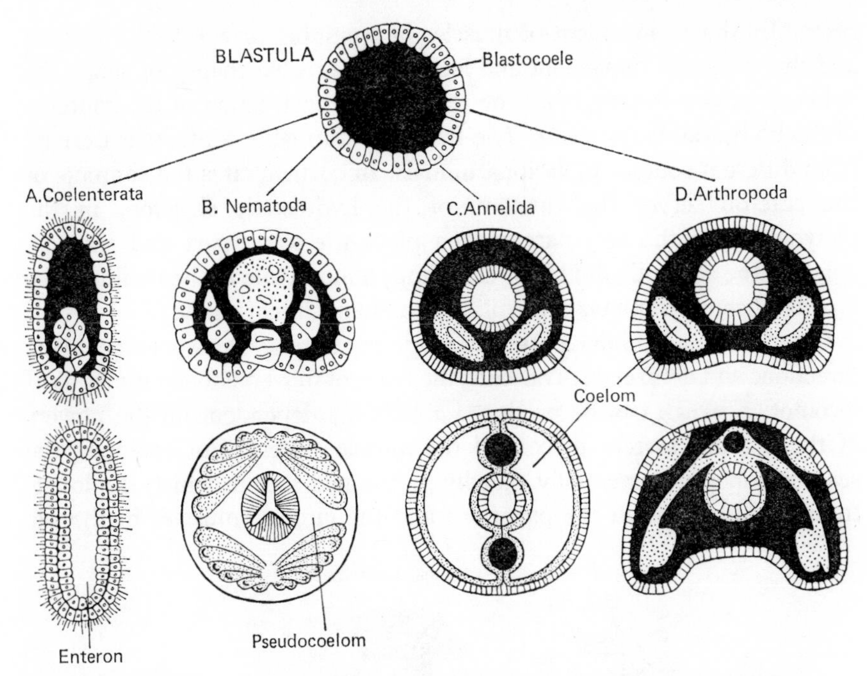

Fig. 9.3 Diagram to illustrate the origin of the body cavities in various animals from an idealised blastula. A, Coelenterata – the blastocoele is obliterated and the functional adult body cavity is the enteron; B, Nematoda – the blastocoele is also obliterated and the functional adult body cavity is a pseudocoelom formed by the vacuolation of large mesoderm cells; C, Annelida – the blastocoele is reduced to the cavities of the blood vessels and the coelom formed by the appearance of a cavity in the mesodermal somites; D, Arthropoda – the blastocoele forms the cavity of the large open blood system and the coelom, although present, is small. The blastocoele is shown black throughout and the mesoderm stippled

required to co-ordinate them sufficiently to produce ordered motion. In fact, such an arrangement is not found and in the course of evolution, a common pattern of longitudinal and circumferential muscle fibres has been adopted by many coelenterates, annelids, nemerteans and members of other phyla. Such an arrangement still demands a high degree of nervous co-ordination since any active muscles are always in antagonism to the others but the system is capable of performing many different movements albeit in a relatively inefficient way compared with a system in which muscles are antagonized individually, such as can take place through a system of levers forming a hard skeleton. Some of the varieties of movement which can be made by a circular and longitudinal muscle system are illustrated in Fig. 9.2. It will be recognized that this simple system resembles the arrangement of muscles in a number of worm-like animals and that although movement and locomotion involve change of shape the volume remains constant and the amount of translocation of the contents of the body wall is not great. The fluid or deformable contents is derived from different sources in various animals. In coelenterates the contents of the enteron serves the function of the hydrostatic skeleton, in the platyhelminths the soft parenchyma plays the same part and in most annelids the coelomic fluid is of chief importance while in the molluscs the blood or haemolymph takes the place of coelomic fluid (Fig. 9.3).

Given that an animal possesses muscles running in two antagonistic directions and deformable contents, the types of movement and methods of locomotion which can be produced are chiefly dependent on the powers of the nervous system to control the muscle system. For example, the sea-anemones are essentially radially partitioned sacs in which the longitudinal muscles lie in the partitions and the circular muscles in the sac

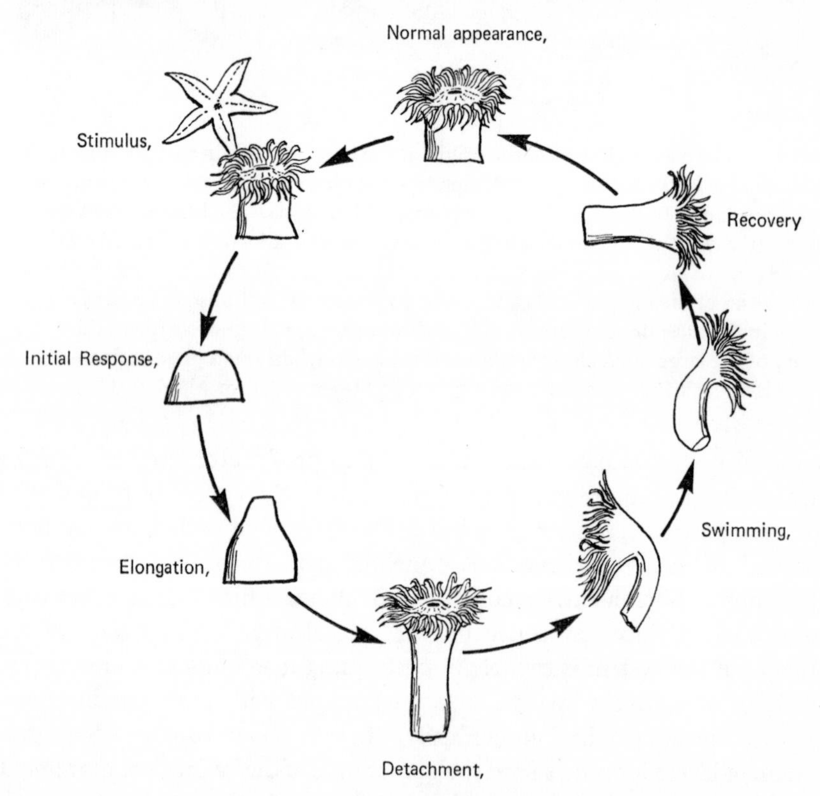

Fig. 9.4 Diagram to illustrate the sequence of movements made when *Stomphia* is stimulated to swim by contact with a starfish, *Dermasterias*

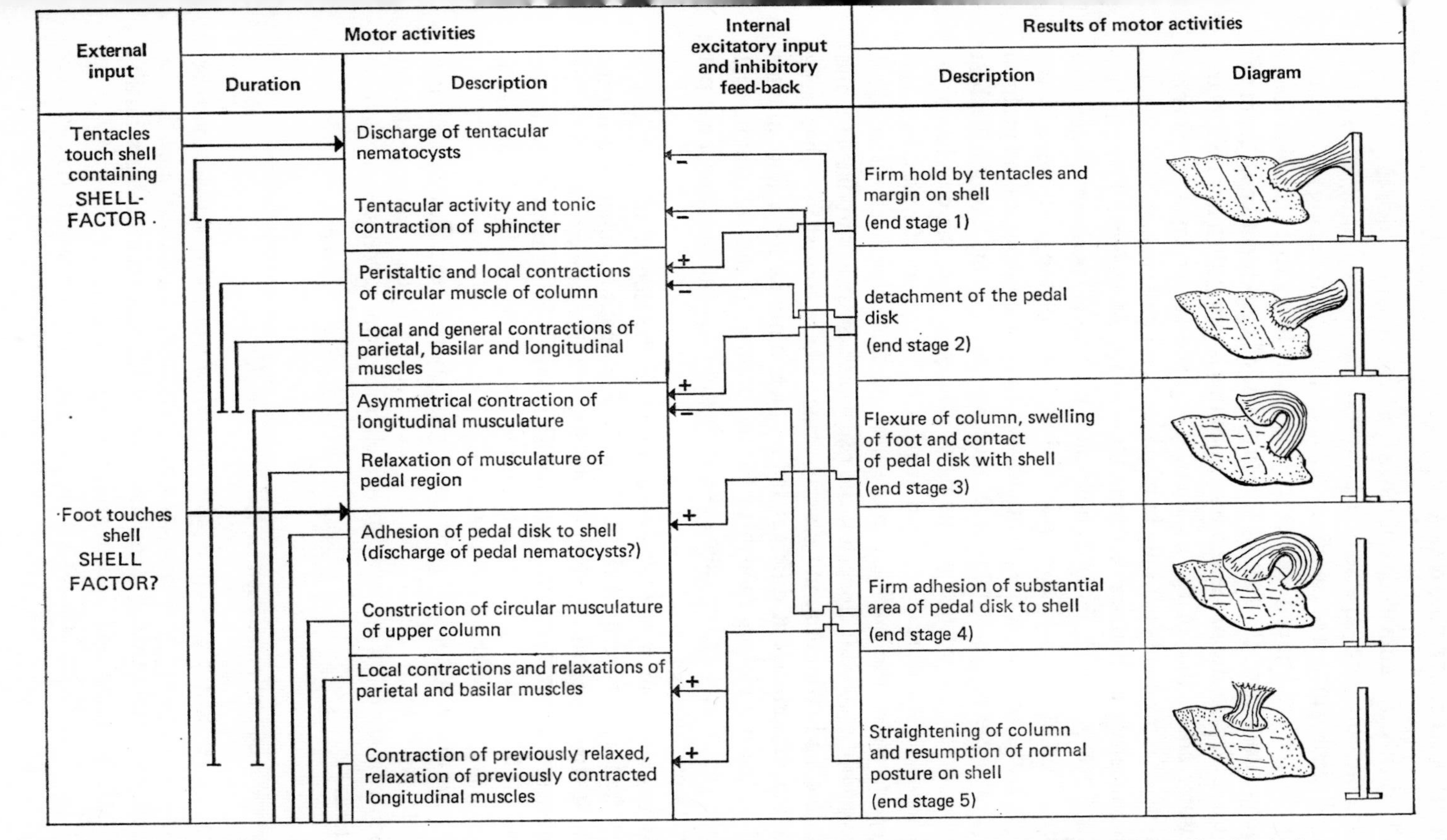

Fig. 9.5 Diagram to show: (a) sequence of motor events in the transference behaviour of *Calliactis parasitica*; (b) the timing and relationships of the excitory (+) and inhibitory (−) influence presumably involved in this sequence. Horizontal divisions in col. 3 separate stages 1 to 5. Each stage begins with the first-named activity out of which the second phase develops

wall. Mostly the animals do not move from the spot on which they settle as larvae but some are capable of a gastropod-like slow creeping locomotion, each little advance resulting from the detachment, stretching and replacement of the leading edge of the foot. It seems likely that the small forces required to extend the foot in the direction of movement come from the hydrostatic pressure in the enteron, since this is generally a few millimetres of water above that of the medium. The fixed sea-anemone performs a variety of movements not only in response to stimuli such as food but also spontaneously as Batham & Pantin showed for *Metridium*. These movements are generally fairly simple and consist of peristaltic waves of contraction of the circular muscle of the column or the contraction of the longitudinal muscles of the partitions. But some sea-anemones can perform much more complex sequences of muscular movement. Two of the best known examples are the swimming of *Stomphia* and the attachment of *Calliactis* to whelk shells. When starfish of the genus *Dermasterias* and *Hippasterias* or the nudibranch *Aeolidia papillosa* approach an attached *Stomphia* the column elongates, the pedal disc constricts in diameter and detaches and swimming movements begin. These consist of rather jerky bending movements at different radii which continue for a few minutes and which are violent enough to lift the animal off the substratum and which can, therefore, be called swimming movements but which are not directional (Fig. 9.4).

When a specimen of *Calliactis* attached to the side of an aquarium is presented with a fresh whelk shell, with or without an inhabiting hermit crab, it attaches to the shell by the discharge of nematocysts which fix the tentacles, the foot is detached, the column bent to bring the foot on to the shell, the foot attached and finally the tentacles released when the anemone is settled on its preferred substratum (Fig. 9.5).

These two examples illustrate and give point to the idea that even a simple sac-like system of muscles, can, with suitable nervous control, produce a series of movements which make up a complex behaviour pattern.

Variations among animals with a hydrostatic skeleton

How far is such a system capable of refinement? The answer may be sought in the power of movement and locomotion of animals higher in the evolutionary scale than coelenterates but which still retain the essentially simple longitudinal and circular arrangement of muscle.

It has been seen how free-living platyhelminths above a certain size crawl by muscular action chiefly by their longitudinal muscles. They also possess circular muscles and often also diagonal fibres. The body is made up of easily deformable parenchyma and quite a high proportion of the total volume can be made up of the fluid contents of the gut. Altogether the mechanical construction of platyhelminths is essentially that of an animal with a hydrostatic skeleton. The flattened shape can be looked upon as an adaptation partly to aid oxygen uptake and partly as providing a large surface/volume ratio which is of benefit in ciliary locomotion. Contact between the worm and the substratum is undoubtedly enhanced by the secretion of mucus, without which it is probable that the muscular crawling type of locomotion would be impossible since the weight in water of a planarian is very small indeed and frictional forces between it and the ground are correspondingly minute.

The movement of parasitic platyhelminths is not very well studied but trematodes such as the liver fluke possess backwardly directed spines in the cuticle which are said to aid their progression along the bile duct. The powerful longitudinal musculature of tapeworms is presumably required to counteract the peristaltic action of the gut of the host, which would otherwise tend to move the worm along together with the gut contents. There is room for further study here which should be possible with the improved techniques now available for culturing parasites.

The nemerteans like the platyhelminths have the space between the gut and body wall filled with easily deformable tissue rather than liquid but some can be even more variable in their shape. The mechanical basis of this ability to undergo great deformation is the arrangement of the inelastic fibres of the basement membrane of the epidermis in two helices of opposed sense. Such an arrangement allows the normal arrangement of circular and longitudinal muscles to bring about such great changes of form that the fully extended length can be up to nine times the fully contracted length. Fig. 9.6 is a graph relating the volume enclosed by a single turn of a geodesic fibre at various angles of inclination to the long axis. Clearly when the fibre is at 0° and at 90° the shape enclosed is a line and a disk respectively and the volume is zero. The greatest volume is enclosed when the angle is about 55°. But, of course, the volume of a worm does not change as its shape changes so that the horizontal line representing the constant volume of a real worm cuts the curve at two points which represent its maximal elongation and its maximal contraction. At these points it is, of course, circular in cross-section but in between, the geodesic inelastic fibres could contain a larger volume than that of the worm which is constant. Hence its cross-section is not circular but elliptical. In the

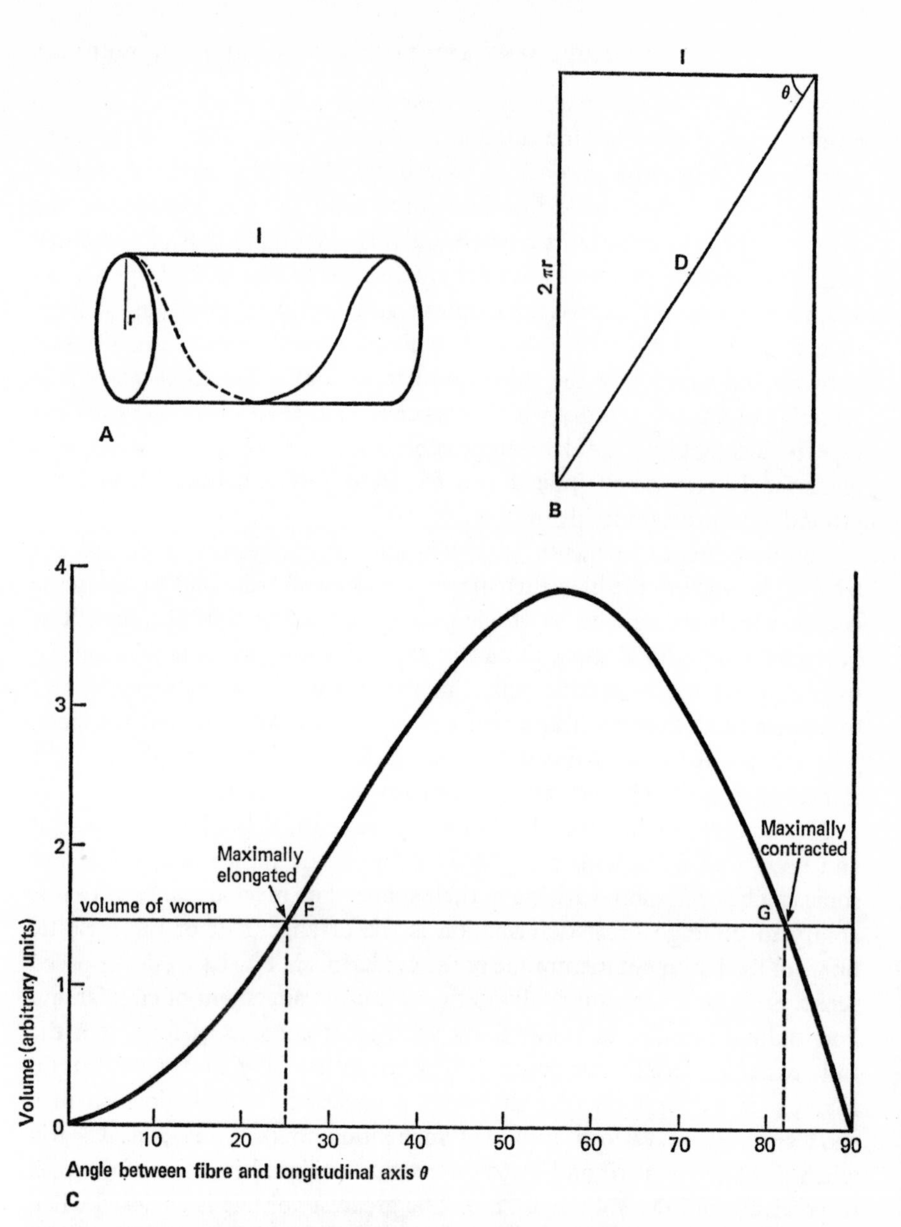

Fig. 9.6 A, unit length of a cylindrical 'worm' bounded by a single turn of the geodesic fibre system (fibres running in the opposite sense have been omitted for clarity); B, the same unit length of worm slit along the top and flattened out; C, the curve represents the theoretical relationship between the volume contained by the fibre system and the inclination of the fibres to the longitudinal axis. The horizontal line represents the actual and constant volume of the nemertine *Amphiporus lactifloreus*. It intersects the curve at F and G which are the limiting positions of elongation and contraction, respectively, for that species

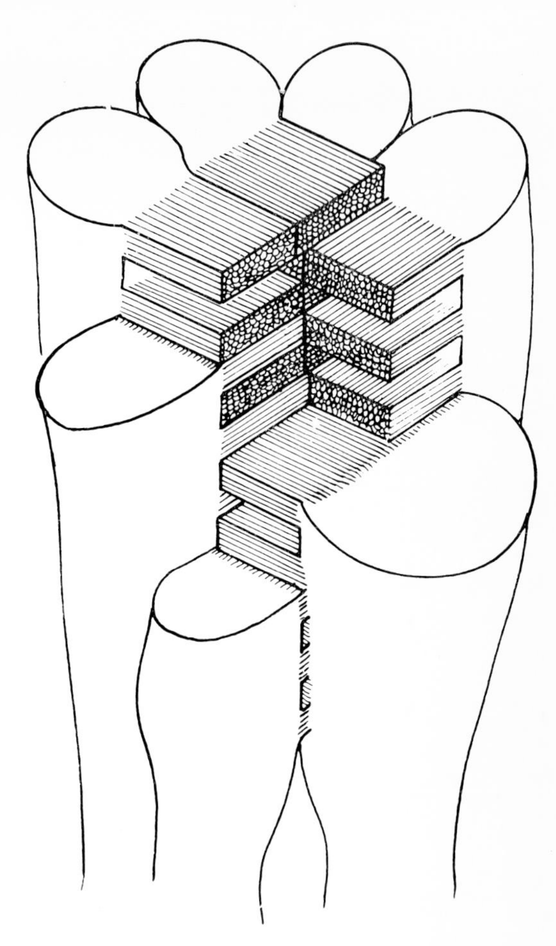

1 (a) Diagram of the arrangement of the cross-plates of tubules in the rhabdom of an eye of a crustacean. (b) Cross-section of the tubules forming the rhabdom of a single ommatidium of the compound eye of the water beetle *Dytiscus*

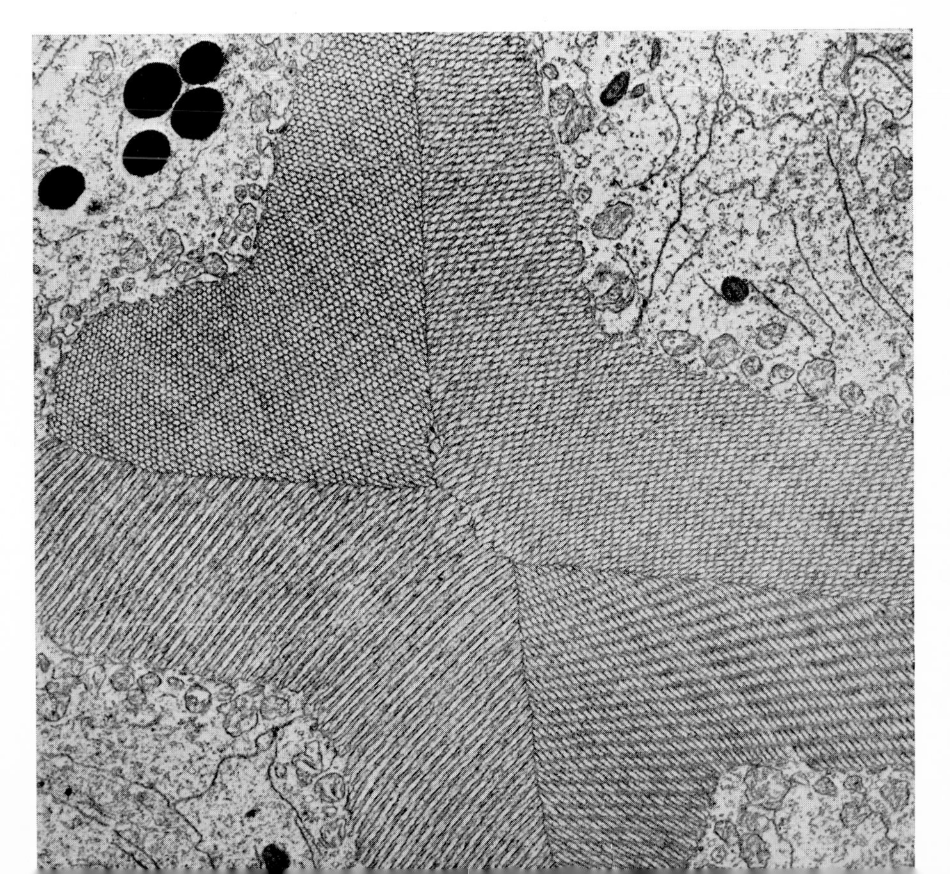

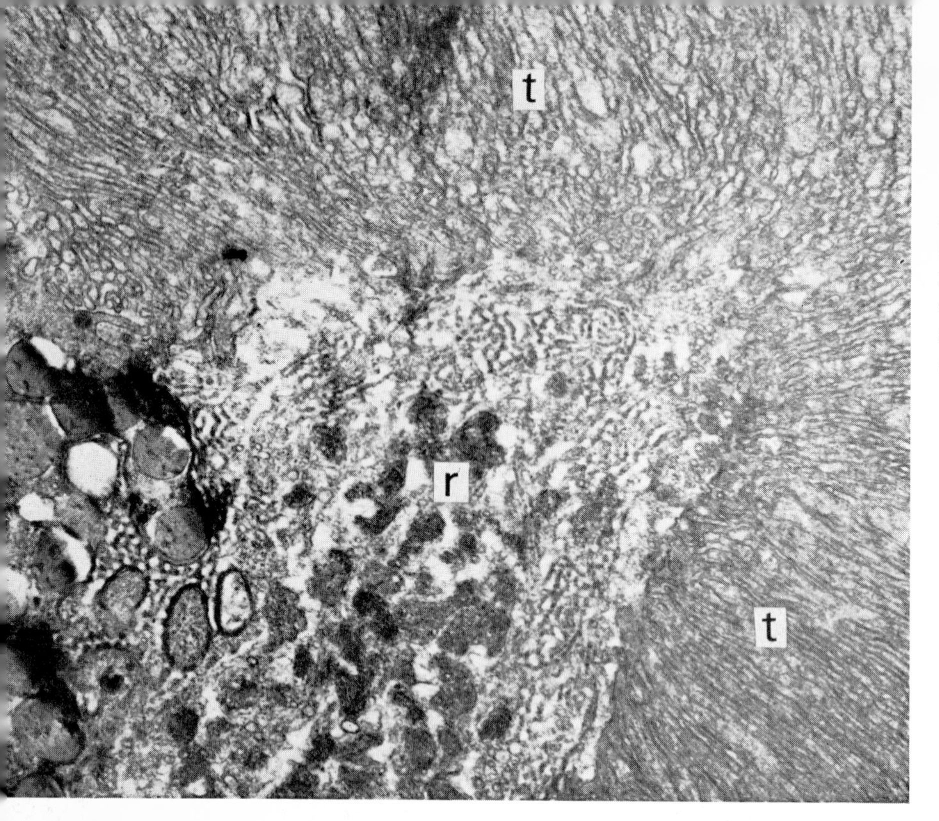

2 The upper ends of retinal cells (r) in the eye of the slug, *Agriolimax*, showing the brush of tubules (t)

3 An Ocypode crab in a threatening posture

4 The pectines of the scorpion *Leiurus quinquestriatus*. Above, the whole organs; below, a scanning electron micrograph of the sensory pegs on one 'tooth' of the pectine

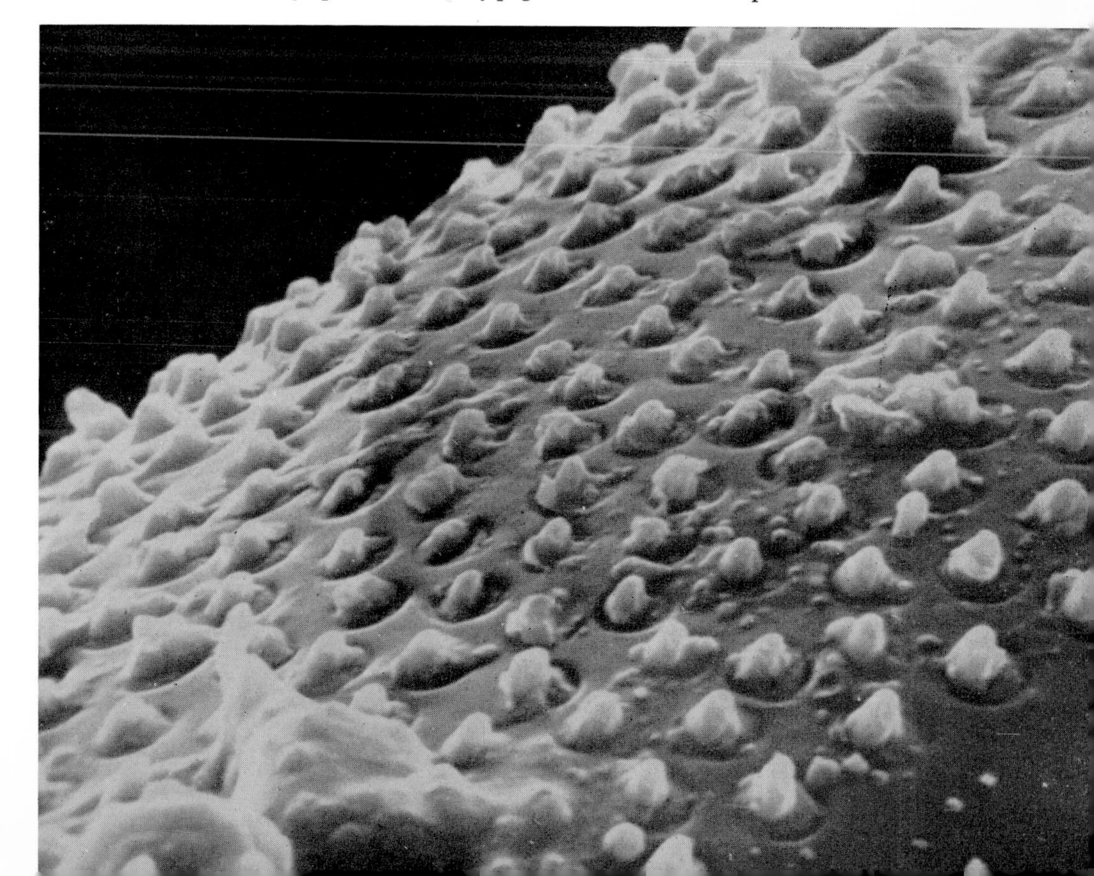

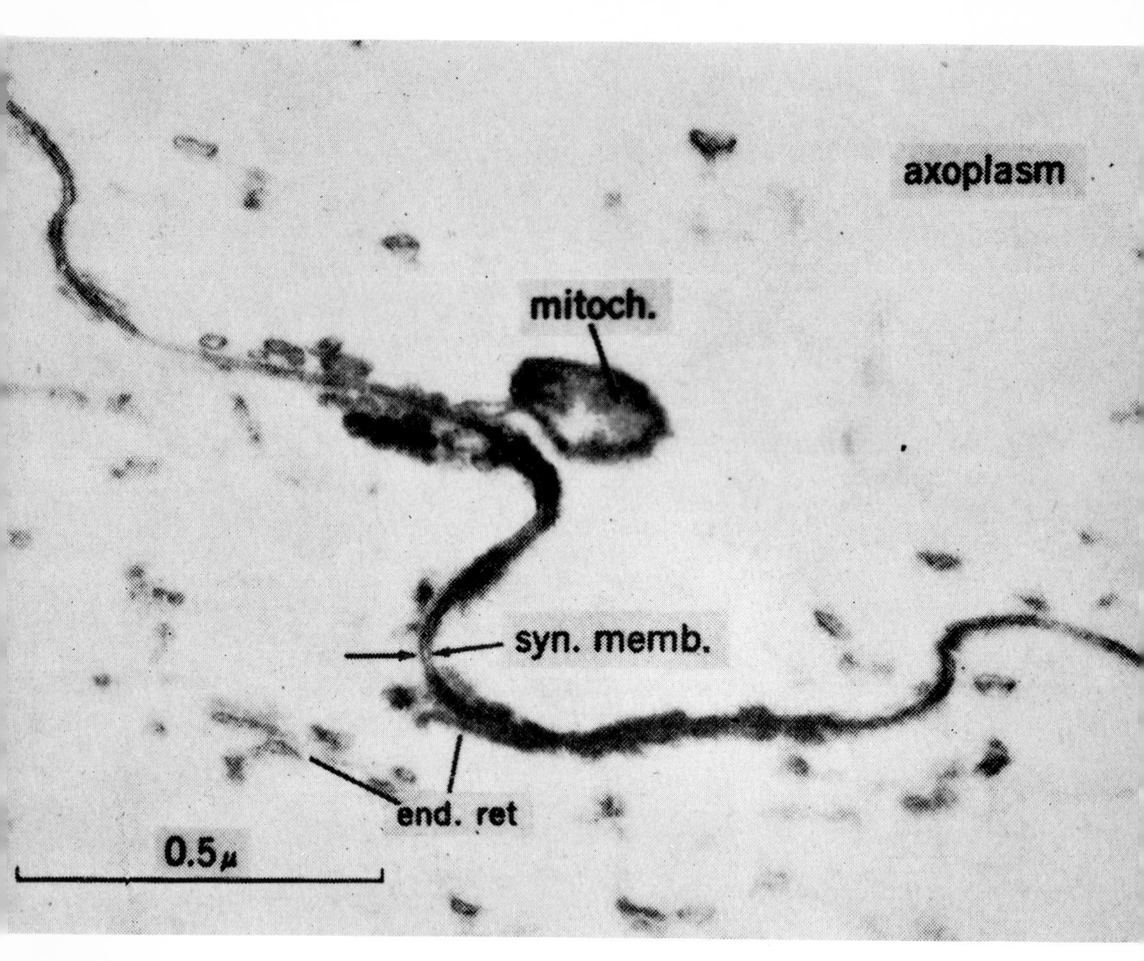

5 Macrosynapse in an earthworm, seen under electron microscopy. There are usually vesicles lined up on each side of the synaptic membrane but they are not abundant in this section. *mitoch.*, mitochondrion; *syn. memb.*, synaptic membrane; *end. ret.*, endoplasmic reticulum

6 *Calanus helgolandicus*

7 The orientation of *Arctosa*. Above, the arrangement of the shield around the dish so that the spider cannot see landmarks. Below, a spider orientating NNW

8 A group of fiddler crabs (*Uca* sp.). The males are distinguished by their large chelae; some are standing at the entrance of their burrows

9 *Arenicola marina*. The casts and head-shaft holes of this burrowing worm can be seen on the surface

10 Dragonfly larva (*Aeschna* sp.) with mask extended

11 Mayfly nymphs (*Ecdyonurus torrentis*). Note the gills along the sides of the abdomen and the nymphs' flattened shape, an adaptation to life in fast-flowing waters

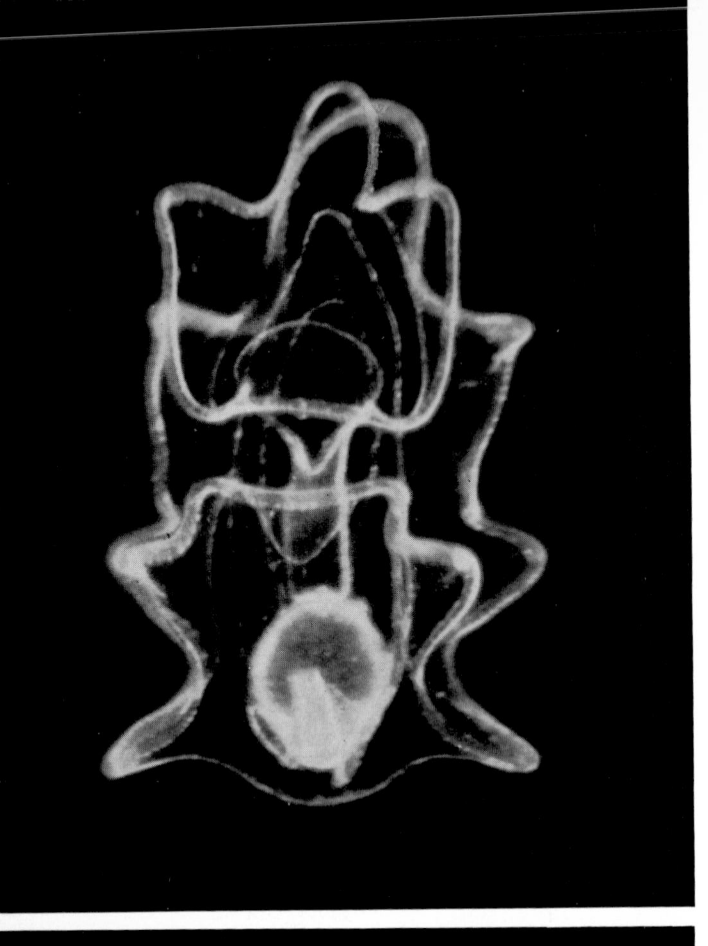

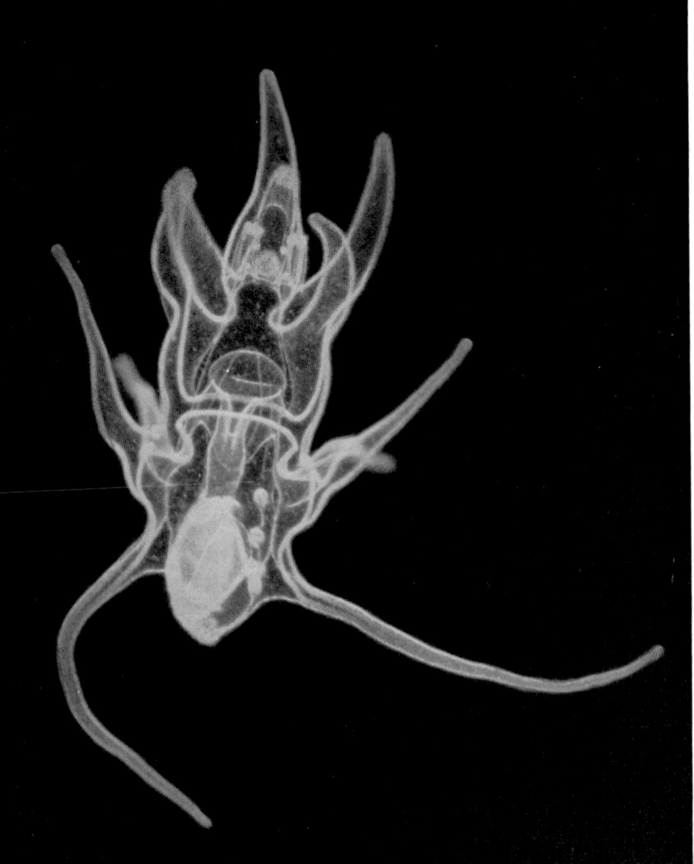

12 Stages in the larval development of *Asterias rubens*:

(i) Bipinnaria larva (ventral view)

(ii) Brachiolaria larva (ventral view)

13 Stages in the development of *Echinus esculentus*:

(i) Early pluteus larva (ventral view)

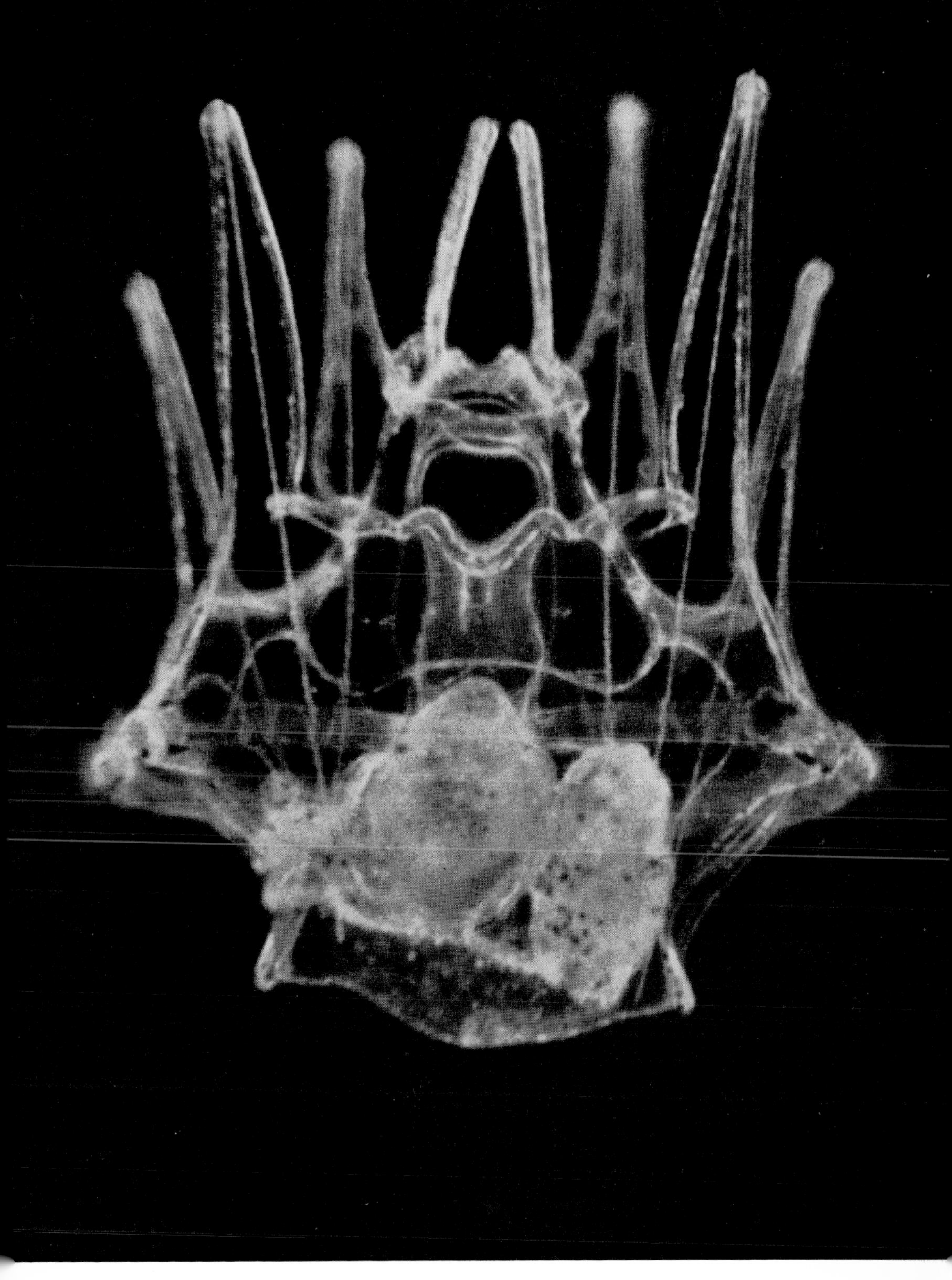

(ii) Later pluteus with eight arms. The rudiment of the future adult can be seen in the right lower part of the larva (ventral view)

continued overleaf

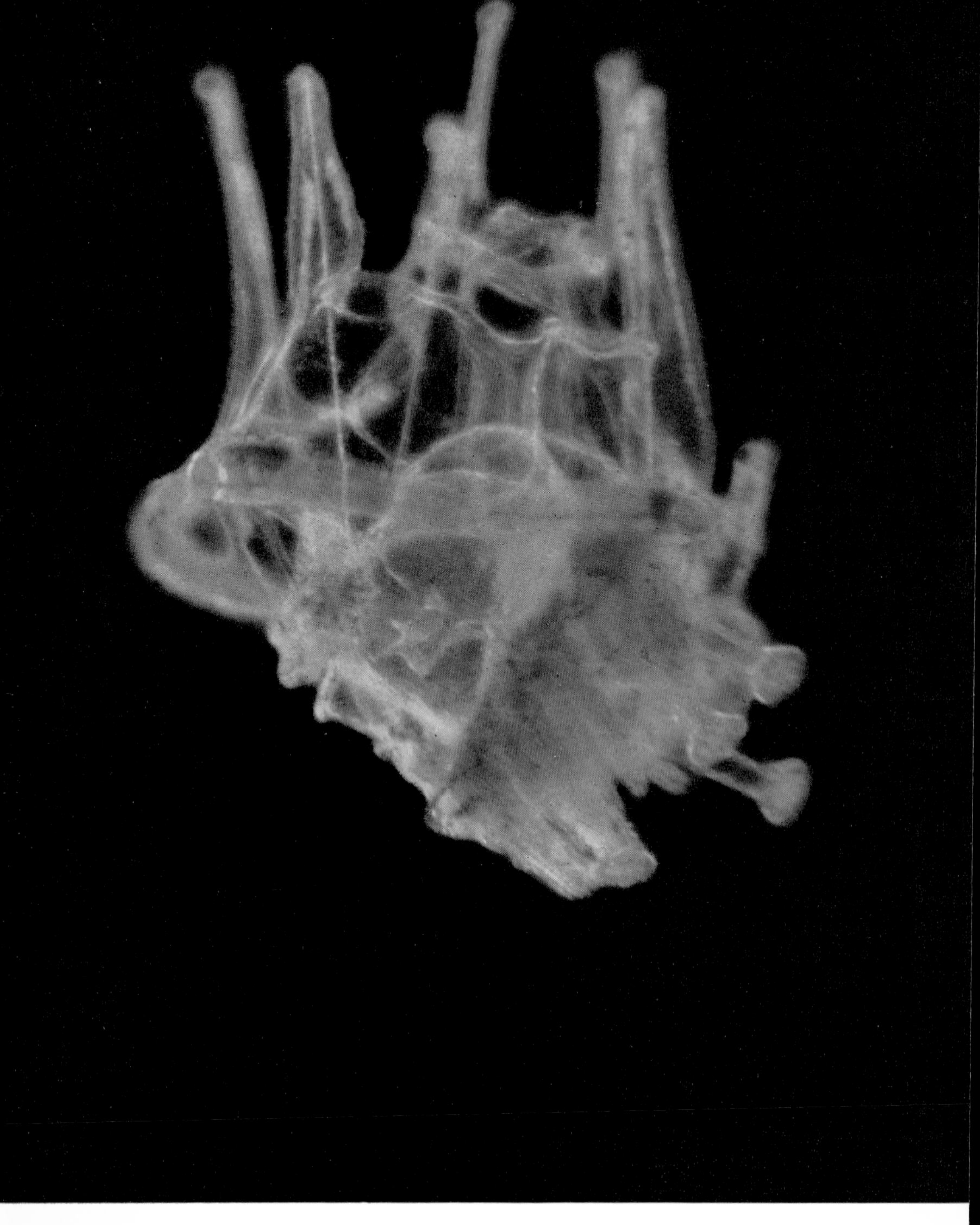

(iii) Late larva, the rudiment has developed spines and tube-feet (ventral view)

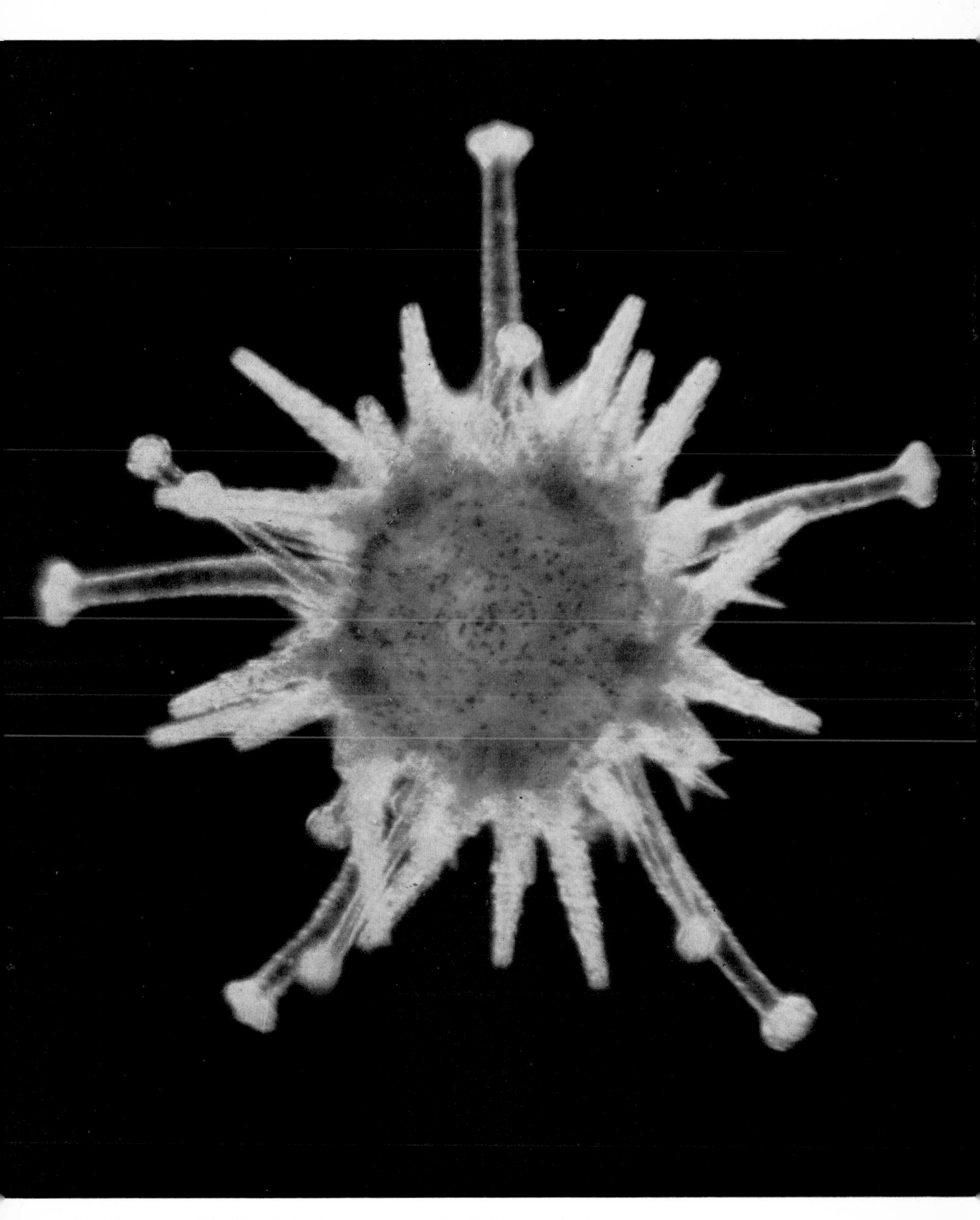

(iv) Young urchin shortly before metamorphosis (aboral view)

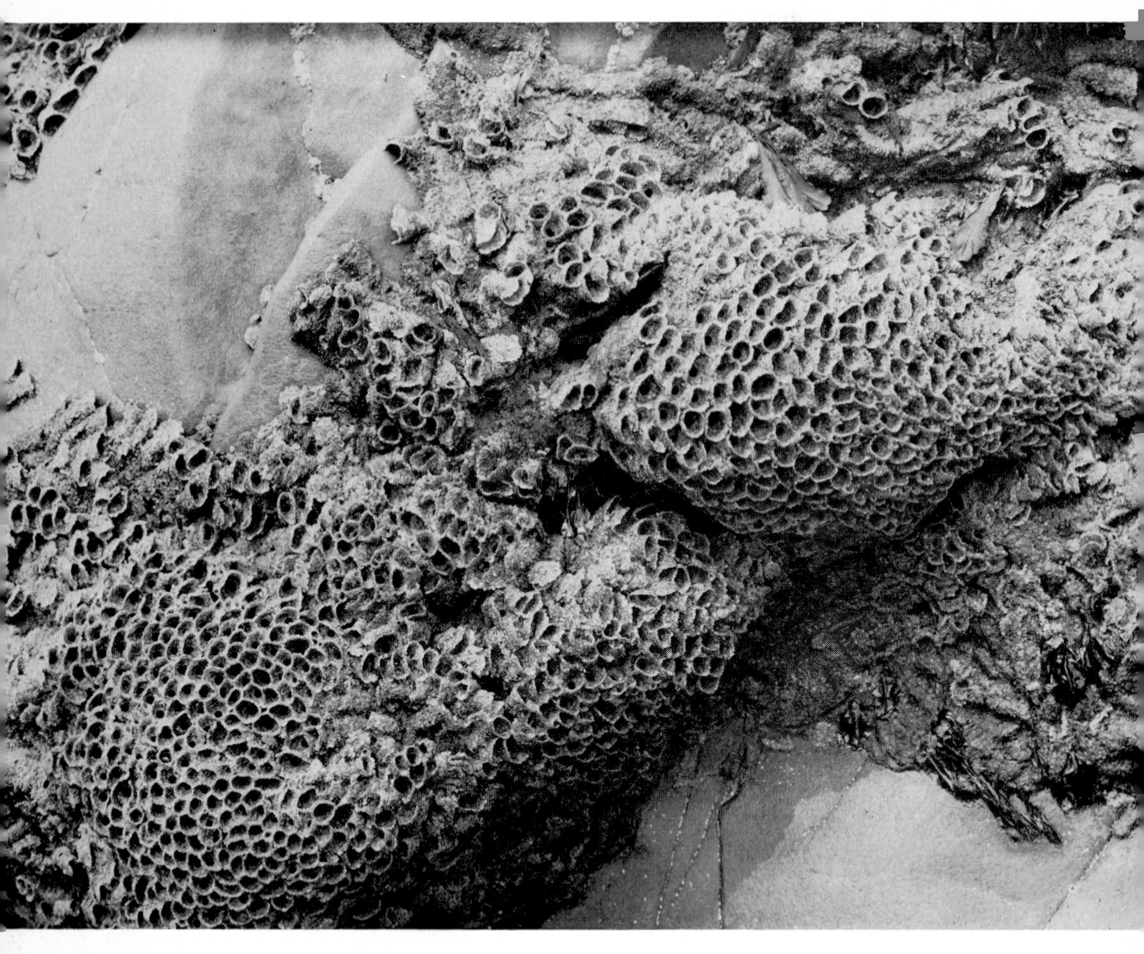

14 Colonies of the honeycomb worm *Sabellaria alveolata*. The tubes are built of sand grains cemented together

15 (a) Late larva of *Sabellaria alveolata* actively crawling, seeking a settlement place

(b) Young worm, completely metamorphosed, in its primary mucoid tube attached to an empty sand tube of another worm

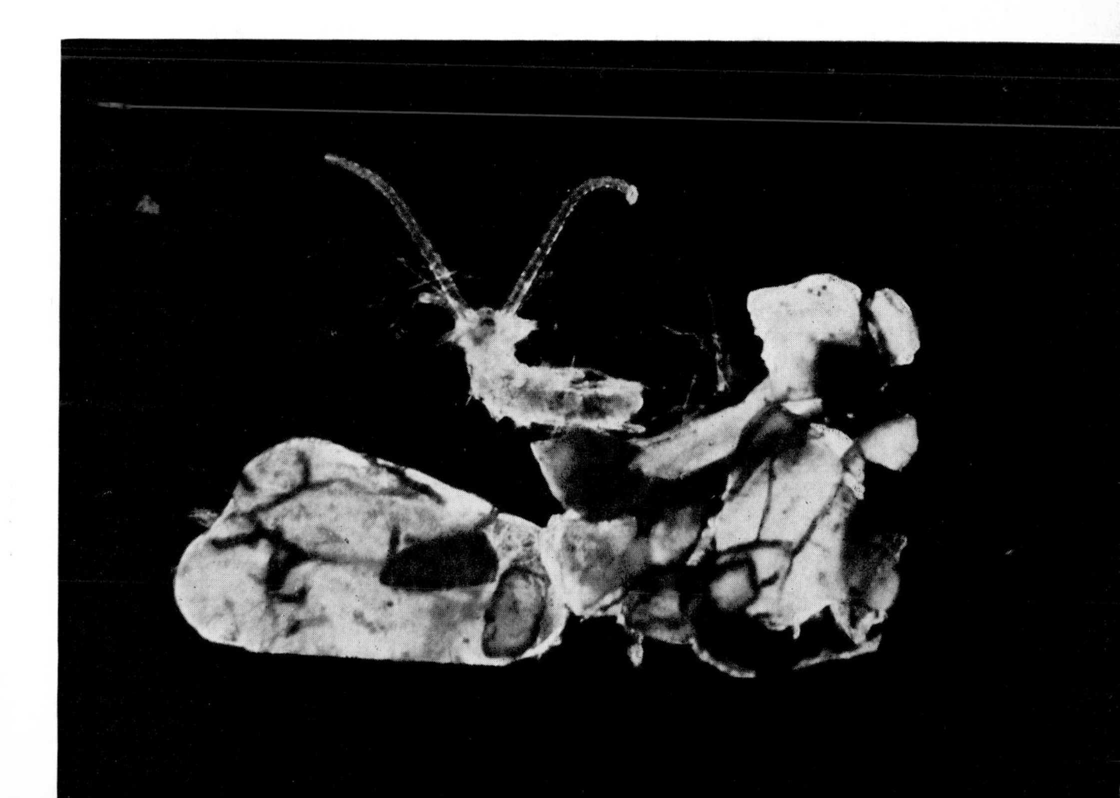

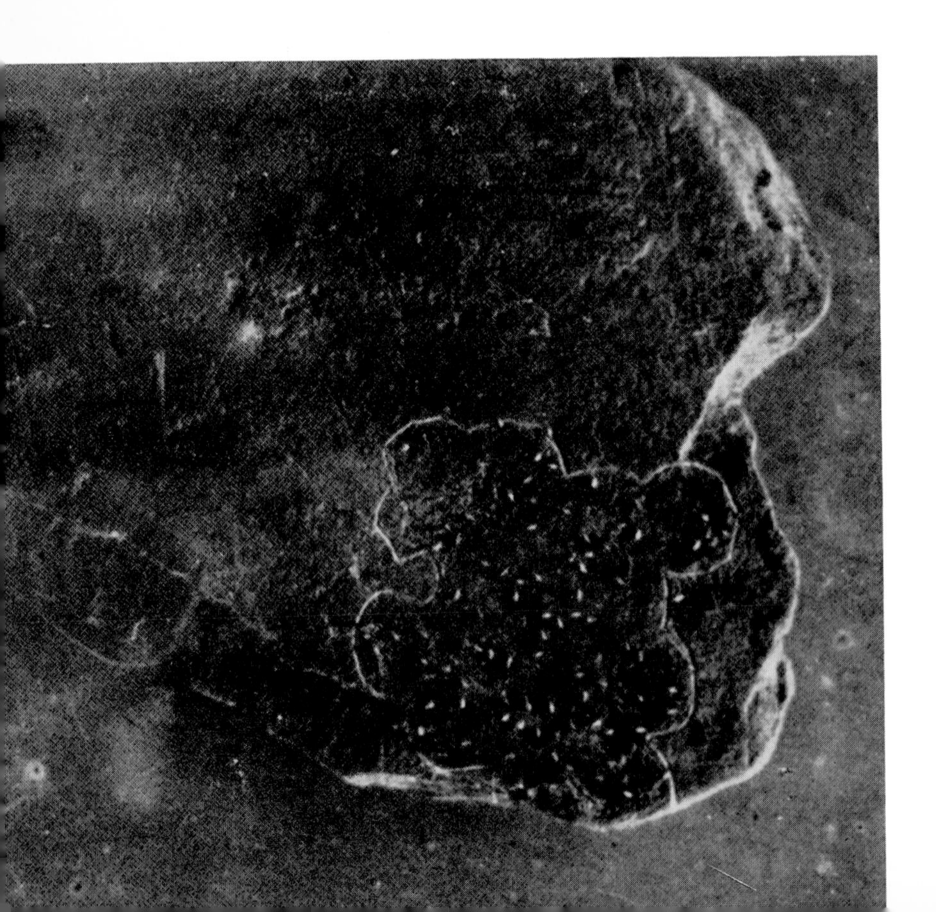

16 Settlement of the barnacle *Balanus balanoides*. The position of adult barnacles (above) is scratched on to the slate before removing them. Cyprid larvae of this species (below) allowed to select where they will attach do so almost exclusively within the area bounded by the scratch

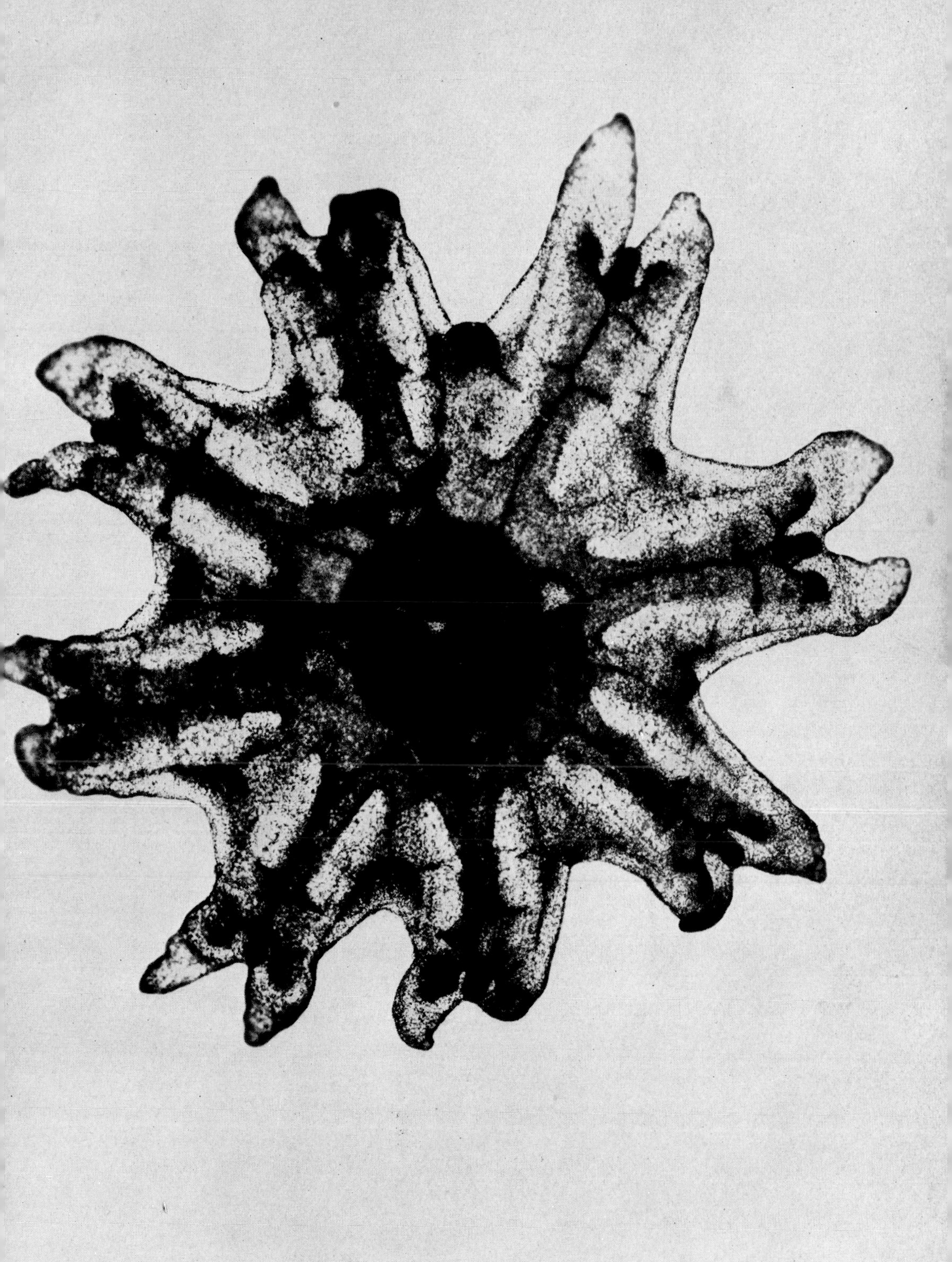

17 An ephyra larva of the jelly-fish *Cyanea*

18 A megalopa larva of the shore crab *Carcinus maenas*

19 A schizopod larva of *Nephrops norvegicus*. The tail processes serve to increase the body area of the animal and thus aid it to float

20 A veliger larva of the mollusc *Aporrhais pes-pelicanae*

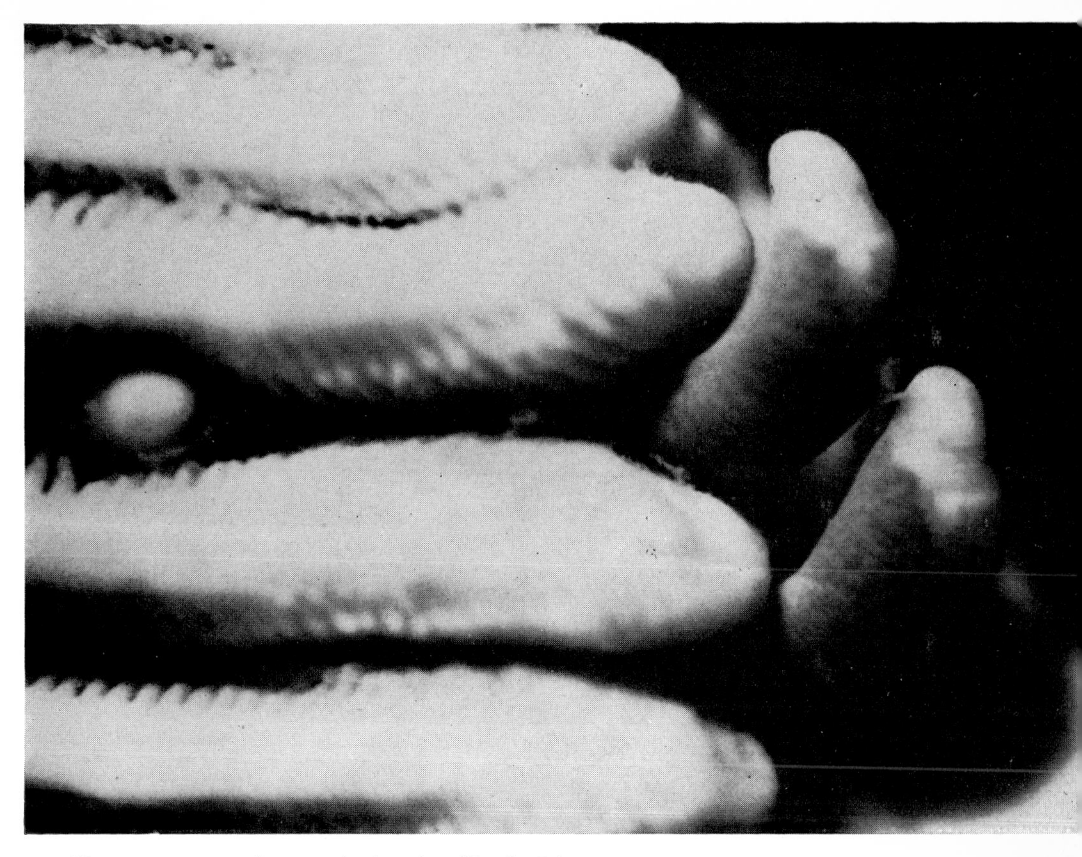

21 Monogenean parasites attached to the gills of a fish

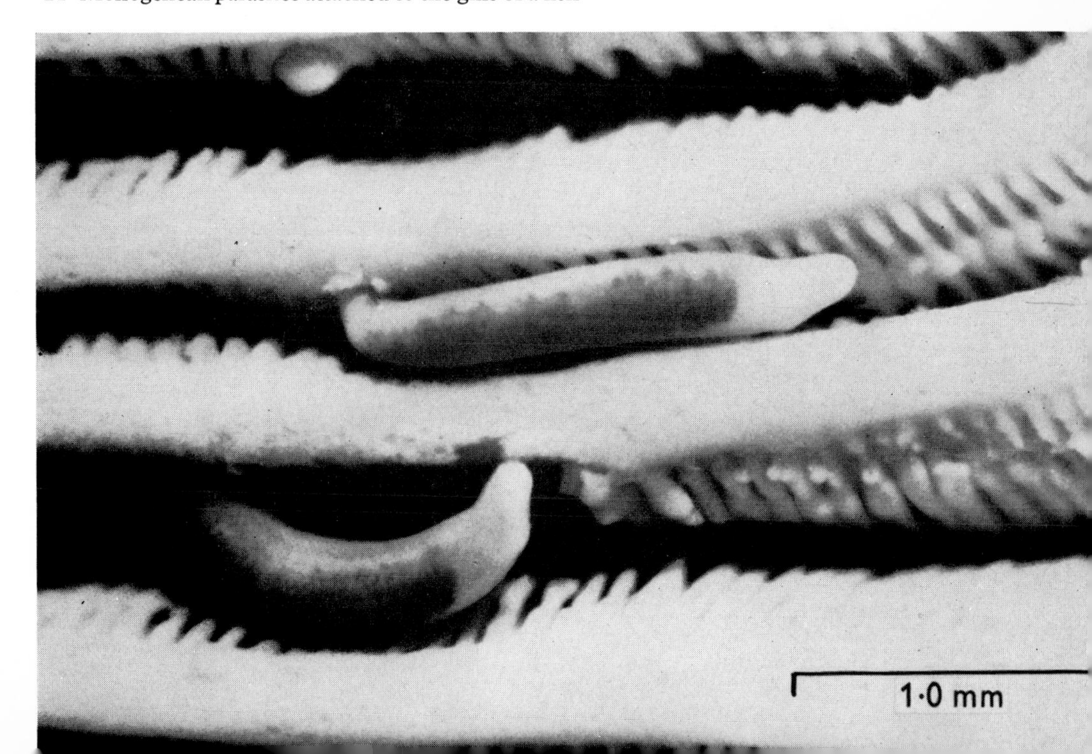

22 The attachment and settling of the anemone *Calliactis parasitica* on a shell occupied by a hermit crab *Pagurus bernhardus*
(a) The anemone is partially attached to the wall of the tank has made contact with the shell
(b) The tentacles adhere to the shell, the column of the anemone twists and the pedal disc begins to lift from the glass (at rear)

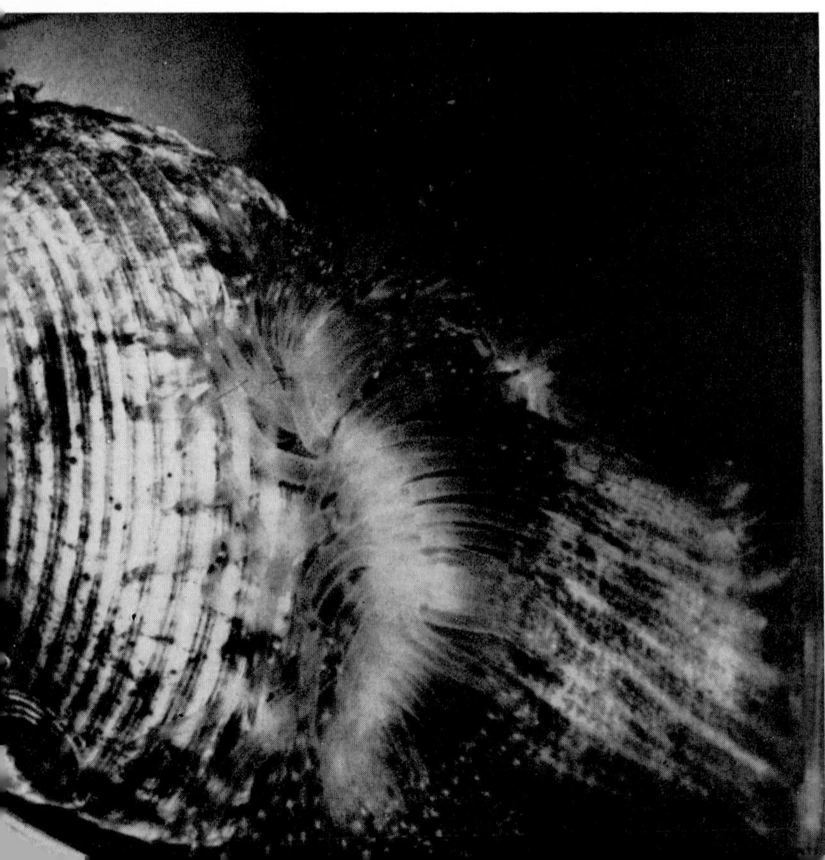

(c) The oral disc bearing the tentacles is expanded to give an increased area of attachment to the shell; the pedal disc is lifted clear and by further twisting of the column is swung round towards the shell
(d) The anemone is fully settled on the shell (from a different sequence of photographs than a, b and c)

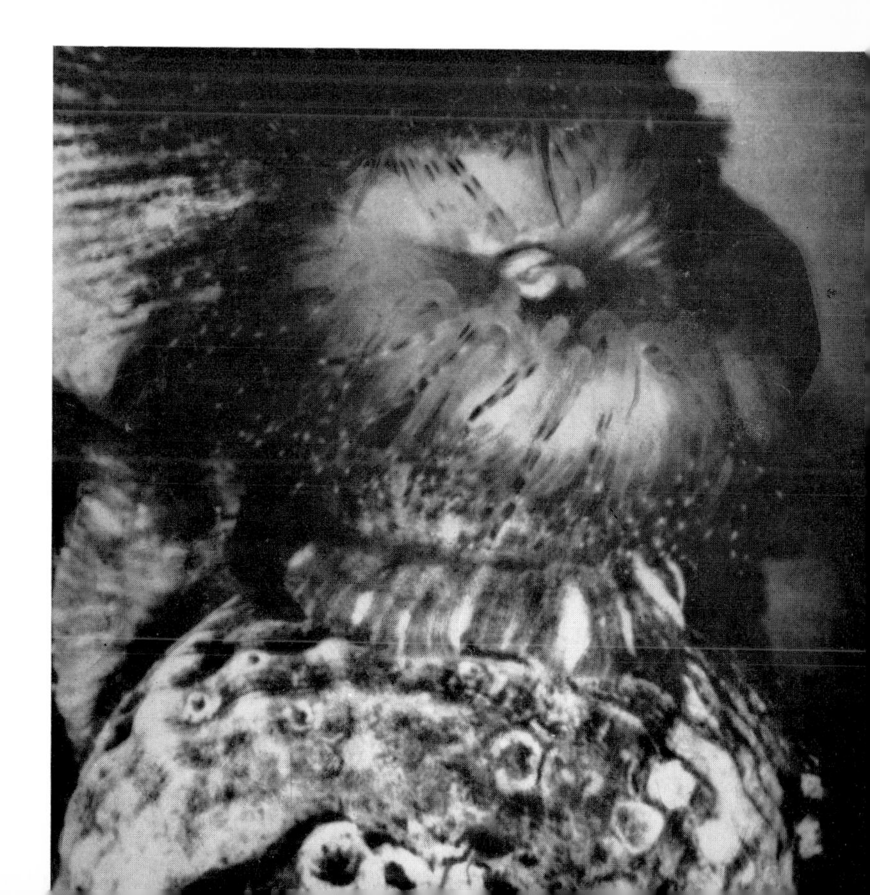

23 Pre-Cambrian fossils from the Ediacara Sandstone of South Australia, one of the first assemblages of recognizable animal remains to appear in the fossil record

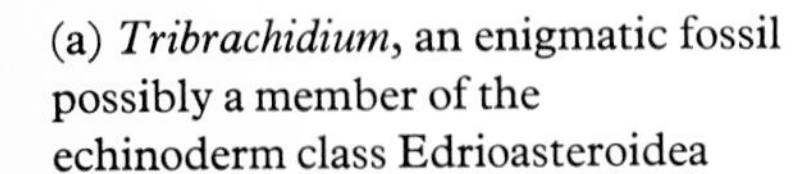

(a) *Tribrachidium*, an enigmatic fossil possibly a member of the echinoderm class Edrioasteroidea

(b) *Pteridinium*, an animal resembling present-day pennatulid coelenterates

(c) *Dickinsonia*, a segmented, annelid-like animal somewhat resembling the present-day polychaete *Spinther*

(d) *Spriggina*, another segmented, annelid-like animal with fewer segments than *Dickinsonia*, somewhat like the present-day pelagic polychaete *Tomopteris*

(e) *Mawsonites*, a medusa-like animal

(f) *Parvancorina*, possibly an arthropod-like animal

24 An artist's restoration of an assemblage of invertebrates of the Middle Cambrian seas, based on the remarkable fossils preserved in the Burgess Shales of British Columbia, Canada. The pipe-like animals are archaeocyathids, and browsing over these and over other organisms are arthropods, trilobites and (bottom centre) an onychophoran. Pelagic trilobites, medusae and primitive arachnids swim in the waters above. Reproduced with permission of the Smithsonian Institution, Washington D.C.

25 An artist's restoration of a Silurian sea-bottom, based on the fossils found in the eastern United States. Predatory eurypterids swim over the weed, among which gastropod molluscs browse. Reproduced by permission of the Chicago Natural History Museum

26 An assemblage of the trilobite *Phacops* from the Devonian shales of Ontario, Canada. In this period of geological time the decline of the trilobites had already begun. Photograph reproduced by permission of the Smithsonian Institution, Washington, D.C.

27 (a) Crinoids from the Carboniferous period. Some rocks of this age are almost entirely composed of crinoid fragments
(b) Restoration of a sea-bottom in Lower Carboniferous times, in which stalked crinoids are most conspicuous. Reproduced by permission of the Smithsonian Institution, Washington, D.C.

28 (a) An accumulation of belemnites, from the Jurassic. (b) Restoration of an Upper Jurassic sea, showing the pelagic belemnites swimming close to the sea-bed. Reproduced by permission of the Chicago Natural History Museum

relaxed or anaesthetized state the contents of the body tend to flatten maximally at which point the angle made by the geodesic fibres is about 55°.

It follows from their shape that flattened worms, such as *Cerebratulus* among the nemerteans, should theoretically be highly extensible. In fact, it is not and its relative inextensibility is due probably chiefly to the relatively inextensible muscle fibres of the longitudinal layer, and also to the presence of dorso-ventral muscles which help to keep the body flattened.

The arrangement of muscles in the body of nemerteans falls into several different categories. In the palaeonemerteans there are inner and outer circular muscle layers with a longitudinal layer in between. In the heteronemerteans the order is reversed and there is a single circular layer lying between inner and outer longitudinal layers, while in the hoplonemerteans there are frequently found two thin layers of diagonal fibres between an outer circular and an inner longitudinal layer. No adequate functional analysis or evolutionary explanation of these variations of the longitudinal and circular muscle pattern has been made since they seem to offer little or no functional advantages over a simple longitudinal and circular system, and they all serve to generate the peristaltic waves which propel the nemerteans over the substratum. The waves resemble those of the earthworm, but, owing to the great relative length of many nemerteans there are often many zones of longitudinal contraction present at the same time and the co-ordination between them is often poor.

Other animals in which the muscles are arranged as circular and longitudinal layers in the body wall include the annelid, echiuroid and sipunculid worms, although among the annelids there is great variation of this basic arrangement chiefly connected with the development of parapodial projections from the body wall in the polychaetes and with the presence of suckers in the Hirudinea.

An alternative to the antagonism of longitudinal muscles by circulars is found in certain hollow parts of molluscs such as the siphons of bivalves and the mantle of cephalopods. Radially arranged muscle fibres diminish the thickness of the wall and hence extend the siphons without blocking the lumen. Similarly the volume enclosed by the contracted mantle of a cephalopod can be increased so that water can be taken in ready to be expelled at the next contraction.

The peristaltic movements of the earthworm are well understood and are illustrated in Fig. 9.7 which shows the external appearance of the locomotory waves and the analysis of the forces affecting a fixed segment. The chaetae can perhaps best be regarded as adding to the frictional

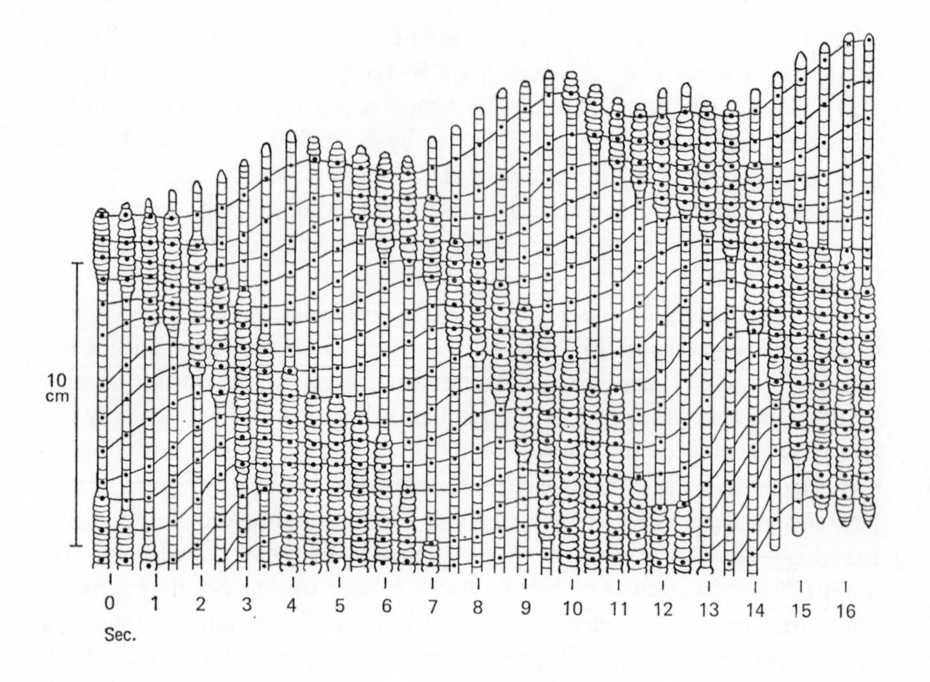

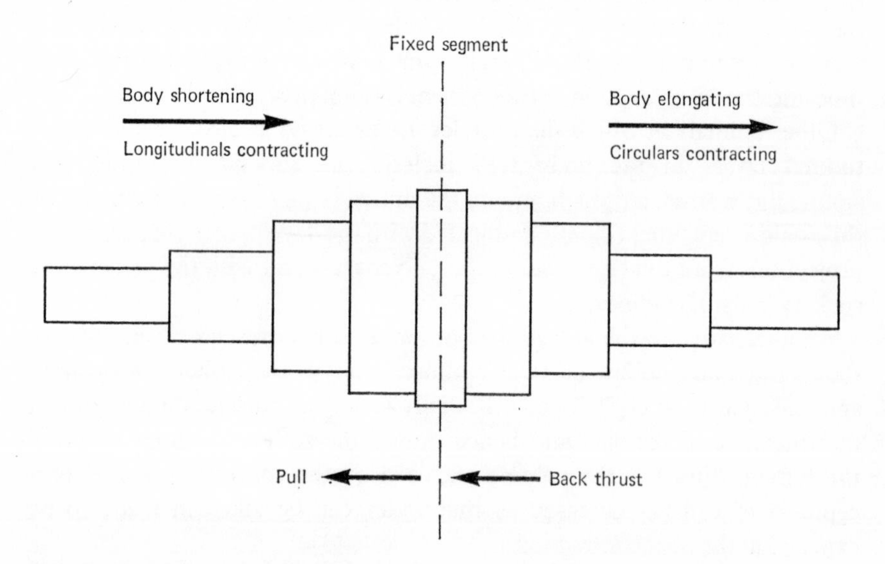

Fig. 9.7 Diagrams to illustrate the peristaltic locomotion of the earthworm. A, the regions of logitudinal contraction (wide with large dots) remain at rest relative to the ground; B, diagram of the forces acting on a stationary segment and the adjacent segments during the passage of a peristaltic wave

resistance of the stationary parts rather than acting as little levers because the animal is capable of crawling over smooth surfaces such as glass which afford no points of application of the individual chaetae and the chaetal musculature is chiefly concerned with protraction and retraction. It is worth noting that an earthworm has only the one locomotory pattern and cannot swim by co-ordinated lateral flexures.

Undulatory propulsion

The means by which any lateral wave brings about forward movement are essentially the same whether the animal is a fish, annelid, nematode or flagellate protozoan. A simplified explanation is that the passage of a wave along a smooth animal presents a backwardly travelling inclined surface to the medium and it is the reaction to this that drives the animal forwards. In undulatory propulsion a wave on one side of the animal is succeeded by one on the other side and hence the lateral components into which the reaction can be resolved cancel one another out leaving only the forward component. This is illustrated in Fig. 9.8. The forward movement of fishes or leeches when swimming in water, the movement of soil nematodes in the surface film or of the parasitic forms in the gut depend on the same mechanical system.

If, however, the animal possesses lateral projections, whether they are movable parapodia or simple fixed struts the case is altered because, when a wave passes backwards down the body, the movement of the tip of the projection on the convex side of the curve is effectively forwards and hence would tend to drive the animal backwards. In animals having projections the undulatory swimming waves pass forwards along the body as they do, for instance, in *Nereis*. This is illustrated in Fig. 9.9.

Undulatory swimming with waves passing backwards down the body is found among polyclad flatworms in which the edges of the body are thrown into undulations. Dorso-ventral undulations are practically the only kind of locomotory movement made by nematodes and are related to their type of muscular organization. Among the annelids the errant polychaetes swim by undulatory waves but these pass forwards along the body. The leeches swim by relatively clumsy retrograde waves when they are not in contact with any solid object. Whilst swimming the dorso-ventral muscles are contracted and the body flattened thereby, in accordance with the dorso-ventral swimming movements. Once the suckers come into contact with a solid object attachment is immediate and swimming stops. Locomotion can then take place by looping, in which the animal is attached by the posterior

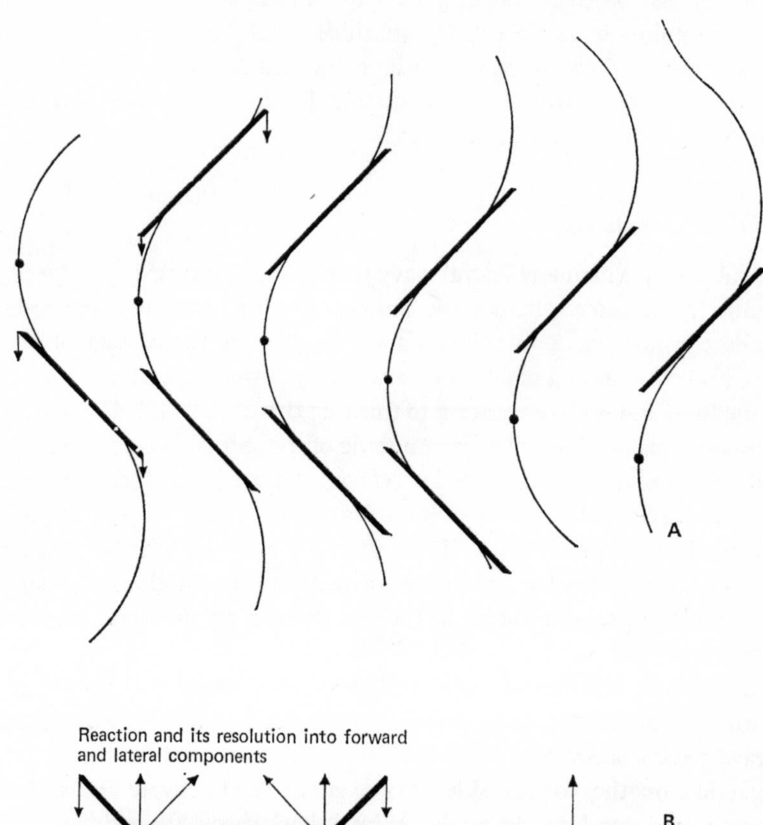

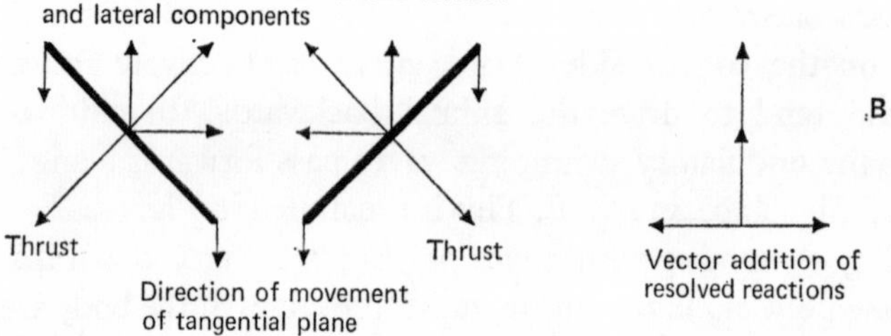

Fig. 9.8 Diagram to illustrate the forces involved in undulatory propulsion. The successive postures taken up during the backward passage of waves of muscular contraction are shown at A. Dots mark one wave. The backward thrust exerted on the water by the passage of the swimming waves is shown as a series of tangents to the convex side of the wave. The resolution of the reaction to the thrust exerted on the water is shown at B. Note that the two lateral components cancel one another

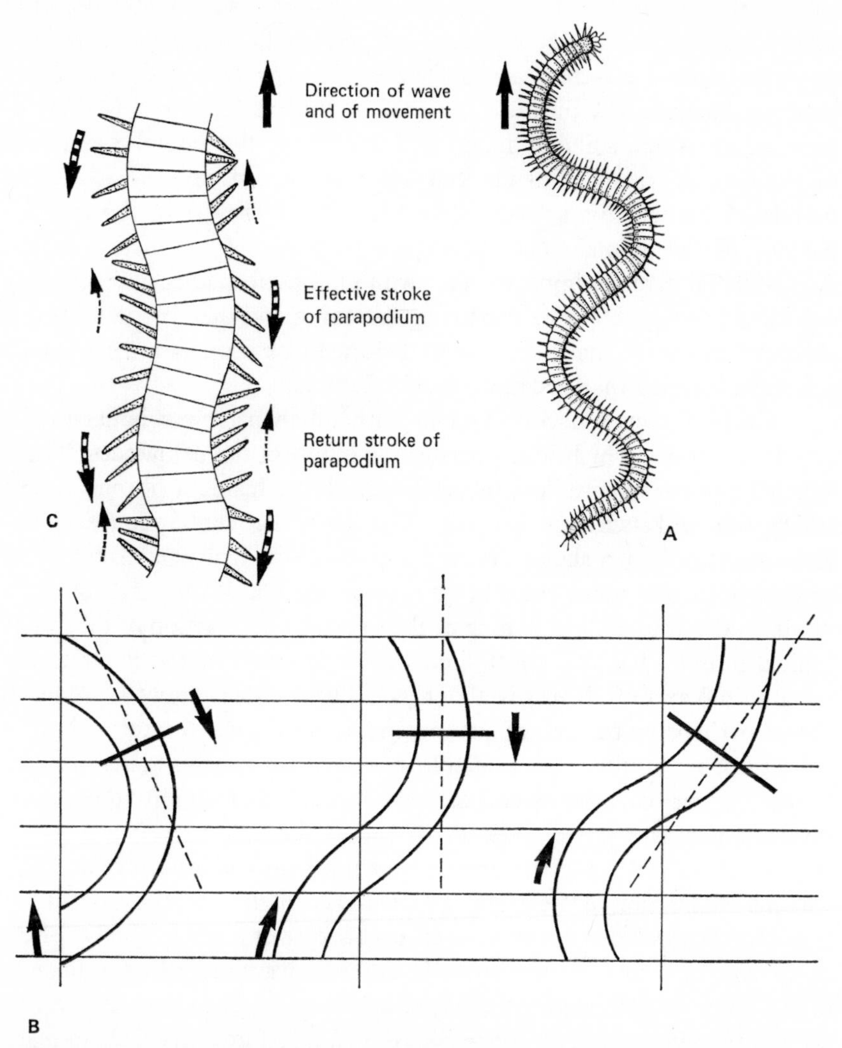

Fig. 9.9 *Nereis*. Diagram illustrating the movements and locomotion of *Nereis*. A, The body form when swimming. B, This illustrates how the movement of lateral projections of the body (parapodia) is brought about by the forward passage of lateral flexures and how the effective backward stroke is made on the convex side of the flexure. Such movements are augmented by movements of the parapodia made by their intrinsic muscles. C, The disposition of the parapodia and the direction of movements which they make during the passage of the lateral flexures from tail to head. The large arrows indicate the propulsive stroke and the small arrows the return stroke. (Based on Gray)

sucker, the anterior end is extended, the anterior sucker attached, the posterior sucker released, the body shortened and bent into a loop and the posterior sucker attached again. These different types of locomotory patterns illustrate how the same muscular system can carry out a variety of movements when suitably innervated and how the lateral swimming movements of many annelids can be transformed into dorso-ventral undulations when this accords better with other aspects of the live behaviour of the animal. The flattening produced by contraction of the dorso-ventral muscles improves the swimming performance and probably enhances looping ability by increasing flattening and thereby permitting a sharper bend to be made, but, nevertheless, these features are essentially refinements rather than necessary design features.

A much simpler muscular system with much narrower limits to the variety of actions which it can perform is found in the nematodes. These animals possess longitudinal muscles only, lying inside a fibrous cuticle which can withstand the pressure that muscular contraction sets up. Reference to Fig. 9.6 shows that, if the angle to the long axis made by the geodesic cuticular fibres exceeds 55° further increase of this angle tends to diminish the volume. Shortening of the worm by contraction of the longitudinal muscles has this effect since the angle made by the fibres to the longitudinal axis of *Ascaris* is about 75°. The volume cannot be reduced except perhaps by the loss of gut contents so an increased internal hydrostatic pressure results and is available to restore the muscles to their resting length. In this way, the nematodes have been able to simplify the muscle system but at the loss of versatility. Almost the only movements which most of them can carry out are dorso-ventral undulatory movements in which a contraction of the dorsal muscle blocks is succeeded by a contraction of the ventral ones as the wave passes backwards down the body. These movements are used in locomotion in different ways according to the size of the worm and the medium in which it is moving. For very small worms such as the vinegar worm, *Anguillula*, the viscous forces available to resist the backward passage of the locomotory wave are sufficient to provide a reaction which propels the animal forwards. For large nematodes this is not so, as can be seen by placing living *Ascaris* in 30 per cent sea water at 37°C. No forward motion results from the muscular waves. Put the animals in a thick suspension or in gut contents and they can rapidly move through it by the same type of muscular waves. Soil nematodes travel in the same way between soil particles or on the water film around them. They also make use of the forces of surface tension and can travel forwards rapidly in a thin film of liquid but only slowly when totally immersed (Fig. 9.10).

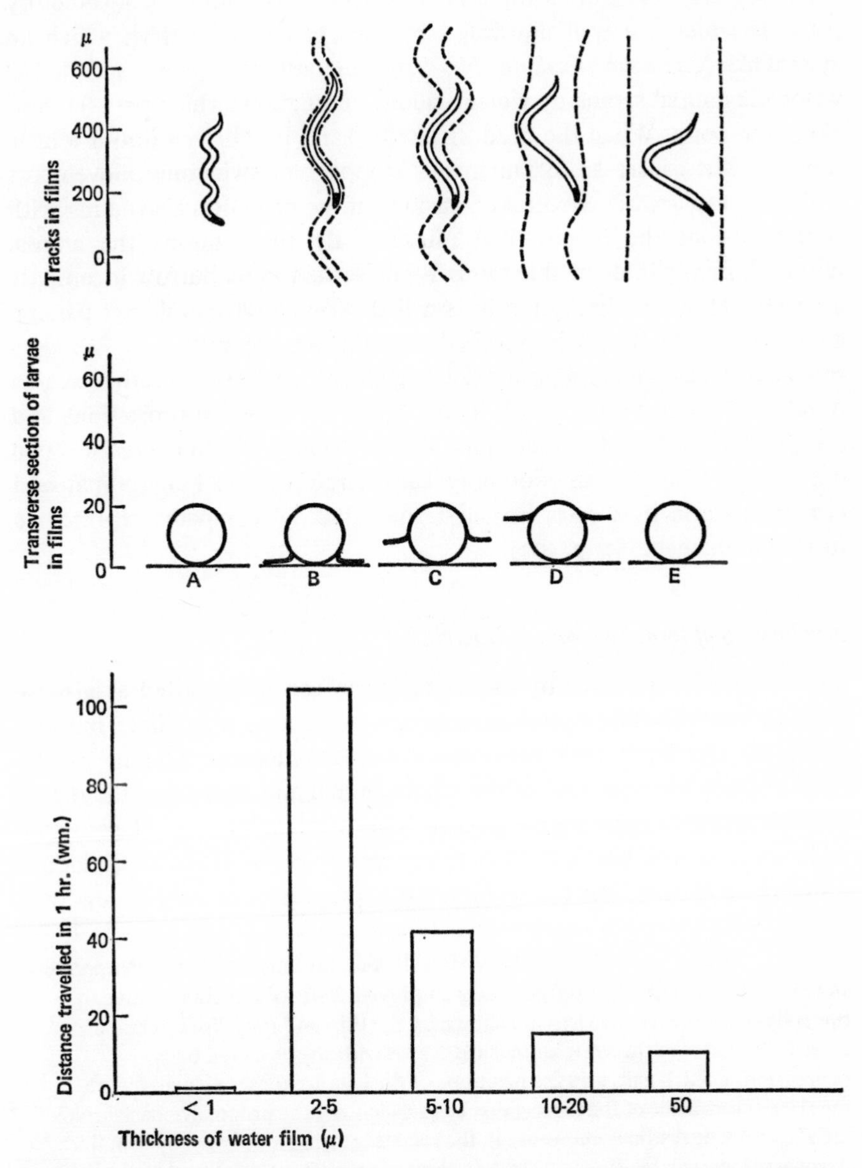

Fig. 9.10 Diagrams illustrating the movement of nematode larvae (*Heterodera*) in water films of different thicknesses. Most rapid progress is made in B where the water film is of such a thickness that a large part of the surface of the nematode is in contact with it. This holds the animal in contact with the substratum and decreases the amount of slip

Among the polychaetes there is considerable versatility of locomotory patterns which are well illustrated by the changes of pattern which an animal like *Nereis diversicolor* or *N. virens* can undergo. When suspended in water the animal swims by violent lateral undulations which pass forward along the body. When the head meets the sand it is driven into it a little way and this causes an instantaneous stoppage of swimming movements which are replaced by forceful extensions of the proboscis alternating with contraction of the longitudinal muscles. By these means the animal rapidly burrows into suitable sand. While settled in its burrow it can pass a current of water through it by small dorso-ventral undulation passing along the body. It will be recalled that the muscle system of *Nereis*, a modified circular and longitudinal arrangement, consists of a rather weakly developed circular layer, much interrupted by parapodial projections and a well developed longitudinal musculature arranged in four blocks – two dorsal and two ventral. Not only can *Nereis* use its longitudinal and circular muscle in a co-ordinated fashion but its parapodia represent a further refinement (see p. 192).

Application of force by means of levers

The application of force by means of levers can be regarded as a more efficient way of using muscular contraction because only those muscles which are exerting tension are in use at any one moment. The inertia of a limb is much less than that of the whole animal and, therefore, the movements made by a limb can be quicker. This can result in rapid locomotion as in cockroaches, but it need not. For example, the limbs of diplopods move quite quickly but the whole animal progresses at only a slow rate,

Fig. 9.11 A summary of the analysis of the locomotory movements of *Peripatopsis* showing the three gaits most frequently employed. The movements relative to the body of two successive legs are shown by the thin and thick lines in respect of time. Silhouettes of the same animal are drawn with the observed body proportions at each gait, and the positions of the legs are those calculated for a regular performance of the three gaits. Legs executing the propulsive backstroke are shown by heavy lines and those in the recovery swing are shown by thin lines. Arrows below each leg shows their directions of movement. (a) The duration of the forward and backward strokes are as (3·7:6·3) the phase difference between successive legs is 0·16, and the duration of the pace is 0·84 sec. (b) The duration of the forward and backward strokes are as (5:5), the phase difference between successive legs is 0·45, and the duration of the pace 0·7 sec. (c) The duration of the forward and backward strokes is as (6·3:3·7), the phase difference between successive legs is 0·16 and the duration of the pace 0·84 sec.

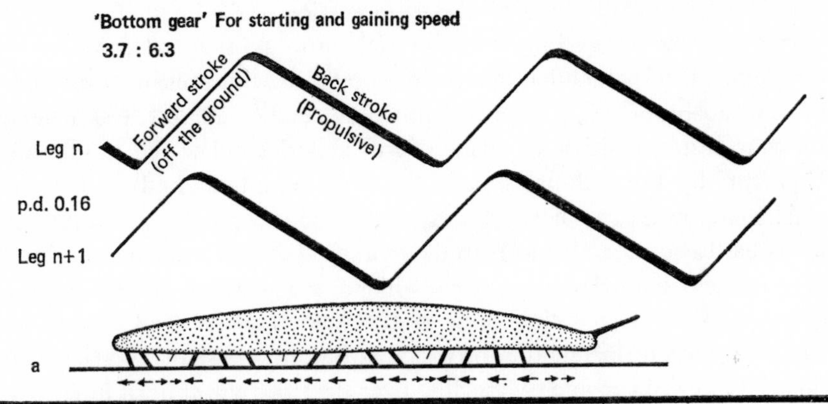
'Bottom gear' For starting and gaining speed
3.7 : 6.3
Forward stroke
(off the ground)
Back stroke
(Propulsive)
Leg n
p.d. 0.16
Leg n+1
a

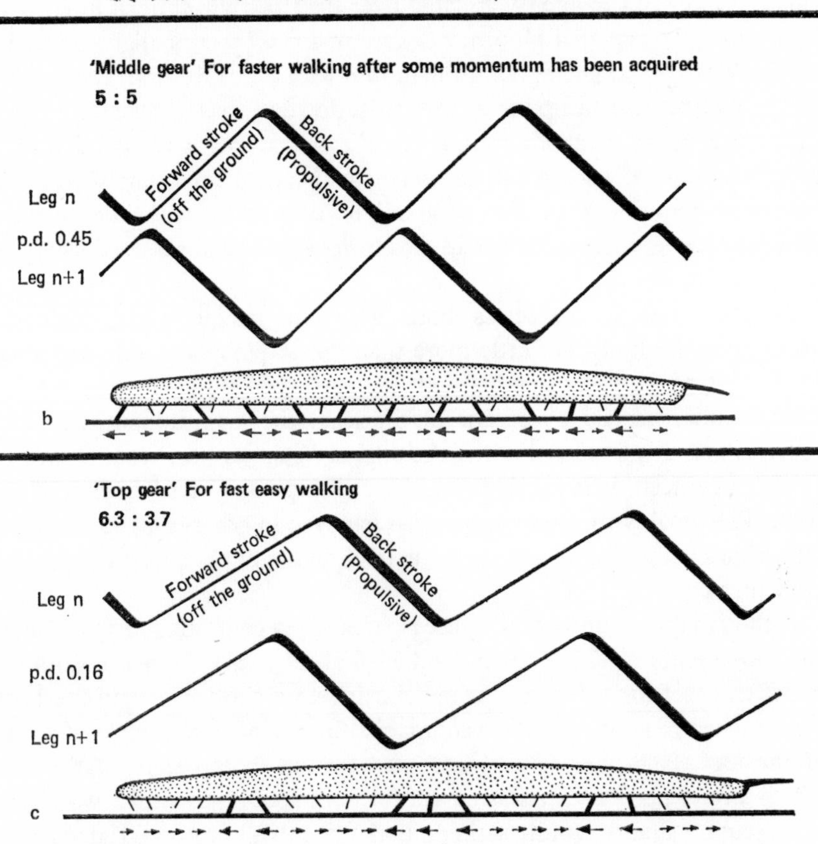
'Middle gear' For faster walking after some momentum has been acquired
5 : 5
Forward stroke
(off the ground)
Back stroke
(Propulsive)
Leg n
p.d. 0.45
Leg n+1
b
'Top gear' For fast easy walking
6.3 : 3.7
Forward stroke
(off the ground)
Back stroke
(Propulsive)
Leg n
p.d. 0.16
Leg n+1
c

similarly the movements of a single podium of a starfish are rapid compared with the speed of locomotion of the whole animal.

It is not difficult to see how the segmentation of the body of an annelid gave a small amount of autonomy to each part, and that this favoured the development of parapodial outgrowths provided with their own muscular system capable of acting independently of the body-wall muscle as a whole. Of course, limbs which are made of soft material and which can only be kept stiff by internal pressure are less efficient than limbs which are rigidly supported by a skeletal system, again because some muscular effort has to be expended to keep the limb rigid so that it can act as a lever. Nevertheless, animals have evolved which make good use of simple turgid appendages for locomotion both in the water and on land. *Nereis* has various modes of locomotion of which swimming has already been mentioned. It is also capable of crawling over the substratum or in its burrow by means of parapodial movements accompanied by only slight muscular waves. The parapodia are not quite in step with each other but each makes its cycle of movement upwards, forwards, downwards and backwards, just in advance of the parapodium in front. Faster crawling is accomplished by a combination of parapodial movement and lateral muscular waves, the backward movement of the parapodium coinciding with the forward passage of the wave, which would also bring about the backward sweep of the parapodium.

Peripatus shares characters both with the annelids and with the arthropods. Its limbs are little more than turgid parapodia although their fluid content is haemolymph and not coelomic fluid. Locomotion is brought about solely by means of the legs: the longitudinal body-wall muscles are only involved when the animal 'changes gear' to a slower or faster gait since this is accompanied by a shortening or lengthening of the body. Presumably the body wall is also involved in generating the pressure which makes the turgid legs strong enough to lift the body off the ground (Fig. 9.11).

Although the duration of the pace is remarkably constant at 0·75 seconds, the movements made by an individual limb can vary from a propulsive back stroke of long duration when starting to a shorter propulsive back stroke when moving quickly accompanied by an increased angle of swing of the legs. High speeds are prevented by the short length of the legs which would need a very short pace for high speed and which would, in turn, require a more rapidly acting muscle than that of the unstriated fibres of *Peripatus*. But the arthropods have, as a whole, evolved very efficient muscles and with this has become possible, first rapid limb movement,

Fig. 9.12 Single segments of a series of Arthropoda are shown in transverse section, with their limbs drawn relative to the ground as seen in the living animals. The scales chosen are those which in each case give a segment of the same volume and the two figures of *Peripatus* represent the same segment employing different gaits. The scorpion and spider show walking leg 3, *Astacus* and the earwig (*Forficula*) walking leg 2, and *Ligia* walking leg 6

then rapid locomotion and finally flight, which makes great demands on the energy output of the locomotory system.

Longer legs, which enable longer steps to be taken, powered by the more rapidly acting striated muscle give rise to other mechanical problems. For example, small animals would be easily upset by air movements if they stood up on the tips of long legs. The majority do, in fact, hang down from their legs (Fig. 9.12). Their centres of gravity remain low and although few legs may be in contact with the ground at any one time, the body weight can be transferred from one leg to the other with minimum risk of the creature being upset. The number of limbs employed in locomotion is closely related to the normal habits and habitat and to the speed of locomotion. Among the myriapods which have a resemblance to *Peripatus* in their shape and in having a relatively large number of limbs, some progress in a manner which is very like that of *Peripatus*, for example *Polyxenus*, whereas others are highly adapted to burrowing by pushing with very many appendages, for example the juliform Diplopoda: others burrow by being able to shorten the body and exert lateral pressure, for example the geophilomorph chilopods. Some can even run fast, like the long-legged *Scutigera* which can travel at 500 mm per second compared with 70 mm for *Peripatus*. Although the fields of movement of the legs of this creature overlap very considerably, this is reduced to a certain extent by fanning out the limbs so that the anterior limb points forwards and the posterior

---

Fig. 9.13 The field of movement of the legs of a series of Arthropoda. The figures of *Peripatus*, *Cormocephalus* and *Scutigera* are drawn to the same segment volume, but the diagrams of the other Arthropoda possess no common scale factor. If the spider-crab and prawn and the spider, *Galeodes*, *Campodea* and *Forficula* were drawn with their bodies of the same weight or volume, their legs would be very much longer and with wider fields of movement than here shown. The heavy vertical lines represent the movements of the tip of the leg relative to the body during the propulsive backstroke. The path of the tip of the limb during the recovery forward stroke is shown by dotted lines for *Cormocephalus* in 'top gear', in *Scutigera* the limb-tips are swung forwards approximately on the common dotted line, and the forward stroke for the other animals is not shown. The two outlines for *Cormocephalus* in 'top gear' represent the extreme positions, about the direction of running, of the body when it undulates. One leg-bearing segment and its right leg are marked, and a dot shows the successive positions of the base of the leg as it moves about the transverse plane. The points A–E show successive positions of the tip of the limb during one stroke. The other animals shown are the spider-crab *Macropodia rostratus*, the prawn *Leander serratus*, the shore crab *Carcinus maenas*, the scorpion, *Buthus australis*, a British lycosid spider progressing at 36 and 250 mm/sec in the two diagrams, *Galeodes arabs*, *Ligia oceanica*, *Astacus fluviatilis*, a British species of *Campodea* and the earwig *Forficula auricularia*

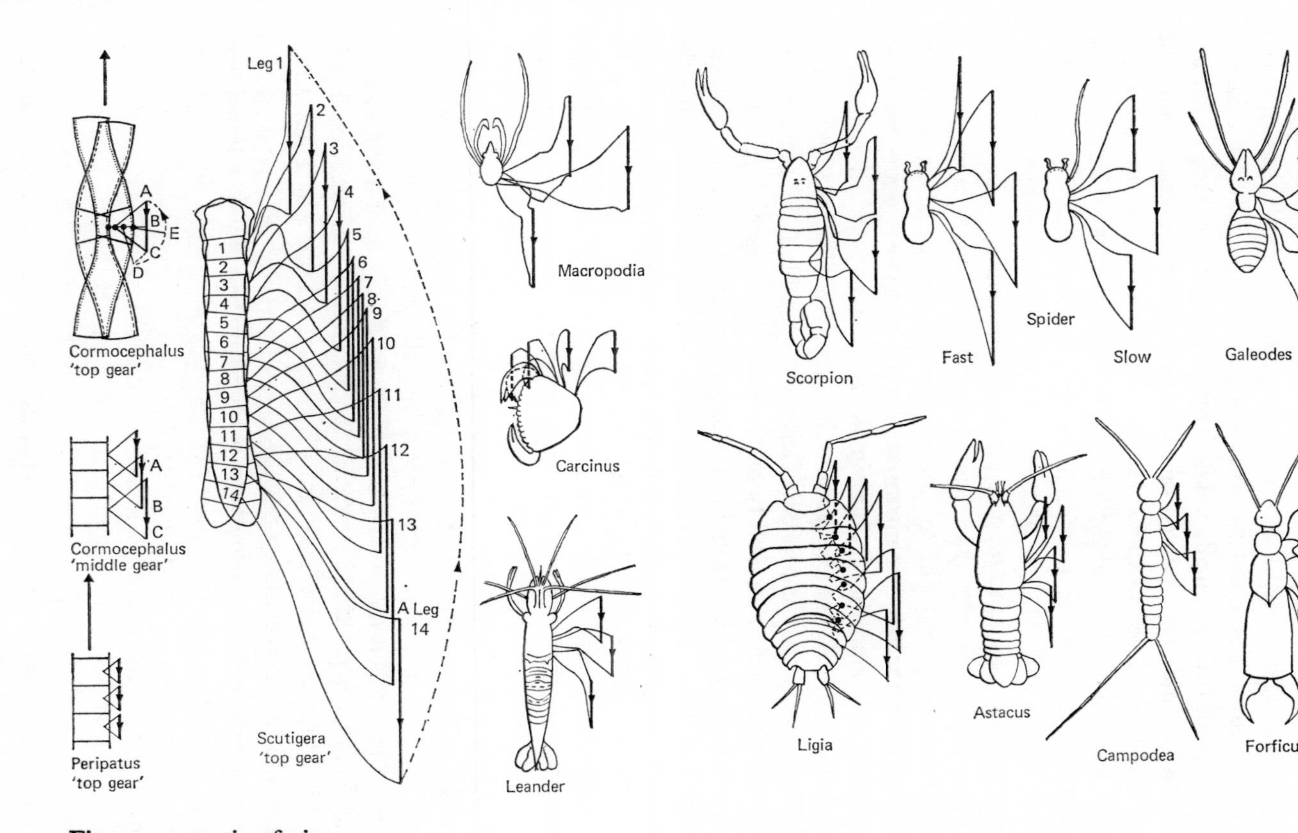

Fig. 9.13 caption facing

backwards instead of all pointing laterally as they do in the longer, slower myriapods. The field of movement of the limbs of various arthropods are shown in Fig. 9.13.

The most highly specialized land arthropods, the insects, have reduced the legs to three pairs and have correspondingly reduced the variations of leg movement that are possible. The basis of ambulation is the support of the body on three legs, two on one side and one on the other while the other three are moved forward. The fore leg serves to pull the body, the middle leg to support it and the hind leg to push it forwards. The centre of gravity only falls outside the triangle towards the end of this phase and the support of the body is then taken over by the other three legs. Although three legs are sufficient to support the insect and to permit the other three to move simultaneously this strict alternation does not always occur; rather the fore leg, opposite middle leg and hind leg are advanced in that order so that in the gait of many insects four or even five legs may be on the ground at the same time. This simple account of insect movement is included to show the progressive simplification which the insect represents and how enormously increased efficiency goes with it in the sense of increased speed and adaptability of the locomotory appendages. This is demonstrated by their adaptation to jumping as in the grasshoppers and locusts and fleas, and to swimming either on the surface as with the Geridae (pond skaters) or floating on the surface with the limbs immersed as in the gyrinid or 'whirligig' beetles. The aquatic bugs (Hemiptera) swim using their hind legs in unison while the beetle *Dytiscus* uses its hind and also its middle legs in the same way. A much fuller discussion of insect movements including flight will be found in the companion volume, *The Life of Insects*, by Sir Vincent Wigglesworth. It is worth while pointing out, however, that it is possible to imagine a series of steps by which the arthropod limb may have been derived from the annelid parapodium. The wings of insects, on the other hand, powered chiefly by indirect muscles which alter their position by altering the position of the thoracic sclerites to which they are articulated, are a structural feature which has arisen during the evolution of insects only.

We have discussed the locomotion of the arthropods as if their limbs were simple movable appendages whereas they are complex structures which have advantages and disadvantages inherent in their method of construction. In general the arthropod limb is a stiff tube inside which the muscles are housed, attached at their origins to the wall of the tube and at their insertions into intucked portions of the outer casing. The muscles are, therefore, in a position of great mechanical advantage and work over only a very short length range compared with the smooth muscles of hollow-bodied invertebrates. They are all striated and quick-acting but each is capable of bringing about movement of one joint in the limb and this is often practically confined to one plane. Each muscle has, of necessity, to be restored by an antagonist. For a limb to be freely movable, therefore, a number of joints, flexing in different planes, are required each activated by at least a pair of muscles (Fig. 9.14). This is the condition of most arthropods but there are some species in which a different mechanism is found. In the trunk limb of many small entomostracans the cuticle of the leaf-like phyllopodia is thin and the limb is extended and their shape maintained by blood pressure. They are flexed by direct muscular action.

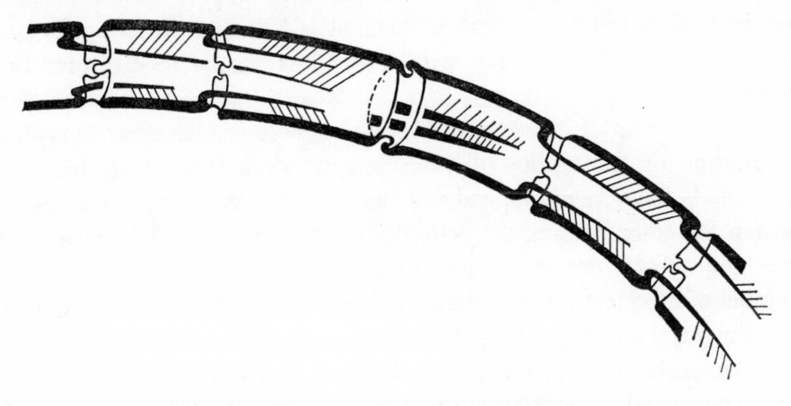

Fig. 9.14 Segments of a brachyuran limb showing diagrammatically the type of joints and the arrangement of muscles. Distal end toward the left. Note that insertions and origins of muscles lie on the same side of the limb; also that because of mechanical relations the positions of extensors and flexors with respect to the direction of bending are the same as in the endoskeletal joints of vertebrates

The thoracic appendages of some of the Cirripedia are also moved in the same way and in the legs of certain spiders extensor muscles are absent. It has been shown that the blood pressure is sufficient to extend the leg and provide sufficient force for the animal's jumps. Since each joint can move in only one plane and each pair of antagonistic muscles activates only one joint, a system of nervous control has evolved which is different from that in many other groups of animals. Whereas little is known about the details of muscular control and tension balance in cylindrical hollow animals, the gradation of contraction in vertebrates is generally brought about by the recruitment of varying numbers of muscle fibres which are called into action by relatively large numbers of motor neurons. The site of control is in the central nervous system. Arthropod muscle fibres on the other hand are innervated by at least two types of nerve elements, the axon from each motor cell branching repeatedly so that all the fibres of a muscle are controlled by very small numbers of neurons. Graded contraction is produced not by recruitment of varying numbers of muscle fibres but by the different effects which the types of innervation have on the muscle. Those from one type of axon cause contraction, while those from the other produce inhibition of contraction. This system seems well adapted to other features of the life of arthropods namely their relatively stereotyped behaviour, the multiplicity of their limbs and the economy of neurons which this demands if the central nervous system is to remain small enough to be contained in a small body. Finally, in the locomotion of arthropods the precision required of an individual limb movement is less than in tetrapods since many limbs are involved but with rather less precise control over their individual effects.

Another group of invertebrates which apply muscular effort to produce locomotion by the agency of levers are the starfishes, which, like some arthropods have many appendages but which have a very much poorer system for co-ordinating the activities of the tube-feet. A starfish moves over the substratum essentially by the simultaneous action of a large number of tube-feet. Each of these performs a stepping cycle of protraction, attachment, bending, detachment, but each foot carries out its cycle independently of others although there is a co-ordinating mechanism which ensures that the steps of the tube-feet on all the arms are in the same general direction (Fig. 9.15). The system seems inefficient in that there is little co-ordination between the tube-feet apart from their general direction of step. It is true that starfishes cannot move very rapidly but perhaps one should marvel that they can move at all in this way. In other classes of

Fig. 9.15 Diagrams to illustrate the mode of action of the tube feet of starfishes. A and B show the opposing muscles of foot and ampulla and C shows the muscles which are used for bending or pointing the feet in different directions. G shows how the movements of the tube feet are not in phase but that the general direction of pointing is the same; this is seen in the aboral view of a moving starfish at D also. E indicates how changes in direction of movement of the whole animal are brought about by changes in the orientation of the stepping direction of the tube feet of all the arms together. In F(b) the three postures X, Y and Z derive from the contraction of the muscles X, Y and Z in F(a) and the relaxation of the muscles whose position is indicated by the corresponding broken letters. Drawings H and J show the stepping cycle of protraction (1), attachment (2), bending (3) and detachment (4) in external view (H) and in section (J) to show the muscles of the tube foot and ampulla which are involved

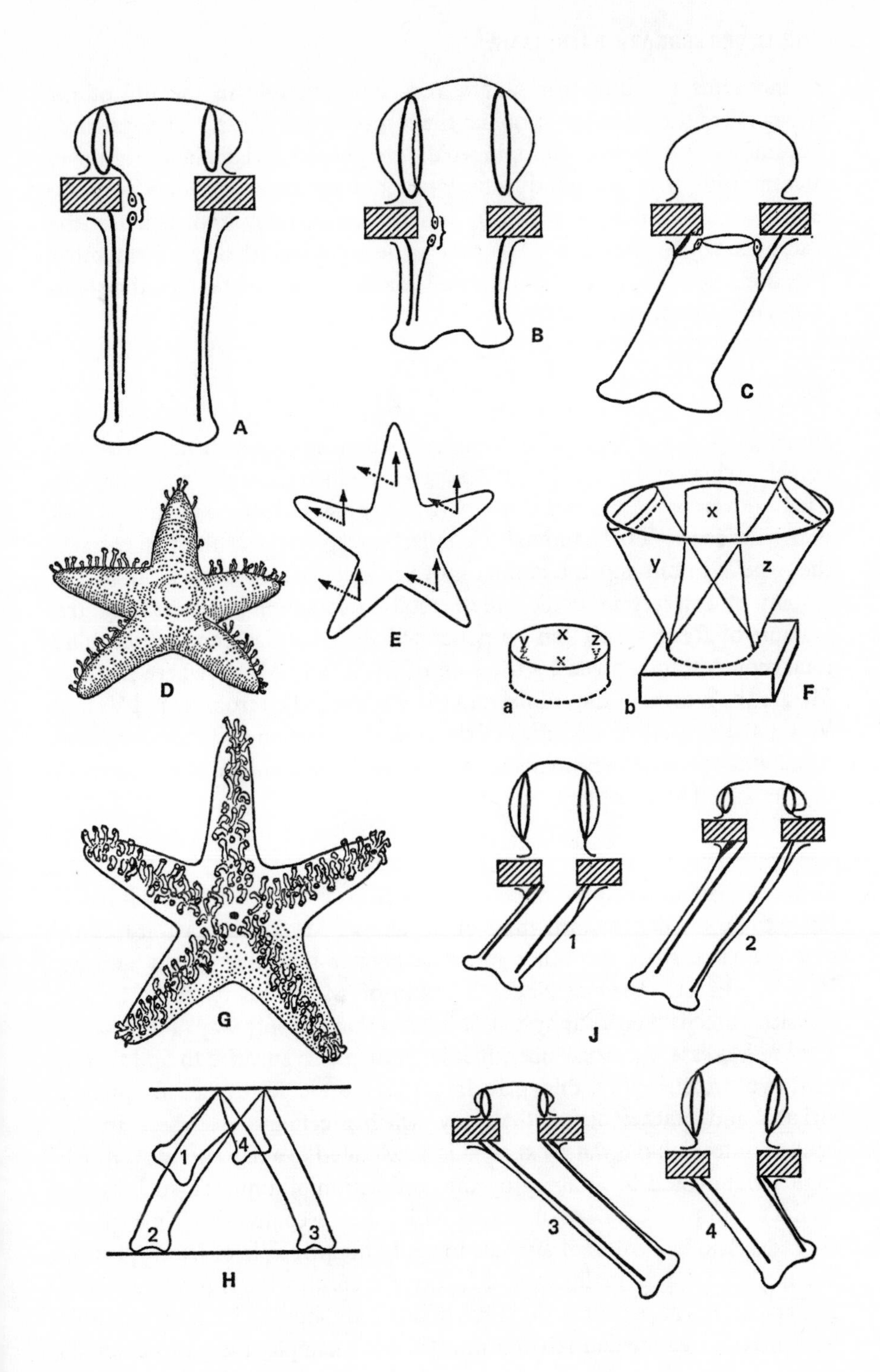
A
B
C
D
E
x
y
z
y
z
x
z
y
x
a
b
F
G
1
2
J
4
1
2
3
H
3
4

echinoderms the tube-feet play a very different role in the life of the animal. In the ancient crinoids they served as feeding and possibly respiratory structures. In ophiuroids they lack suckers and in many holothuroids they are poorly developed but in some echinoids they are active and locomotory while in others considerably reduced. In the asteroids where they are important in locomotion, they perform other functions as well, in particular they all enable the animal to open the shells of the bivalves on which it feeds.

Jet propulsion

Finally, one more type of locomotory method is found in invertebrates, namely, what can be called jet propulsion in which muscular contraction is used to overcome the inertia of a partially enclosed volume of fluid. In so doing the inertia of the animal is similarly overcome and the two masses, the expelled fluid and the animal move in opposite directions. In such a system of interacting bodies their total momentum is limited by the amount of force which can be generated by muscular contraction. The total momentum is given by the sum of M_aV_a and M_wV_w where M_a and M_w are the masses of the animal and the water expelled from it, and V_a and V_w are the respective velocities of these masses. Action and reaction being equal and opposite the momentum imparted to the animal is the same as that imparted to the water:

$$M_aV_a = M_wV_w.$$

However, the quantity of water of mass M_w expelled from an animal is likely to be smaller than the mass of the animal M_a. If $M_a > M_w$ it follows that the velocity of the water must be greater than that of the animal, $V_w > V_a$. In other words since the ratio of M_a to M_w is fixed by the physical dimensions of the system it follows that the only way of increasing V_a is to increase V_w. Consequently it is, perhaps, no surprise to find that in medusae which apply this principle the muscle fibres are frequently striated and quicker acting than any which are found elsewhere in the coelenterates, i.e. the water expelled is accelerated to a high velocity, much higher than could be attained by the contraction of a muscle such as the sphincter muscle of a sea-anemone which has a contraction time of 5·0 seconds. Another group of animals to apply the principle of jet propulsion are the cephalopod molluscs and in these creatures the muscles are also fast acting, having a contraction time of 0·068 seconds, which is comparable with that of mammalian striated muscle. For example, the gastrocnemius of the rat has a contraction time of 0·034 seconds. The speed of exit of the water is also increased by narrowing the aperture through which it is ejected. This purpose is served by the siphon of the cephalopod, the velum of the hydromedusae and similar structures which occur in scyphozoans. Swimming by the rapid closure of the shell valves has also been developed by some bivalves and in *Pecten* the striated, fast-acting portion of the adductor muscle which brings about this type of movement has a contraction time of 0·046 seconds as against 2·28 seconds for the non-striated portion.

Movement by one or other of the many methods discussed in this chapter is one of the most striking features of animal life. Indeed the perfection of a muscular mechanism for the transduction of chemical into kinetic energy on a substantial scale distinguishes animals from plants although motile systems at the cell level may be found in most groups of organisms. The evolution of powers of movement requires the application of a control system of great complexity and precision since it has the task not only of co-ordinating an animal's movements but also of relating these to conditions in the environment and within the animal.

Chapter 10

Sensory equipment and perception

IN KEEPING with the diversity of their habits and the variety of their structure, invertebrates have achieved the reception of information from the environment by a number of different morphological paths which are not always very similar to those used by vertebrates. The soft-bodied members with their epithelia in direct contact with the surroundings are in a simpler situation than the arthropods covered as they are by a hard exoskeleton which, at the same time that it guards them against the environment, excludes stimuli which might excite sensory cells in their hypodermis.

To us, as animals who depend so much upon vision, light as a stimulus seems of paramount importance, with sound a close second. If the elaborate structure of many invertebrate eyes is any measure, vision is predominant in the repertoire of their senses as well.

However, in coelenterates, apart from some medusae, it is not always possible to see spots of pigment or anything which looks like an eye and yet many show responses to light. This is an example of the generalized light sensitivity or dermal light sense which is found, or suspected, in various degrees even in arthropods and lower vertebrates. Vision in eyes is always dependent upon a change in a light-sensitive pigment caused by the energy of the light waves striking it. This change is essential to the process of initiation of nerve impulses in the sense cells. However, thus far, evidence for light-sensitive pigments in areas of animals' bodies which show generalized light sensitivity is slim. Indeed, it may well be that this sensitivity is the result of the direct action of light energy on nerve endings. For example, shading a light shone on the radial nerve of the sea-urchin, *Diadema antillarum*, produces visible responses including the jerking of neighbouring spines. As the direction of the jerking is orientated to the light source, the result is that dark-adapted animals move out of the light and light-adapted ones leave the shadow. This may well be the mechanism behind the observed movements of *Diadema setosum* which, in the Gulf of Suakim, moves under cover by day but emerges at night to feed. If the

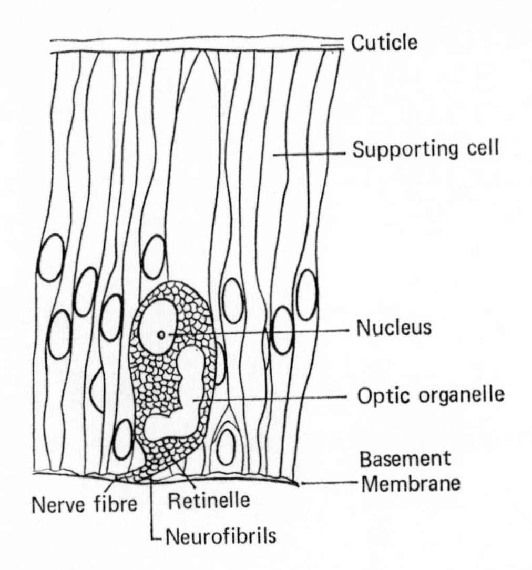

Fig. 10.1 Photoreceptor cell in first segment of *Lumbricus*

reception of the light in these cases does not involve a sensitive pigment, this visual sense is different from that found in all eyes and therefore it would be unlikely that vision had originated in such a generalized sense.

In the epidermis of earthworms specialized light receptors cells can be seen (Fig. 10.1). They are more common in the anterior part of a worm such as *Lumbricus terrestris*, and it is this part of the body which must be stimulated to bring about the responses of the worm. Light shone on the front end of a worm extended from its burrow causes it to withdraw rapidly back into the ground, a response mediated by giant fibres (see p. 236). Though earthworms (*L. terrestris*) are most sensitive to light of a wave-length of 483 mμ there is no evidence that they can discriminate between coloured lights on a basis of their wave-length.

But for directional movements some way of limiting the direction from which light reaches the eye is important. The shadow of the body is sometimes sufficient; a blowfly larva has 'eyes' which consist of a small group of sensitive cells in each of two pockets, one on each side of the internal skeleton of the head (Fig. 10.2). Light coming from behind the larva does not stimulate the cells because the bulk of the body shields them. This is very important in orientating the animal's movement with respect to the light source.

A cup of black pigment serves the same purpose in the eye of a planarian (Fig. 10.3). Here again light can only reach the sensitive cells from a limited sector. These worms can direct their movements towards or away from a light source. Moreover, stimulation of particular cells in the eye

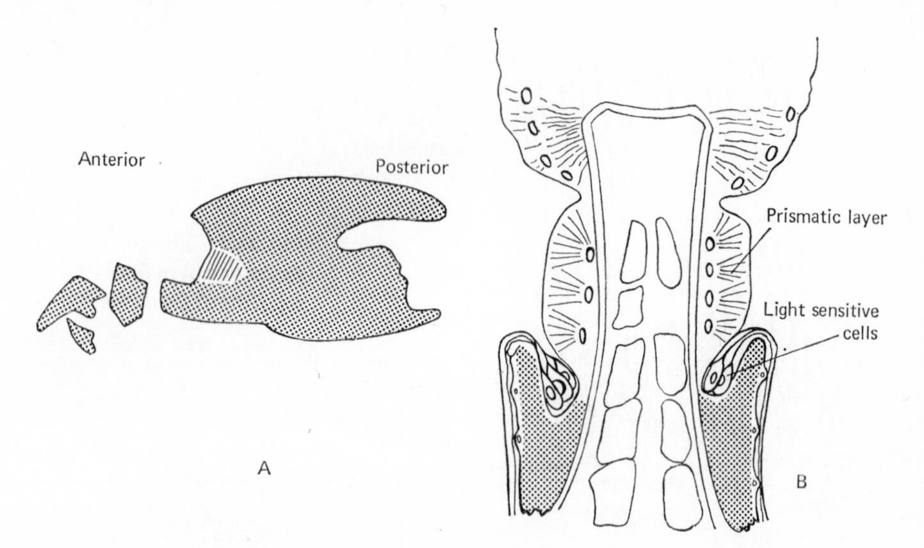

Fig. 10.2 The light receptors of a blowfly larva. A, the pharyngeal skeleton of *Musca*; the pocket in which the light-sensitive cells lie is shaded. B, horizontal section of the head and skeleton showing the light-sensitive cells and the prismatic layer

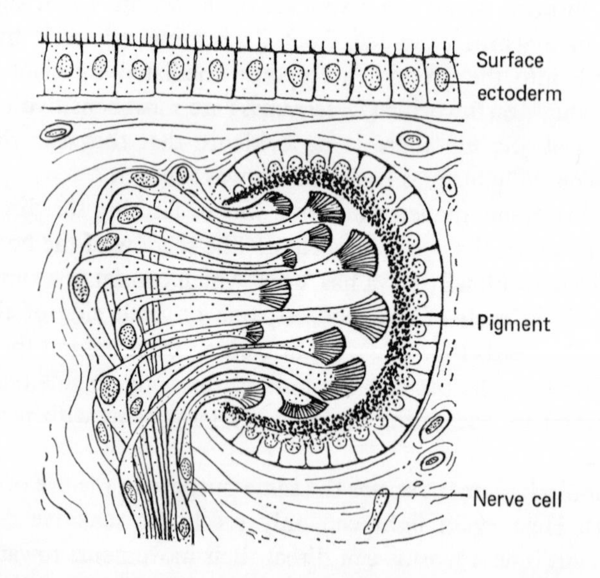

Fig. 10.3 Section through the eye of the flatworm *Dugesia gonocephala*

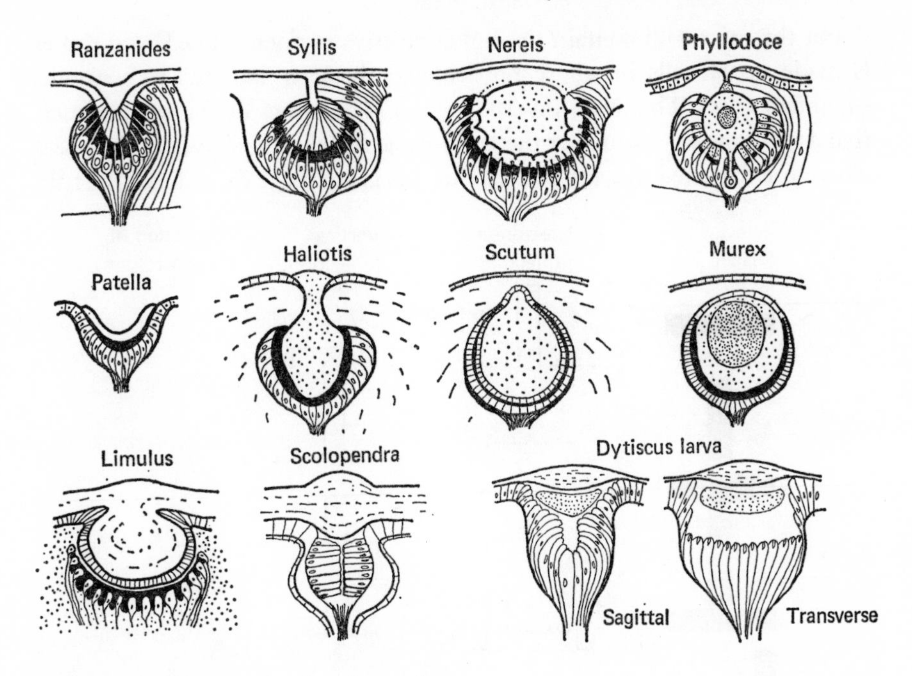

Fig. 10.4 The eyes of various invertebrates

leads to different responses of the whole worm. Light shone on posterior and ventral parts causes the animal to turn towards the same side, but stimulation of the other parts produces a turn in the opposite sense.

The directionality of an eye is also increased if it has a lens. In this case light from certain directions may be brought to a focus on a sensitive retinal surface. In addition this has the effect of concentrating the light increasing the amount of energy falling on the retinal cells. One supposes that this permits such eyes to function in lower light intensities than if they had no lenses.

A number of invertebrates in many different phyla have eyes with lenses (Fig. 10.4). Many of the lenses are spherical, as would be expected for aquatic animals, but the spatial arrangements of the retina with respect to the optical properties of the lenses show that many of these more simple lensed eyes cannot be image-forming. The focal points of the lenses seem to lie behind the retinal layer. In the eye of *Pecten*, for example, this is so. The eyes of this scallop are arranged along the edges of its mantle. Each has a double retina, containing about five thousand receptors. The proximal retinal cells have a roughly conical extension from which arises a brush of microvilli, a type of visual cells also found in slugs (Plate 2); no doubt the microvilli contain the light-sensitive pigment. The distal retina is made up of cells in whose border next the lens flattened sacs extend (modified cilia). This is an inverted retina. The optics of the eye are such that a real image is formed at the level of the distal retinal layer at the place where the flattened sacs are found. But this is an image formed secondarily

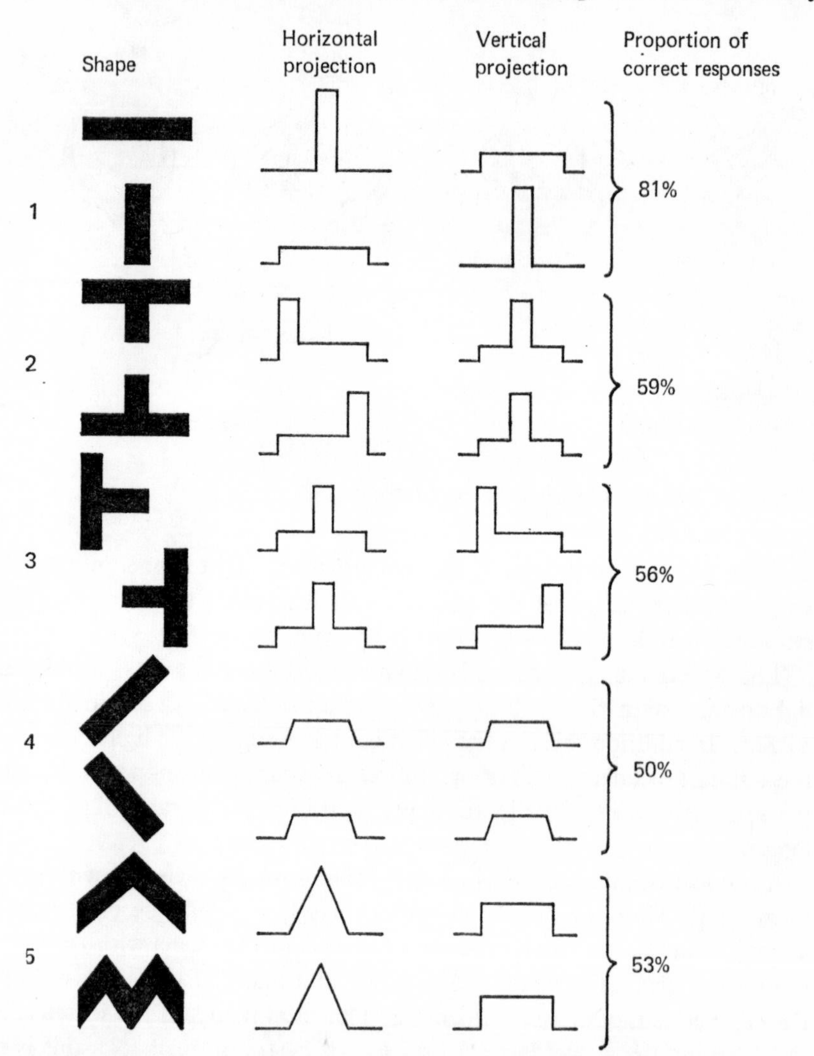

Fig. 10.5 Pairs of shapes that octopus can and cannot learn to distinguish. The projections are those which would be derived from a theoretical grid composed of horizontal lines and vertical columns. The percentage figures show the proportion of correct responses made in the first 60 trials (pairs 1–4) or 240 trials (pair 5) of training

by reflection from the silvery layer of guanin crystals lying at the back of the eye. At the level of the proximal retina the light rays are not brought to a focus by the lens.

The scallop responds to shadows by swimming away. It seems likely that the distal cells, being able to respond to a change of light intensity in a very small part of the environment, bring about the reaction to the abrupt reduction in illumination which accompanies a shadow. The proximal cells on the other hand probably direct movement, for the pupil of the eyes restricts the light rays entering the eye to those in a cone of 50°, small enough to give a fair accuracy for directing movement.

Amongst the molluscs the cephalopods show the greatest extreme of morphological complication and their eyes are no exceptions. Not only do they have lenses, but the focal length of the lens can be altered by a process of accommodation. By this means distant objects and those close to will cast focused sharp images on the retina. *Octopus* can learn to distinguish between various shapes (p. 277). Given the task of reacting differentially to horizontal and vertical rectangles, these animals can adjust their behaviour but they prove unable to distinguish between a rectangle sloping to the left and a similar one sloping to the right. This suggested that the shape might be analysed in the brain from a series of inputs from a regular grid of columns and rows of sensory units (Fig. 10.5). Now the retinal cells of this eye are arranged in a regular pattern (Fig. 10.6). Moreover, if the spatial arrangement of an octopus's eye is disturbed, its ability to tell the difference between a vertical and a horizontal rectangle is impaired. Normally, the pupil of the eye is horizontal but if the statocysts of an octopus are destroyed the eye may be orientated in any way. Coincidently with this change in the eye comes the breakdown of discrimination between vertical and horizontal. Thus a fixed spatial orientation of the retina is important in the analysis of images.

But invertebrates do not only have such vertebrate-like eyes. The problem of good form vision has been solved in evolution by the development of the compound eye, found in a majority of arthropods, indeed in all insects. These eyes are composed of numbers of units, the ommatidia (Fig. 10.7). Each one is essentially an elongated, self-contained light-sensitive unit with a lens and sensitive cells called retinulae. The retinula cells each have a thick fringe of micro-tubules arising from their inner sides, within which the sensitive pigment is deposited (Plate 1). The micro-tubules from the group of retinulae in one ommatidium are packed closely together giving rise to a rhabdom which appears under the light microscope like a rod.

The highly ordered structure of the rhabdom (Plate 1) is one which probably forms an analyser for polarized light. Coupled with independent evidence that the analyser must be within the ommatidium, it is likely that this typically arthropod structure is the reason for the almost complete uniqueness of the arthropod ability to analyse directly the plane of polarization of light. *Octopus* also has this ability and the micro-tubules arising from its retinal cells are similarly close-packed and neatly arranged.

Whether the system of lenses acts as such or are merely wave guides is not clear; however that may be, no image is detected by an individual element though images can be seen at various levels behind the lens of an insect's eye. The ommatidia signal only the average intensity of light falling on their rhabdoms. In this way an image is built up centrally from a series of 'dots'. The greater the number of these, the more detail there is in the image. Insects with poor vision have small numbers of ommatidia in their eyes, but a dragonfly, or a honey-bee, both flying insects with acute vision, have thousands.

Acuity of vision in Crustacea seems to be aided by movements of sleeves of pigment which surround each ommatidium. These are in two

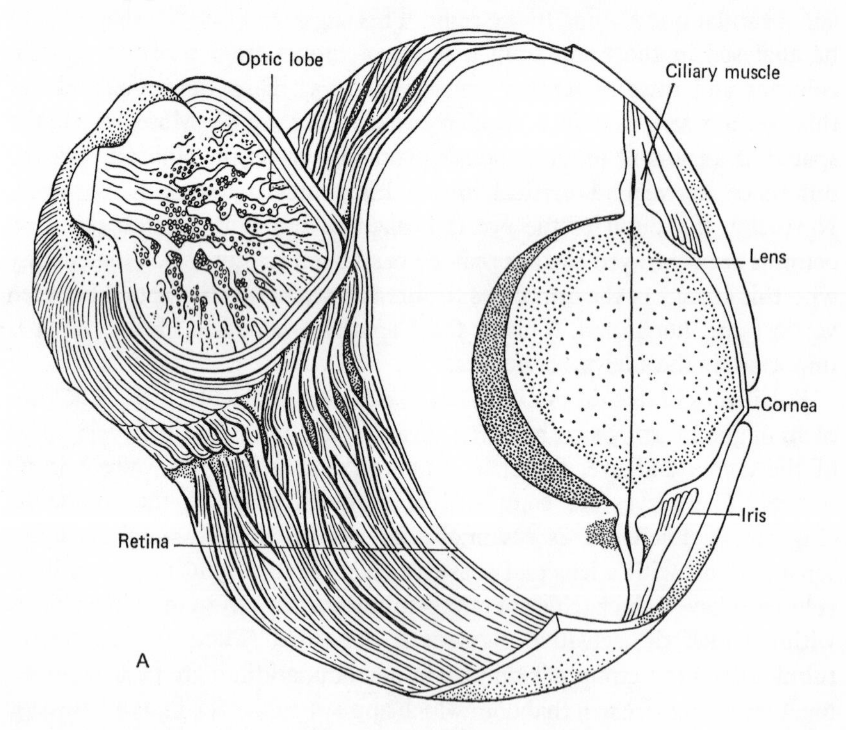

Fig. 10.6 (A), a dissection of the eye and optic lobe of *Octopus*

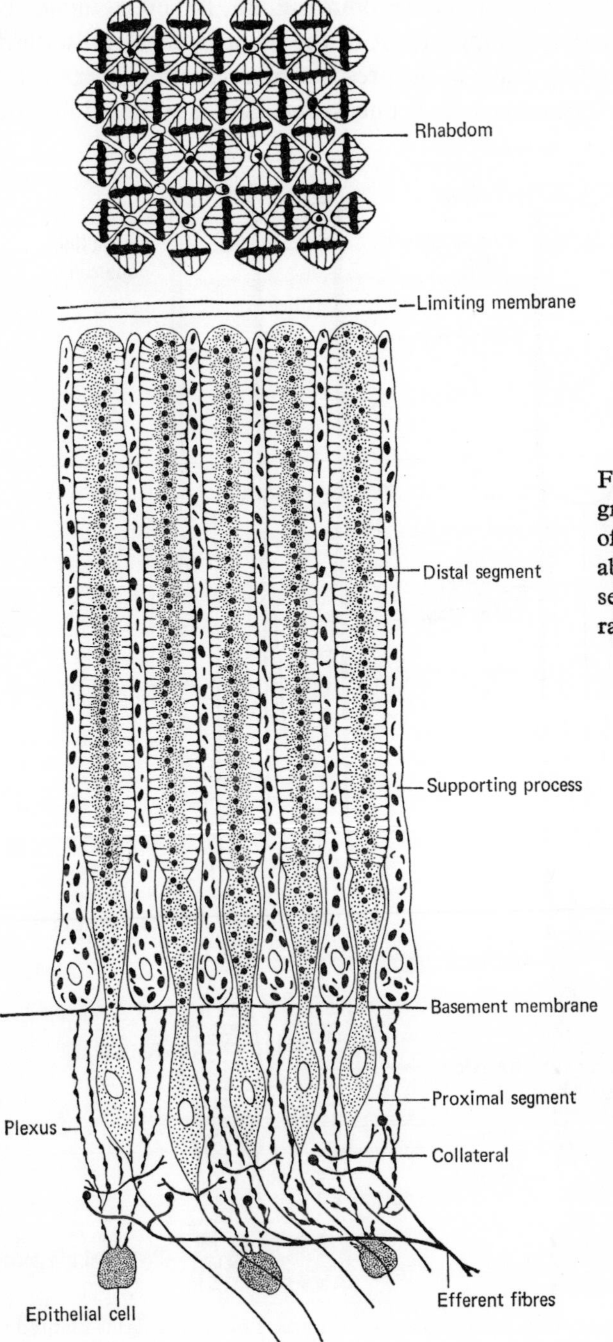

Fig. 10.6 (B), diagrams of the structure of the retina of *Octopus*, above, in tangential section and below, in radial section

parts, one round the inner end of the ommatidium, the other round the outer dioptric part. It is the movement of the inner pigment cells which affects acuity. The effect is due to the presence of a reflecting layer behind the retinulae cells. When this layer is uncovered, light is scattered into the

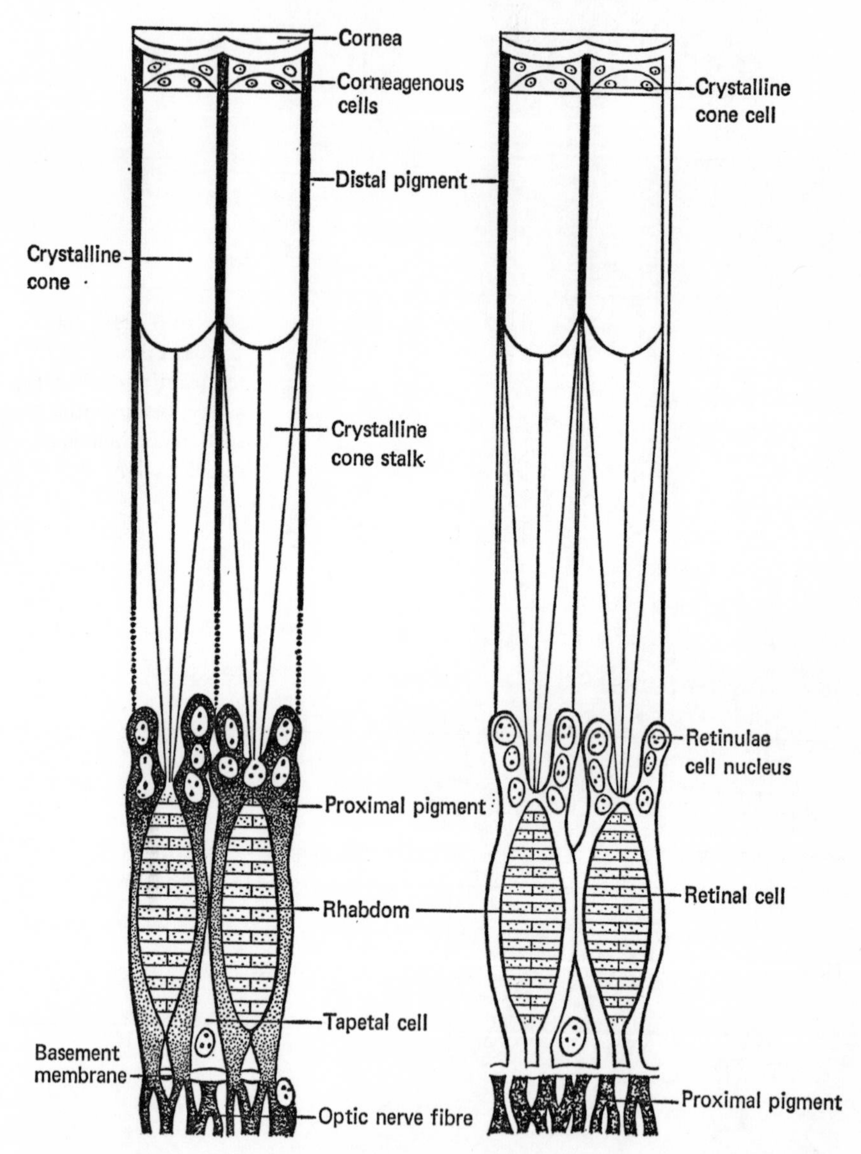

Fig. 10.7 The structure of the ommatidium of *Astacus*. Left, in the light-adapted condition; right, dark-adapted

sensitive cells by reflection producing a diffuse image, but when the pigment obscures the reflecting layer, light which has passed down one ommatidium will not be scattered into neighbouring ones and detail in the image will be preserved. Prawns with their proximal pigment withdrawn react to broader stripes than when the pigment is extended to prevent light scattering.

Colour vision

It is often difficult to prove that an animal has colour vision because a true ability to distinguish between lights of different wave-lengths can easily be confused with an ability to pick out colour by its brightness in comparison with others. Von Frisch's classic experiments with honey-bees eliminated the possibility by offering the bees paper of various shades of grey inter-

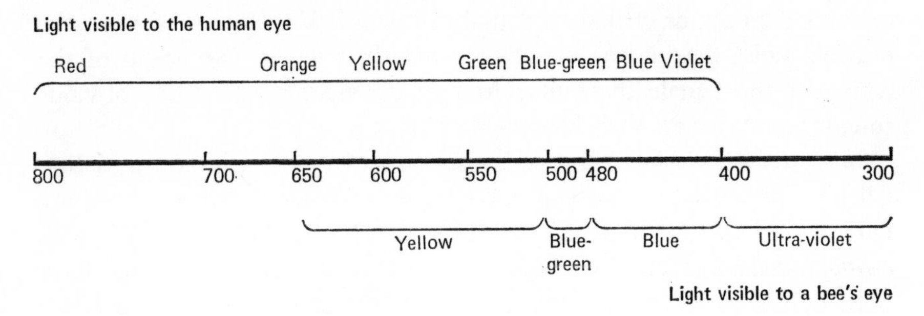

Fig. 10.8 The spectral sensitivity of a honeybee's eye compared with that of a human

mingled with coloured papers. When the bees continued to go to the papers of one colour only, ignoring the grey papers with equivalent brightness, it was plain that they could distinguish colour. Indeed the honey-bee's spectral sensitivity is as broad as that of a human (Fig. 10.8) but it extends into the ultra-violet and does not include the red end of the spectrum.

Some crustaceans are colour sensitive. Light of wave-length less than 500μ causes *Daphnia* to swim downwards, while yellow light of wave-length more than 500μ stimulates them to swim upwards. In red light, however, they tend to oscillate up and down with little horizontal movement. These cladocerans feed on phytoplankton and their reactions to light will tend to keep them in the area where their food is, for the green phytoplankton has the effect of filtering off the shorter wave-lengths.

Behavioural responses of earthworms may indicate two visual systems, each with maximal sensitivity in different parts of the spectrum. Their responses to shadow seem to be most sensitive in the yellow region of the spectrum; the receptors for this appear to be scattered uniformly over the body. The responses of changes of speed of movement show best when the light is blue; the receptors seem mostly to be gathered at both ends of the body.

Hearing

Hearing, in the sense of reaction to air- or water-borne sound, plays little part in the behaviour of invertebrates other than the arthropods; but this may be a measure of our ignorance rather than the true state of affairs. Among insects there are many which make noises in one way or another, for example by rubbing their hind legs against their folded wings as grasshoppers do, or clicking a drumhead, which is the mechanism of the cicada's noise-making, or merely beating their wings, the origin of the whine of the female mosquito. And in correlation with these obvious sounds insects have several kinds of hearing organs.

But even among the arthropods this seems largely an insect speciality. Clicking prawns are known to make noises with their chelae but we are ignorant of the function of the sounds. The ocypode crab, *Ocypode ceratophthalmus* (Plate 3), makes noises when in its burrow. They have been described as 'rasps', 'rattles' and 'burbling gargles', to which are added sharp knocking sounds on occasion. Perhaps these are warnings of a burrow's occupation. Nevertheless, unless crustacean statocysts (p. 214) act as hearing organs, which is not entirely impossible, structures which can be interpreted as ears are absent in this group.

Important information can come from noises carried in the ground from one animal to another. Insects are known to pass information in this way, and indeed, the sound produced by a honey-bee during its dance may convey important information to another bee and be detected through its legs. Spiders are the only other arthropods proved to be able to react to vibration in the frequencies which we call sound. On their legs lyriform organs (Fig. 10.9) are found, they appear as a set of minute slits. To the thin membrane in each is attached a sensory ending. They appear to be stimulated when a strain on the cuticle of the leg causes the slits to be opened. Such a group of organs at the tarsal-metatarsal leg joint of the web-building spider, *Archearanea*, will react to vibration of the leg in response to sound between 20 and 45,000 cycles per second. It seems that

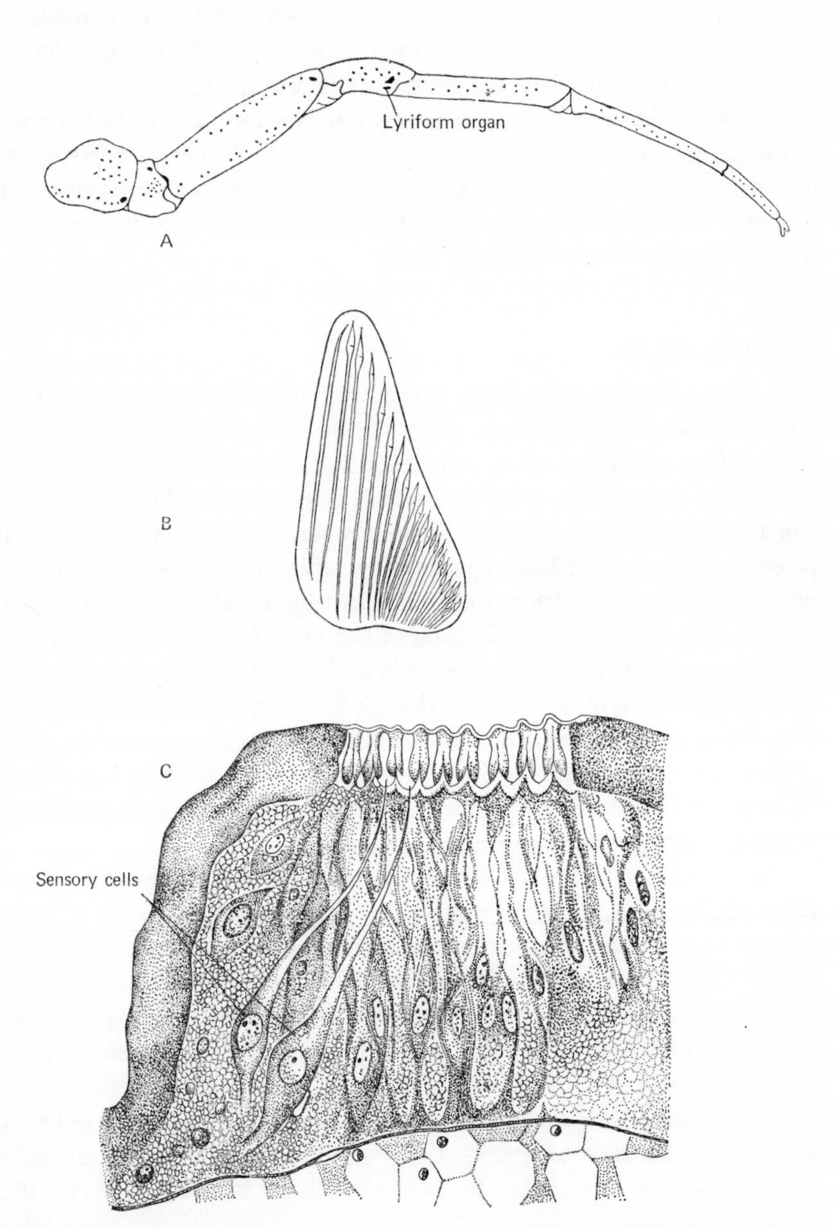

Fig. 10.9 A, the distribution of lyriform organs on the leg of a spider *Aranea ovigera*; B, an organ enlarged; C, cross-section of an organ showing two sensory cells

the different sensory endings are tuned to different ranges of frequencies which means that discrimination of frequency can take place. Normally the vibration is picked up from the thread of the web through the spider's leg resting on it. This species will only move out over the web to attack sources vibrating at 400–700 cycles per second, ignoring higher and lower frequencies. Plainly by this means the spider can discriminate between a trapped prey (this is roughly the wing-beat frequency of a fly) and other disturbances not due to potential food.

Postural senses

Part of the information which any animal requires in determining its response is that which informs about the position of the animal in space, whether it is right way up or not. Though many animals have receptors sensitive to the pull of gravity, they are not essential to an ability to swim or run the right way up. Indeed, all insects and many crustaceans do not possess such an organ. Instead, they show typical dorsal light responses in which they turn their backs to the incident light maintaining their body axes at right angles to the direction of the rays. Leeches such as *Hirudo* will swim with their backs upwards when illuminated from above, turning over only on being lit from below (though they correct their position later). The same happens with the tadpole shrimp, *Triops cancriformis*. Even *Artemia salina*, the brine shrimp, loops the loop when a light is switched on below it, though in this case it is a ventral light response for its normal position is ventral side uppermost (Fig. 10.10).

The dorsal light response is most important in determining the correct flight path of a dragonfly as well as the correct swimming path of many aquatic insects. Often this light sense is combined with some sensitivity to gravity though insects, remarkably enough, lack a true statocyst of the type found in vertebrates, decapod crustaceans and cephalopods. This consists of a fluid-filled cavity with sensory cells forming part of its wall. On these cells rests a statolith, a body of higher density than water, usually formed of calcium carbonate (though it may consist of sand grains in some prawns). Under the influence of gravity this statolith is moved as the animal's posture changes, stimulating the sensory cells by bending their hair-like sensory processes.

Like most decapod crustaceans, the crayfish *Astacus astacus*, has a statocyst in the base of each of its antennules (Fig. 10.11). It is possible to remove the statolith from these without damaging the sensory cells. If, then, the sensory hairs are bent towards the midline by using a jet of water

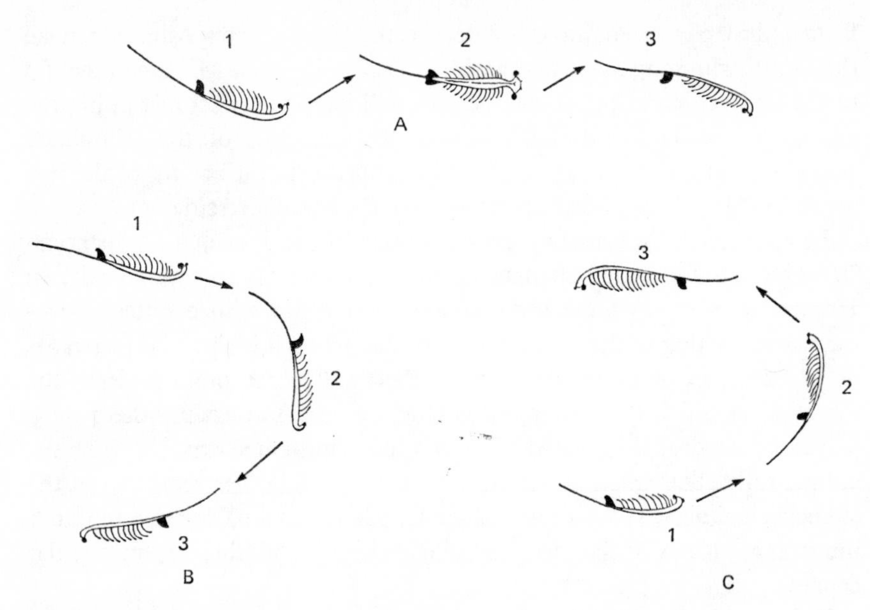

Fig. 10.10 Three ways in which brine shrimps *Artemia salina* may turn over when lit from below

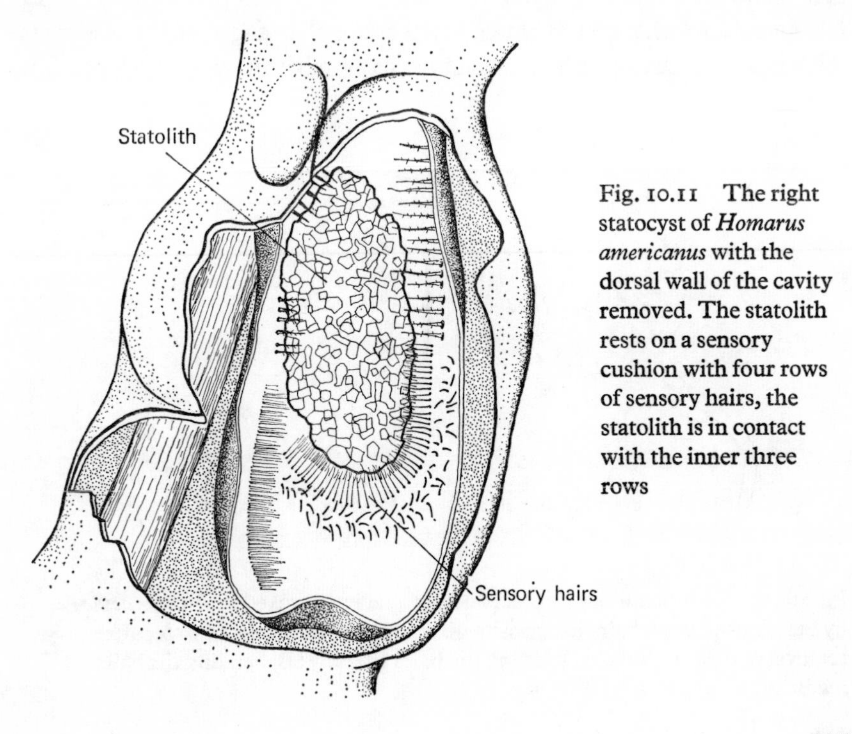

Fig. 10.11 The right statocyst of *Homarus americanus* with the dorsal wall of the cavity removed. The statolith rests on a sensory cushion with four rows of sensory hairs, the statolith is in contact with the inner three rows

from a pipette inserted into the statocyst, the legs of the opposite side make thrusting swimming movements while those on the same side are drawn up to the body (Fig. 10.12). These actions will have the effect of tipping the animal about its longitudinal axis in the direction of the stimulated statocyst. But if the sensory hairs are bent outwards, the actions of the legs are reversed and the animal tips away from the stimulated side.

In addition to information about orientation in space, the crustacean statocyst can also signal changes in posture about the transverse axis. In *Homarus americanus* there are receptors which fire with a different frequency according to the position of the head when it is tipped downwards or upwards, in other words, as it pitches. Other receptors indicate the direction in which the movement is taking place. And again, others show a distinct response for movements about the longitudinal axis, i.e. when the animal rolls. The result of the activity of these and other receptors in the statocyst is almost as much postural information as comes from a vertebrate's inner ear with utricular and saccular maculae and three semi-circular canals.

This vertebrate plan is closely mimicked in the structure of the statocyst of cephalopods. Again this is a fluid-filled sac, one being placed on each side below the brain in a cavity in the cartilage. Within this sac (Fig. 10.13) is a macula, of what one might call conventional arrangement, and a crista. This is a long narrow strip of sensory cells all of them with long hairs. The

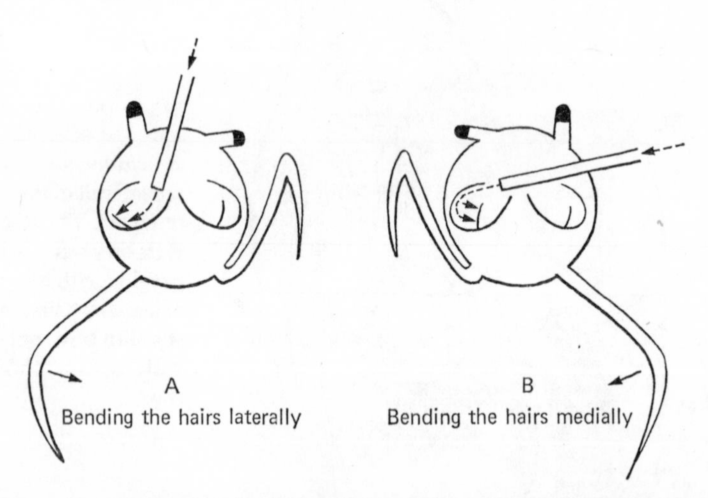

Fig. 10.12 Compensatory eyestalk and limb reflexes evoked in *Astacus astacus* by bending statocyst hairs in opposite directions with a fine jet of water after removal of the statoliths. A, bending the hairs laterally; B, bending the hairs medially.

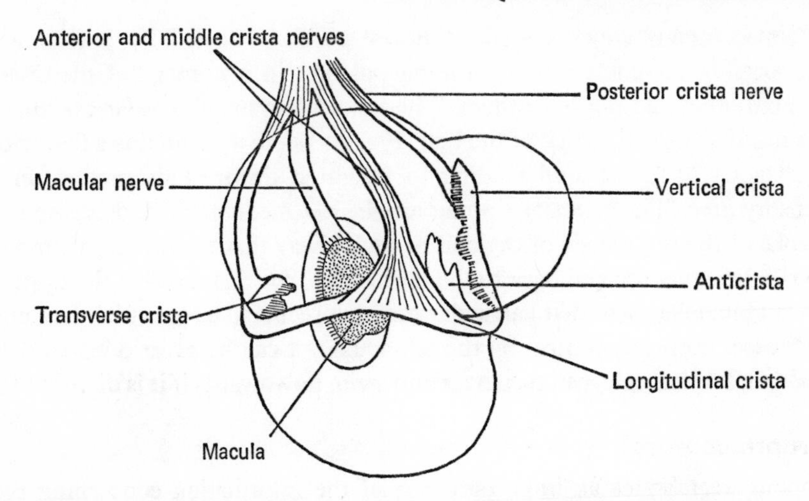

Fig. 10.13 The structure of the statocyst of *Octopus*. This is from the animal's left side and is viewed from the same side. Note the cristae arranged at right-angles to each other

strip is in three sections, in planes at right angles to each other. As a separate nerve supplies each part, they are strongly reminiscent of the vertebrate semi-circular canals arranged at right angles to each other. The hairs of the cells are enclosed to form a flap, similar to the cupulae of the ampullae of the vertebrate inner ear.

Destruction of these statocysts causes the expected disturbances in the animal's posture and movement, and disorientates the eyes as mentioned above (p. 207). But the exact way in which the parts function is not yet known in the detail with which we know the working of the decapod statocyst or the vertebrate labyrinth.

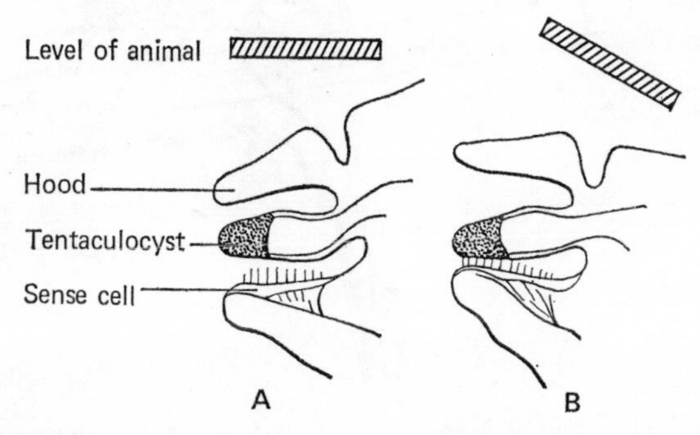

Fig. 10.14 The tentaculocyst of *Aurelia*. In the horizontal position (A) and in the tilted position (B) when the tentaculocyst presses down on the sensory epithelium

Sense organs which contain a heavy portion affected by gravity are responsible for control of swimming position in a number of medusae, though others do not have them. The tentaculocysts of *Aurelia aurita* are arranged around the edge of the bell (Fig. 10.14). Each contains a flap with inclusions within its end tending to weight it down. This presses on a sensory area. The effect of tipping an *Aurelia* medusae is to decrease the stroke of the upper side of the bell in such a way that the lower part tends to rise bringing the bell back to the horizontal. For this reaction the uppermost tentaculocyst at that particular moment seems to be essential. In some medusae, however, control by the tentaculocyst can be overridden so that *Pelagia*, for example, can turn over and swim downwards if it is disturbed.

Proprioceptors

Among vertebrates an important part of the information concerning the body's position comes from sensory endings in the tendons and muscles which signal the state of tension or contraction of these tissues. Until 1951 when Alexandrowicz described stretch receptors in the muscles of decapod

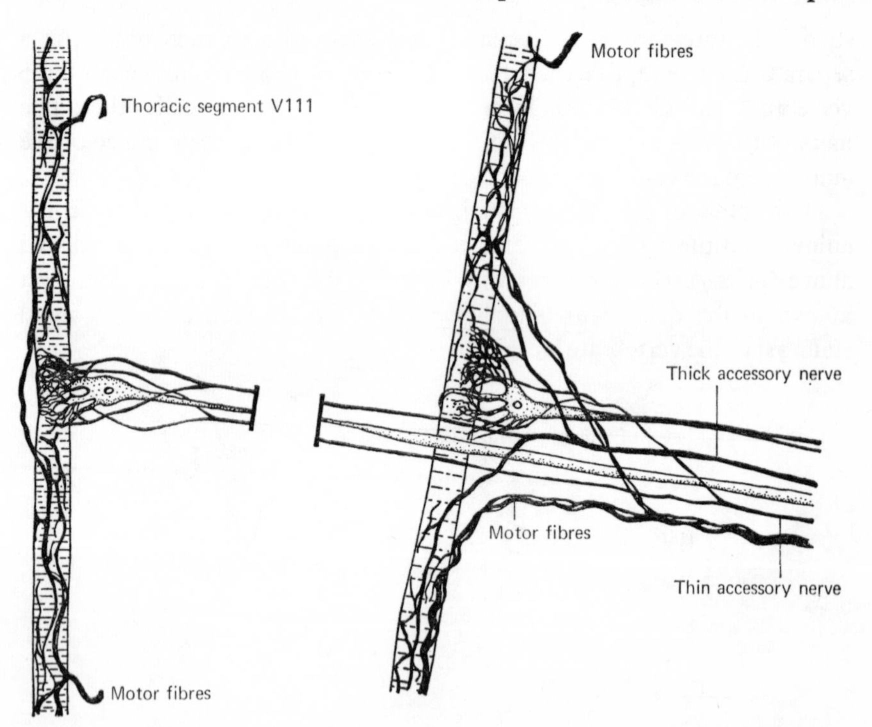

Fig. 10.15 The receptor muscles on the right side of the eighth thoracic segment in *Homarus vulgaris*

crustaceans this 'inner sensitivity' of proprioception was believed to have little in the way of counterpart in invertebrates. The ones found in *Homarus vulgaris* consist of modified muscle fibres in which part of the fibre is replaced by a region of connective tissue (Fig. 10.15). Beside this area lies a sensory nerve cell whose fine dendrites ramify into the modified fibre. Contraction causes a discharge in the nerve, evoked apparently by the stretching of the nerve cell endings.

Similar kinds of endings have also been demonstrated in insects, though they have, in addition, sense organs, the campaniform organs, which are stimulated by stresses set up in the cuticle by movement, thus supplying further information for the control of posture and movement.

Muscle-endings are well developed in the muscles of molluscs, and nerves which end diffusely on the muscles of turbellarians are suspected of having a proprioceptive function. Such endings may be difficult to stain and it may be almost impossible to prove their function but they may be expected to occur anywhere where elaborate movements take place. They have been identified in annelids, for example, and their function has been confirmed by the discharge in the afferent nerve from a segment when the skin is stretched. Reflexes based upon such mechanical responses play an important part in the control of annelid movement (p. 235).

Touch and chemical sensitivity

In soft-bodied invertebrates numbers of what on morphological grounds appear to be sensory cells lie in the epidermis, often with characteristic cilia-like extensions. It is often impossible to be certain whether they are stimulated by touch or by chemicals, indeed it is not unlikely that a simple sensory cell of this sort stimulated directly may respond to stimuli of different classes producing a different pattern of impulses for the different modalities.

However that may be, there is some evidence of specialization in the skin receptors of turbellarians (Fig. 10.16). There are three main types which can be distinguished morphologically. The 'tactile' receptors have a few branches which spread widely; these cells are found in the skin of the whole body. But the 'rheoreceptors' (touch receptors sensitive to water currents) lie in the anterior third of the body. They are large cells, occurring in pairs, each of which has many long bristles projecting from it, penetrating the epidermal cells and extending beyond the cilia covering the body. The last group, the chemoreceptors, occur only on the head of *Mesostoma*. Each has a single dendrite only from which arises a bunch of short pegs which, like the rheoreceptors, penetrate between the epidermal cells.

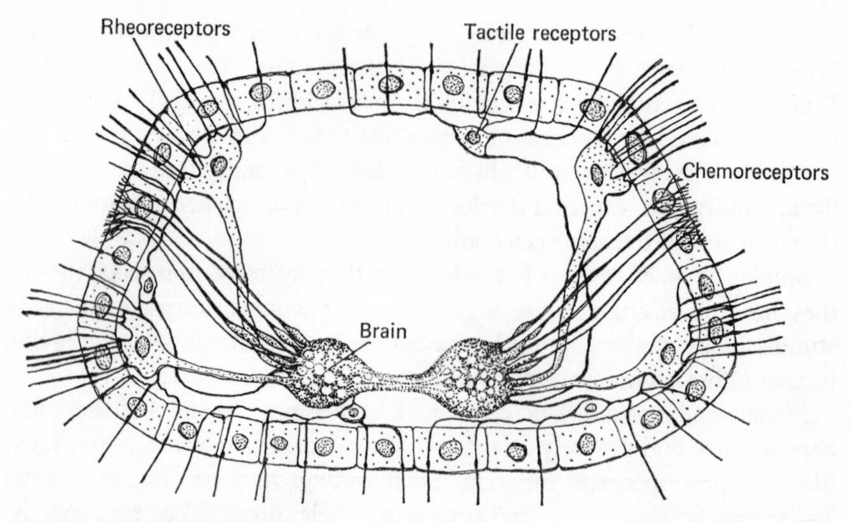

Fig. 10.16 A section through the head of the flatworm *Mesostoma* to show four of the eight rheoreceptors, the two groups of chemoreceptors and four tactile receptors

The functions attributed on morphological grounds appear to be correct, for removal of long-bristle groups and the peg groups causes loss of response to water currents and to chemicals respectively. No proof is yet possible of the function of the numerous sensory cells scattered through the endoderm and ectoderm of coelenterates. These are single cells with a long process and one or more nerve fibres. In all probability these are sensitive to touch as well as to chemicals, but at the moment it is not possible to do more than to demonstrate them and note their distribution.

Near the base of the gills of many gastropods lies the osphradium. This structure is strategically placed in the path of the water currents circulating through the mantle cavity. The prosobranch gastropod, *Buccinum*, has a well-developed osphradium which has been proved to be a chemosensitive organ. The whelk draws in water through a mobile siphon which can be used to sample the surroundings. Response to chemical substances in the water allows the whelk to find its food. Doubt as to whether this organ is always chemosensory is raised by the failure of all tests of chemicals to show any sensitivity of this sort in the osphradial epithelium of *Planorbarius corneus*. Possibly it may have a dual role, functioning also as a touch receptor reacting to sediment carried into the mantle cavity. A quick reversal of the water current would then prevent the gills becoming clogged. Whether this organ is responsible or not there is no doubt that *Nassarius reticulatus* can detect chemicals in the water entering its siphon.

It reacts positively for example to soluble starch, glycogen, aminobenzoic acid and skatol and, in nature, detects decaying flesh at a distance. Lactic acid in concentrations below 0·055 mg/ 100 cc is approached while higher concentrations cause the snail to retreat from the area.

Where the body of an invertebrate is encased in an impervious hard exoskeleton, special sense organs have to be developed to 'look out' on the environment through chinks in this armour. The sensory cells of the epidermis of the types described above are replaced by sensory hairs. According to their structure they may be touch-sensitive or chemosensitive or both.

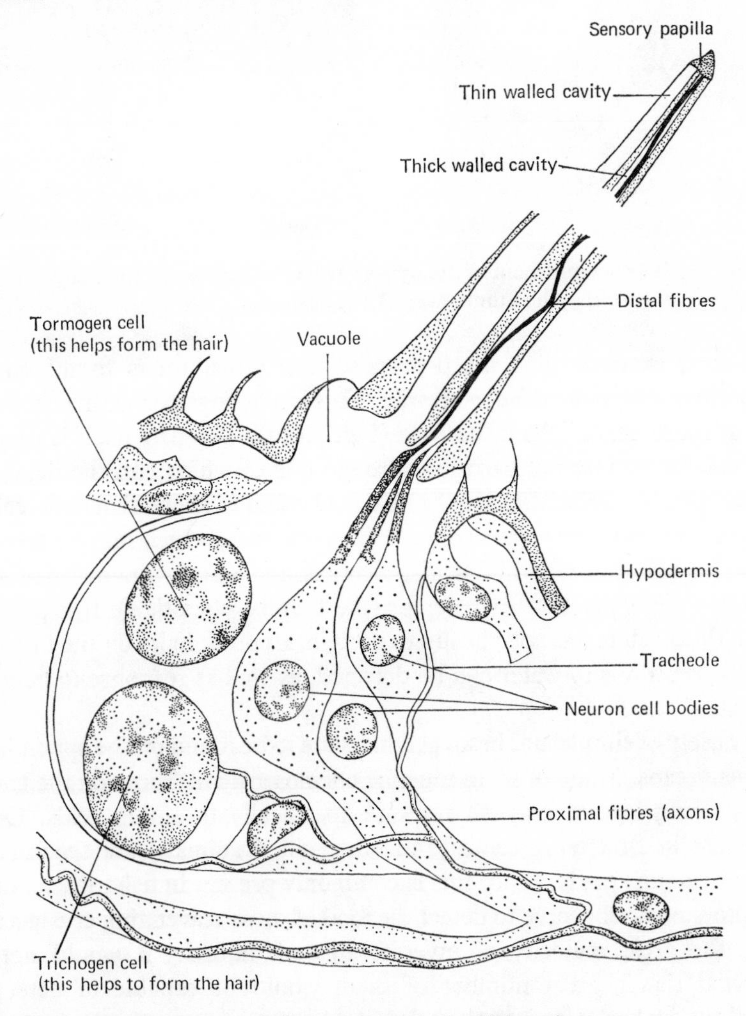

Fig. 10.17 Triply innervated chemosensory hair of *Phormia*

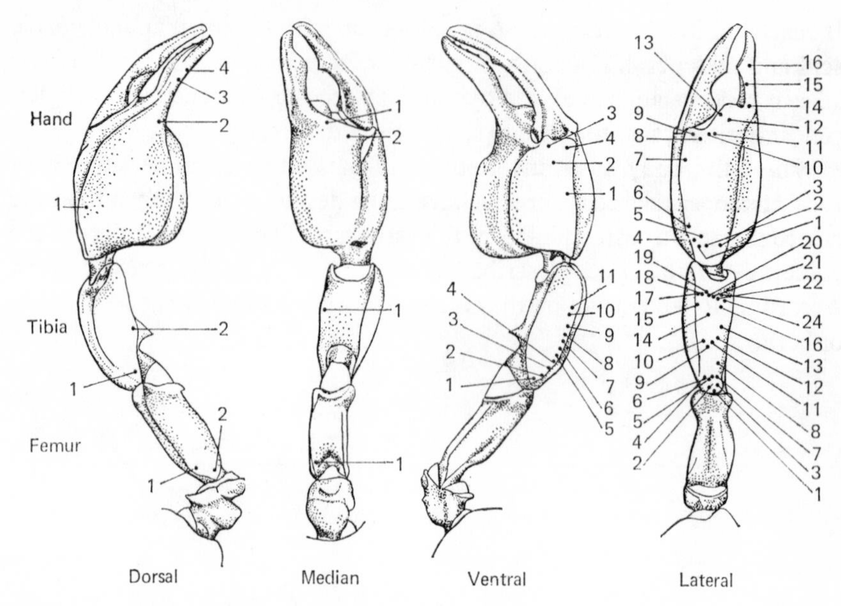

Fig. 10.18 The arrangement of the sensory trichobothria on the pedipalps of *Euscorpius carpathicus*; the numbers mark the positions

A good example of a multi-purpose hair sensillum is found on the labellum of *Phormia regina*, a blowfly. This is a hollow hair with a cluster of two or more nerve cells at its base (Fig. 10.17). The hair itself is divided into two by an internal partition through one of which run the dendrites of the sensory cells. Electrophysiological studies show that one cell is stimulated mainly by sugars (the S spike potentials) which may be taken as attractive substances, while acids, alcohols and salts – unattractive substances – stimulate another receptor (the L spike potentials). It is possible that a third cell reacts specifically to protein. In other hairs on the tarsus of this fly, reactions to water can be detected, as well as response to bending the hair.

A variety of simple and branched hairs are to be found on the appendages of crustaceans. Many of them must be chemosensitive; those on the chelae and walking legs of crayfish respond specifically to amino acids. Other receptors in Brachyura (crabs and other allied animals) are sensitive to trimethylamine oxide. Since this is commonly present in fish muscle, these receptors no doubt serve to detect the food of these scavenging crustaceans.

As the great majority of invertebrates are aquatic, it would not be expected that a great number of them would be capable of detecting chemicals in low concentrations in air, that is, endowed with a sense of smell. But insects with their thoroughgoing adaptation to life in air are often able to react to scent; indeed, for many, including most of the social insects, it is their dominant sense. Again, thin-walled pegs, mainly on the antennae, are responsible. The dendritic extensions of the sensory cells extend into the peg branching to end as naked endings in very small pores in the walls. It is receptors such as these which, for example, permit locusts to detect the scent of grass in the air and a honey-bee the scent of its own hive.

Hairs may react to vibrations in the air as do those on the cerci of cockroaches. The long trichobothria on the pedipalps of the scorpion, *Euscorpius carpathicus* (Fig. 10.18), are moved by small air currents. As each one is mounted on the cuticle in such a way that it can be moved in one plane only, and the various trichobothria have different planes of movement, these sensitive mechanoreceptors are well suited to determine the direction of a source of air disturbance. Probably the movements of prey are detected in this way.

The pectines of scorpions like *Leiurus quinquestriatus*, living in dry, sandy places, bear sensory fields of stubby pegs (Plate 4) on the teeth which are ranged along the appendage. These are mechanosensitive being stimulated when they are distorted. A possible use for them is the measurement of the size of sand grains on which the scorpion comes to rest for these animals show distinct preferences for sand with grains of 0·5 mm diameter or smaller.

With the range of sense organs which the average invertebrate possesses these animals can hardly be described as insensate beings. Just as it is unwise to underestimate their sensory powers so it is foolish to ignore the complexity of their behaviour, as will be seen in Chapter 12.

Chapter 11

Nervous system and co-ordination

WE ARE familiar with the idea that nervous systems are the essential co-ordinating systems in animals responsible for the patterning of muscular responses into integrated movements of all or part of the body. They also act, through the sense organs, as the collectors of information about the external and internal environment, information which is then used to modify and initiate activity in the muscles. It is difficult to conceive of nervous systems without systems of contractile elements through which their integrative function can be expressed. Indeed, in searching for the origin of the nervous system, Parker postulated as the first stage in the evolution of a nervous system the occurrence of contractile elements which are independent effectors. Citing the myocytes of sponges as contractile cells which each responded separately to stimulation of itself, Parker proposed that near such contractile elements, epidermal cells evolved into sensory elements capable of receiving the environmental information, a contact between a contractile and a sensory element constituting an excitation-transmitting synapse.

But there are a number of objections to this idea, and perhaps a more acceptable theory is one which arises from the discovery of pacemakers responsible for initiating spontaneous rhythmic activity in *Hydra*. Passano suggests that the contractile elements which were, everyone agrees, the first to arise, had amongst them pacemakers which evoked activity in them. From this point the contractile cells became more specialized as muscle, while the pacemakers evolved into nerve cells, at first becoming sensitive to the environment, so that stimuli from without affect their rhythmic activity modifying the responses produced by the contractile elements. Later the nerve cell becomes adapted for conduction, though its first (evolutionary) function was that of the repetitive initiation of activity.

Cells have been described in sponges which are identified as nerve cells but there is little evidence, either morphological or physiological, to support this supposition. Therefore it is in coelenterates that the first organized nervous system is encountered. Even there it is in the form of a diffuse

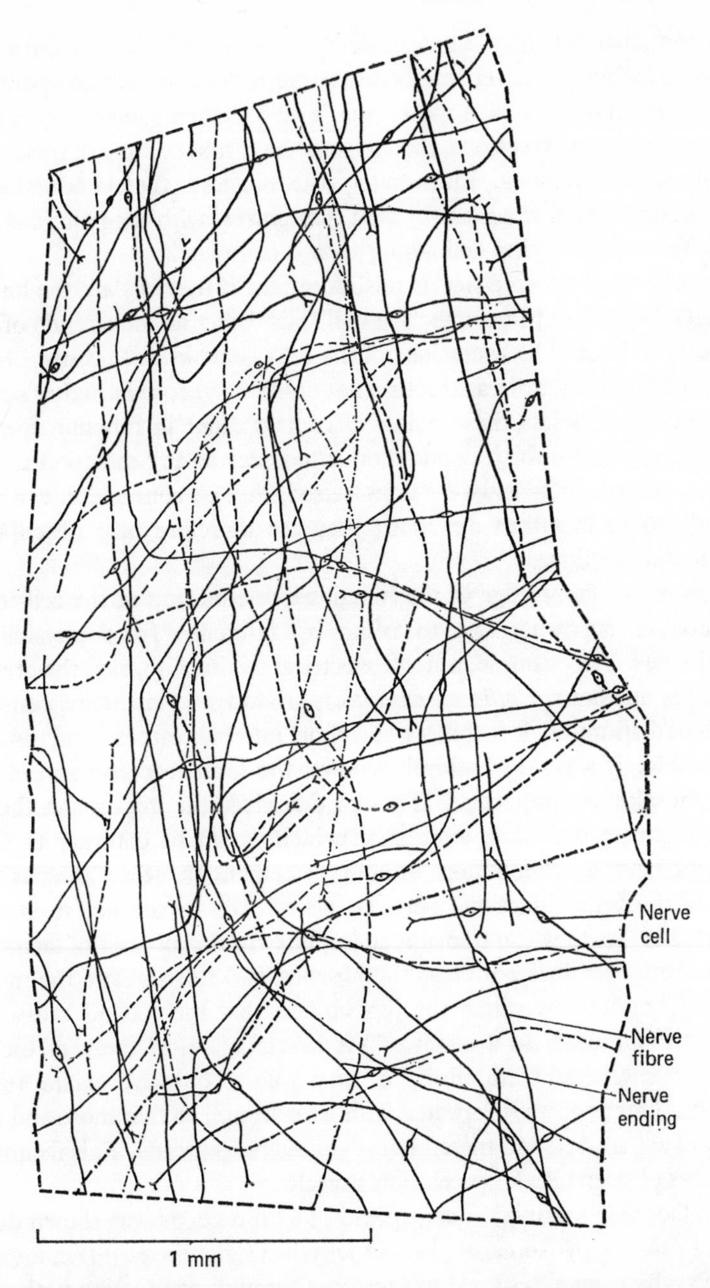

Fig. 11.1 A plan of the nerve net on one face of the mesentery of an anemone (*Metridium senile*). Since the method of staining does not necessarily colour all of each nerve and because some will go out of the plane of focus of the microscope each cell does not necessarily appear to connect with another

intra-epithelial net (Fig. 11.1) made up of cells, which sometimes form linear tracts but are never bundled together to form nerves comparable to those of the more complex metazoans. These elements of the net are cells with two or three extensions, though some have more – each of them seems equally well developed. Immediately this suggests that conduction can occur in any direction across the cell; the contrast with the polarized nerve cell with short dendrites and a long axon is complete.

The expression 'nerve-net' is misleading for if the cells are the knots of the net, they are not in connection with each other as the strands of a net join knot to knot. The extensions of the cells come in close contact but do not fuse; thus they form a structure not unlike a synapse in higher nervous systems. Nevertheless these points of contact differ in function from the vertebrate synapse, for if conduction is to occur in any direction through the net, the synapses must be unpolarized, that is, conduction can occur through them in either direction; they are therefore best described as interneural junctions.

Most of our knowledge of the functional relationship of the cells of the coelenterate nerve-net has to come by inference from physiological experiments. For example, a single electrical stimulus to, say, the oral disc of the sea-anemone, *Calliactis parasitica*, produces no visible response but the second stimulus, if it follows at a short interval, causes local muscular contraction. If a train of stimuli follows, the local response spreads and more muscles are implicated. Though this gives the impression that the nerve impulses die away, a situation which would be contrary to all we know of nervous conduction without decrement in other animals, it is explainable if the junctions between nerve cells in the net need to be activated before they can transmit an impulse. In this process of facilitation the first impulse does not cross the junction and so stimulate the muscle. It does, however, sensitize the junction so that the second crosses the synapse more easily as a result of the previous arrival there of the first impulse; subsequent impulses, if they follow quickly, excite further neurons. This process is repeated with later stimuli so that more and more synapses are crossed and therefore the conduction spreads. This facilitation also takes place between net cells and muscles.

The fact that conduction can spread in all directions was shown during the last century by Romanes. He cut jelly-fish (*Aurelia aurita*) in a variety of ways which were designed to sever any through-conducting paths (Fig. 11.2). Nevertheless a propagated wave of muscular contraction would pass through the tissue provided that the parts were connected by bridges at least 1 mm wide. Similar sorts of result were obtained by slicing *Calliactis*

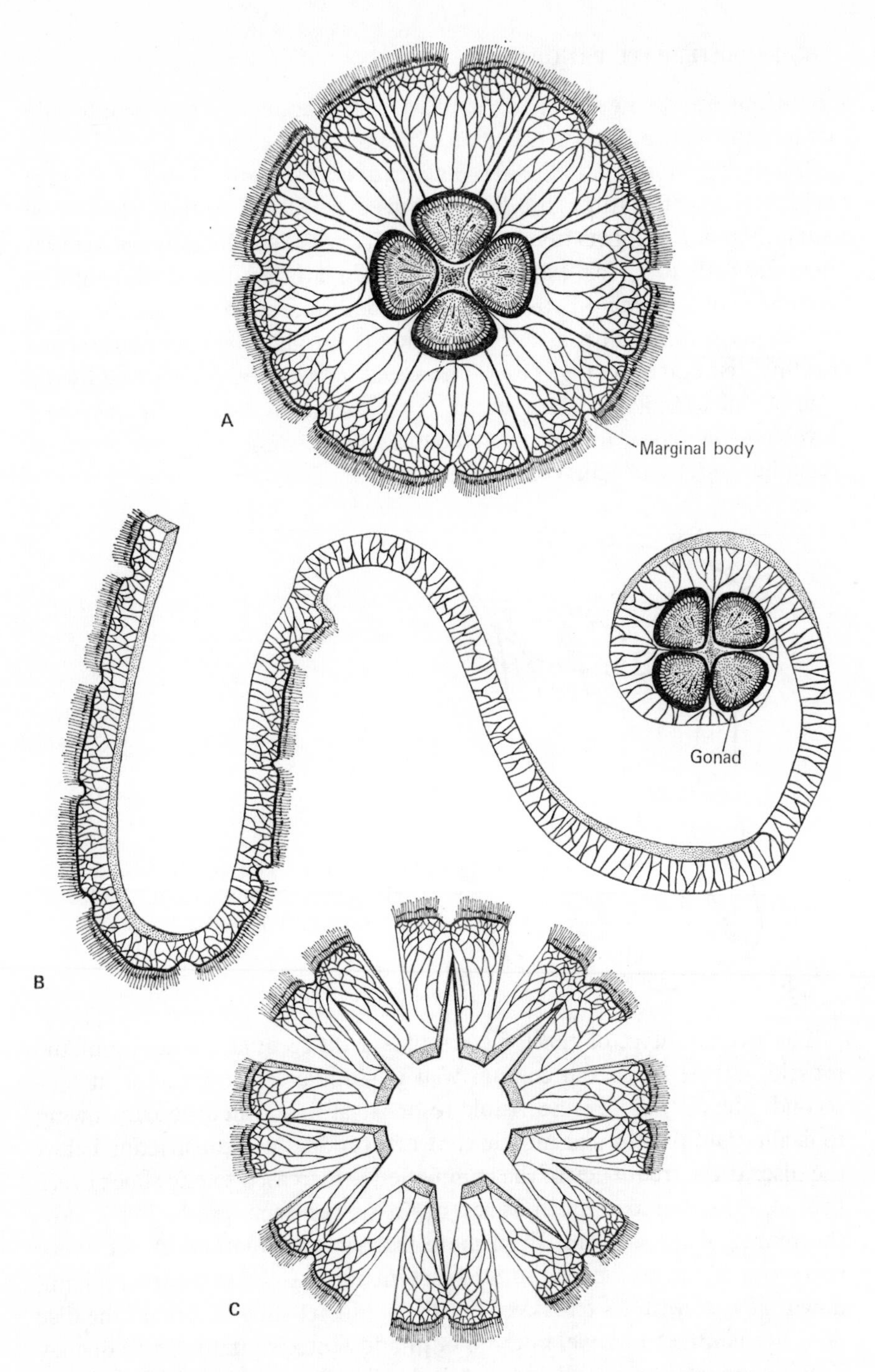

Fig. 11.2 A, *Aurelia aurita* viewed from above. After the cuts had been made, which are shown below (B and C), conduction still occurred through the tissue

in various ways, stimulation at the end of a strip connected to the body only by its end still produced contraction of the whole anemone.

Though conduction in the nerve-net may occur in any direction, there is evidence of paths through which impulses travel faster than elsewhere; often there are two nets involved each with special conduction characteristics. In *Calliactis*, for example, conduction around the surface of the column is of the order of 10–20 cm per second but round the disc the speed is 1 m per second. The fastest conduction is through the mesenteries connecting disc and foot (Fig. 11.3). The speed seems to be determined by the number of synapses to be crossed. There seem in fact to be two nets involved, one co-ordinating slow contractions lasting many minutes and the other fast, twitch-like contractions.

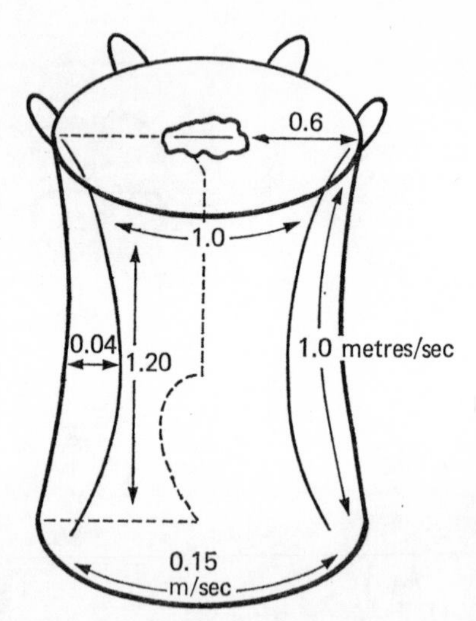

Fig. 11.3 The speeds of conduction in metres/sec along various paths through *Calliactis*

The overall effect of these factors plus a differential sensitivity of the muscle, is that when the column wall is stimulated with shocks at ten-second intervals, there is no visible response at first. But gradually, owing to facilitation, the circular muscle contracts producing a constriction below the disc. If the frequency of the stimulation is increased to one shock every five seconds, the parietal muscles begin to contract, producing a slow shortening of the column. Even more frequent stimulation (at one every two seconds) causes the longitudinal mesenteric muscles to contract pulling down the disc with its tentacles; but the sphincter muscle around the disc does not contract to envelop the mouth and tentacles until the frequency rises to one per second. Thus, with increasing frequency of stimulation, a sequence of contractions take place which are the same as those which produce the whole closing response of an anemone when it is violently stimulated.

The division of the nervous system into two nets is particularly well exemplified by the ephyra larva of *Aurelia* (Fig. 14.1). This eight-armed medusa has a feeding response in which one arm at a time catches food and bends inwards towards the manubrium which in its turn swings over to it. More than one arm may be catching food at the same time, the arm movements occurring simultaneously with swimming movements or while the larva is floating quietly. To swim the whole animal shows rhythmic symmetrical contractions. There are two nets which can be distinguished morphologically. One, which has been called the giant fibre system, is composed of bipolar cells rather larger than usual. Being found only on the circular and radial muscle fibres, it is immediately suspected to be the controller of the swimming beat which involves these muscles. Woven through the epithelial spaces of the ectodermal layer is a net of finer extensions of smaller, often multipolar cells. These appear to be responsible for the feeding reactions.

Analysis of nervous functioning also comes from experiments in which cuts are made at strategic places. For example, cutting the circular muscles and their giant fibre system at the base of each arm, causes the muscles to contract in an uncoordinated manner, though the whole animal can go into a spasmodic contraction, which, like the feeding reaction, seems to be mediated by the fine net.

The co-ordination of the many individuals in various colonial forms can also be explained in terms of these properties of the nerve-net. Facilitation seems unnecessary in the organ-pipe coral, *Tubipora*, for one stimulus immediately spreads throughout this alcyonarian colony; in *Heteroxenia* as the number of stimuli increases, so the effect spreads further until all the individuals are involved. Though the first shock is effective in the reef-forming corals *Acropora* and *Porites*, made up of many polyps, further spread will only occur at a frequency of one per second, lower frequencies being ineffective. But in *Palythoa* there is little spread with further stimuli, a frequency of eight per second bringing in only about ten polyps, while such a frequency will involve all the polyps of *Heteroxenia* or *Acropora*.

Despite the apparent simplicity of construction of the nerve-net, such relatively complex movements as those of an anemone selecting and mounting a whelk shell (p. 180) are possible. It is probably the limited range of their sense organs and the absence of a central co-ordinating and controlling system which impose the limit on the behavioural possibilities of coelenterates.

Segmentation and cephalization

The differentiation of the nerve-net in the coelenterates so that through-conduction can occur foreshadows the condensation of the nervous system which is found in other invertebrates. The nerve cells (Fig. 11.4) become larger and cease to conduct in any direction; at least as they are connected in the nervous system each tends to receive impulses through dendrites or on to the cell surface and conduct along one elongated extension, the axon. Thus, impulses are conducted for much greater distances without having to cross synapses and as a result facilitation becomes less important as a factor determining the pattern of response of an animal.

Instead, variability comes about by the very much increased variety of cross-connections which can be made. Nerves coming from and supplying various parts of the body feed into long nerve cords in which there are many nerve cell bodies. The occurrence of connecting or internuncial cells whose sole function is to make connections between cells within the cord means that responses will depend upon the particular cross-connections that are brought into operation. The more internuncials there are the greater is the number of possible paths by which incoming impulses can flow up or down the cord. Indeed the routeing within the cord becomes a more and more important aspect of the organization of the central nervous system.

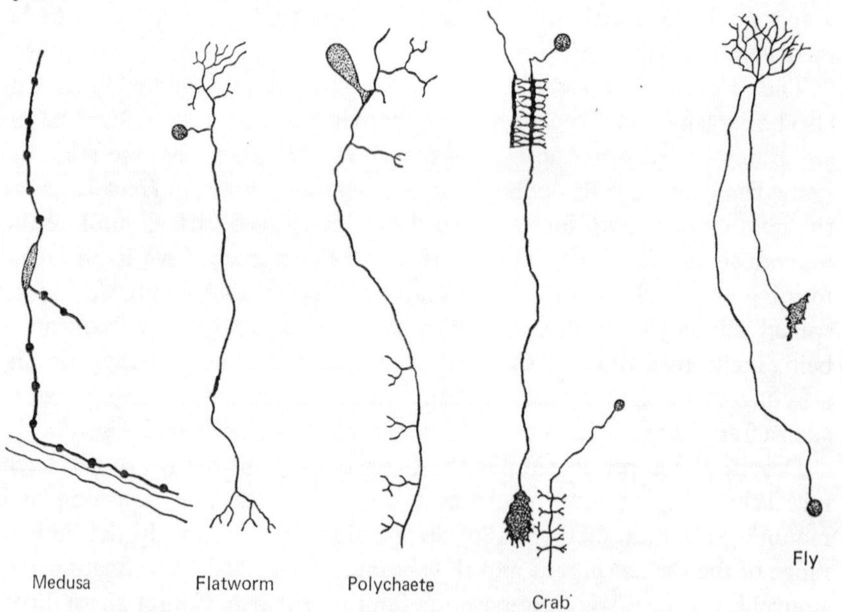

Fig. 11.4 The different morphology of neurons from selected invertebrates

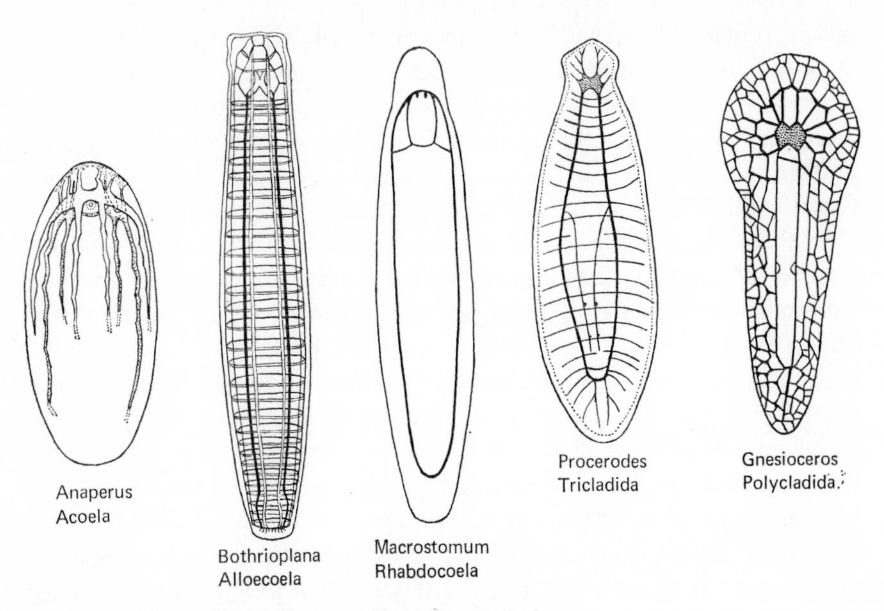

Fig. 11.5 Plans of the nervous system of turbellarians

The majority of coelenterates are sessile animals but most metazoans can move freely, many of them actively. And since their movement tends to be in one direction, the evolution of a condensation of sense organs at the leading end is clearly advantageous, for in this way the environment ahead is sensed and the movement directed. The input from the sense organ is an essential part of the determination of the responses and behaviour of an animal and it is not surprising, therefore, to find that there is a trend to a condensation of nerve cells at the head end to form a brain.

These tendencies – to the formation of a brain and a concentrated nervous system – can be found in the flatworms. Though the nervous system of a turbellarian may still retain the diffuse nature of a nerve-net in part, some of the strands of nervous tissue stretching back along the body are distinct enough to be called nerve cords (Fig. 11.5). Experiments involving the removal of the brain of a planarian show that this collection of nerve cells is essential for behaviour to be properly directed, though even then co-ordinated movement is still possible. The nerve-net found in peripheral parts of the animals is still capable of organizing activity independently when the central nervous system is removed. For example, the proboscis of a planarian is capable of showing responses to food juices even when severed from the body.

With the appearance of metameric segmentation, each segment of the body is provided, in its early development at least, with its own complement of ectodermally-derived nervous tissue. In the annelids and most of the arthropods, the segmentation of the central nervous system is evident; in insects and crustaceans there may be fusion of segmental ganglia in groups to form ganglionic masses but the derivation of these often remains obvious from the relationships of the nerves leading from the mass.

In these animals, the central nervous system is in the form of a clearly defined cord running the length of the body, which in section is seen to be two solid cords fused together, though its double nature remains outwardly obvious in arthropods. Once again cell bodies are confined to the cords apart from the bipolar sensory cells found in the skin and elsewhere. At the head end are two much enlarged pairs of ganglia, one, the sub-oesophageal lying beneath the gut, joined by two connectives passing round the oesophagus to a second enlarged ganglion, the supra-oesophageal. In general the sense organs, the eyes, and appendages bearing sensory hairs of various kinds, are connected to the supra-oesophageal ganglion, while the sub-oesophageal mass is a condensation of the ganglia of a few of the succeeding anterior segments.

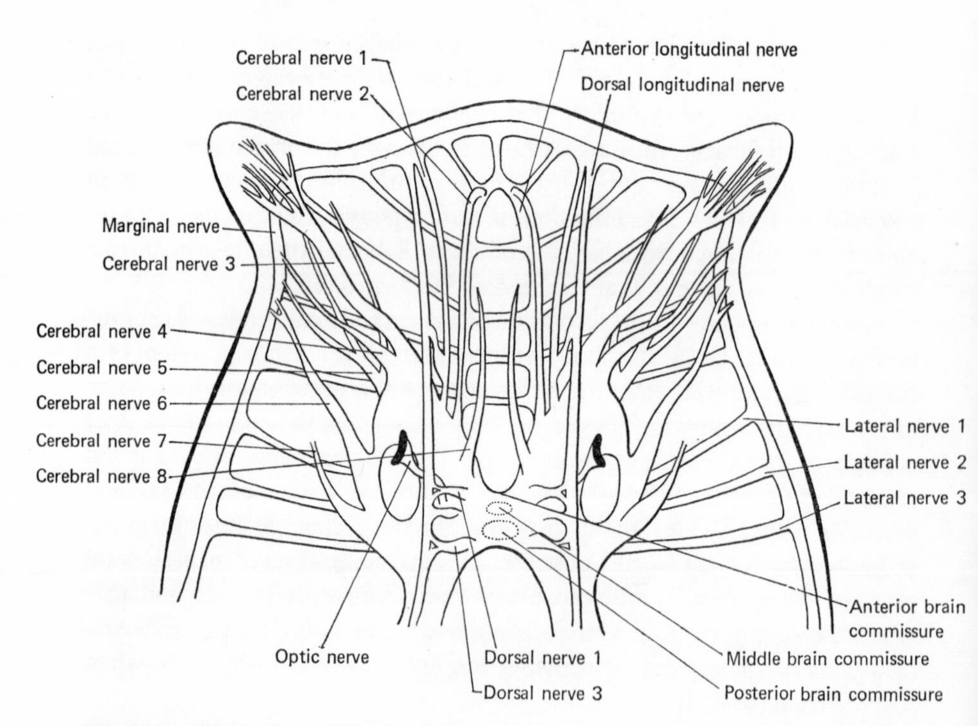

Fig. 11.6 The brain of *Planaria alpina*

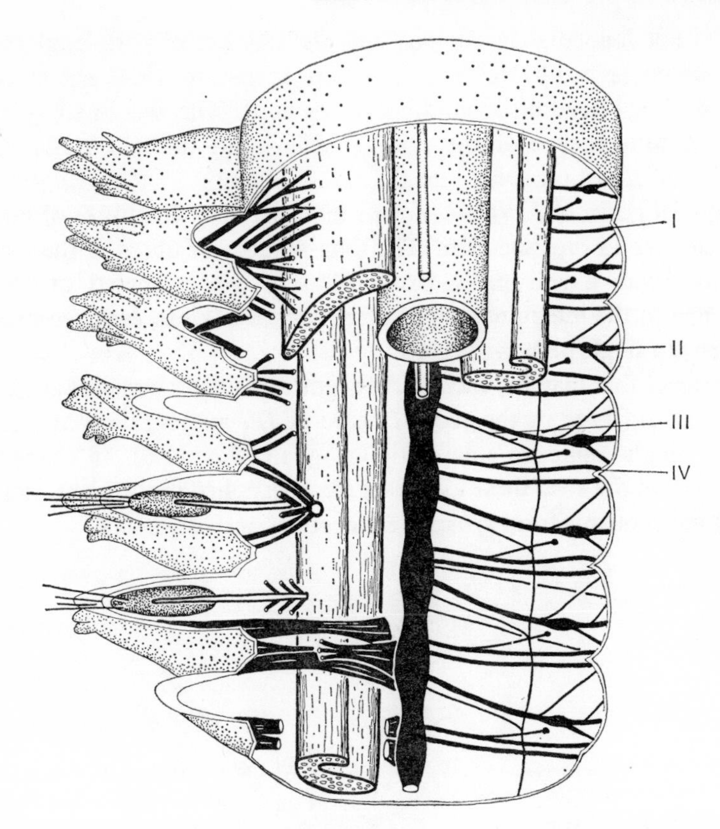

Fig. 11.7 The arrangement of the segmental nerves (i–iv) of *Nereis virens*

Being fully equipped, as it were, with sensory and motor elements the segmental nerve system is potentially autonomous. The organization of the polychaete segment is a good example to take. In nereid polychaetes, each ganglion gives rise to four pairs of segmental nerves which supply the body musculature and epidermis in that segment (Fig. 11.7). The first nerve arises from behind the septum separating the segment from the one anterior to it, the second and third arise closely together near the posterior end of the ganglion at about the middle of the segment. The second nerve is the larger. The fourth nerve leaves the cord just in front of the posterior septum from the ganglion of the succeeding segment which pushes forwards slightly into the segment in front. Thus any one ganglion has nerves IV, I, II, and III arising from it in that order. Nerves IV and I supply the dorsal parts of the integument, II passes to the parapodium, while III links with neighbouring segments through a lateral nerve.

It is not immediately obvious how the number of peripheral receptor cells are connected to the central nervous system for these nerves contain relatively few large fibres running to the cord. This would suggest some kind of convergence of the sensory cells input through a diminishing number of fibres on entry into the central nervous system. However, the parapodial nerve of *Harmothoë* contains many very fine fibres of less than 1μ diameter in cross-section. The total number of fibres, if these are included, bears a reasonable relationship to the estimated numbers of receptors in the integument and on the muscles. Thus, each receptor may well have a single path to the cord.

However that may be, each of the segmental nerves contains at least one motor axon (three in the second nerve of *Harmothoë*) derived from cell bodies on the opposite side of the ganglion (Fig. 11.8). In crossing the substance of the cord these axons are closely applied to the giant fibres (see later) and probably form synaptic connections at these points.

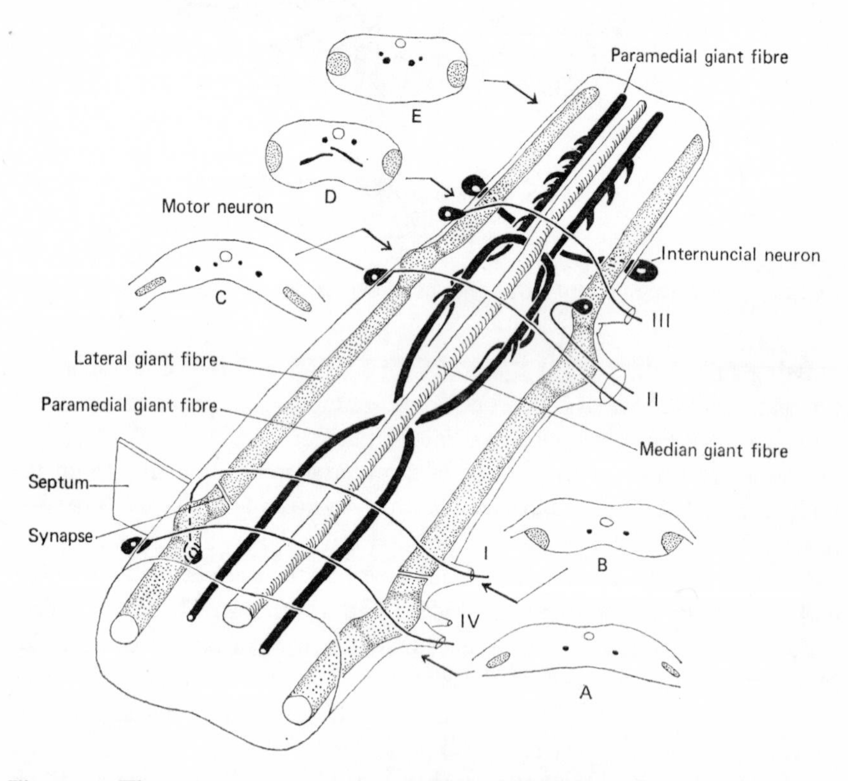

Fig. 11.8 The main motor nerves in a segment of *Nereis diversicolor* and their relationship to the giant fibres. The synapses between the giant fibres can be seen

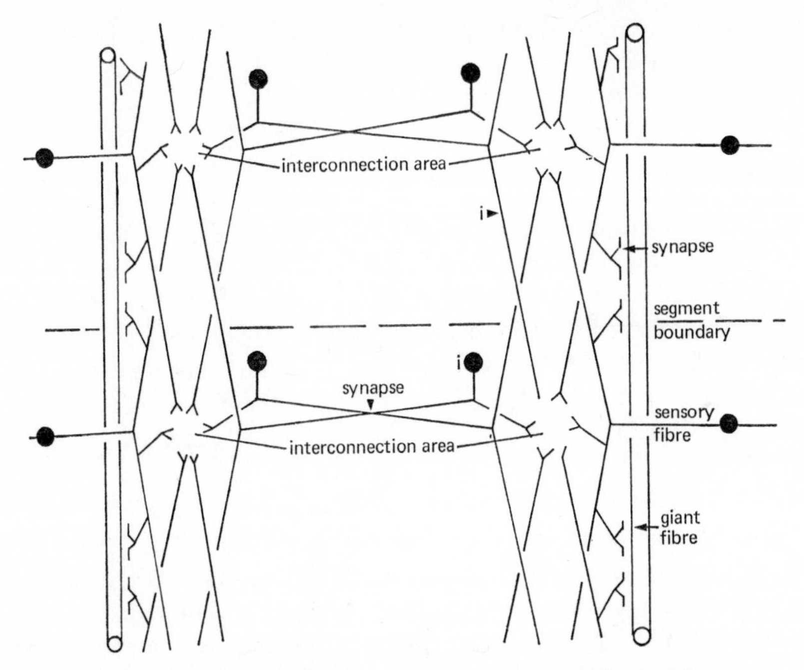

Fig. 11.9 A possible plan of the connections between sensory nerves between and within segments of *Harmothoë*

The sensory nerves of *Harmothoë* seem to be intricately interconnected within the cord, for stimulation of the bristle receptors in any segment causes impulses to emerge in sensory fibres of nerve II in other segments and indeed in the circum-oesophageal connectives. Thus there must be connections up and down the cord, not only to neighbouring segments, but for greater distances as well (Fig. 11.9). This arrangement is strangely reminiscent of a nerve-net.

It can be shown that locomotion in annelids utilizes local reflexes. Waves of contraction will pass along an earthworm's body even when it is severed completely apart from a nerve cord connection. Thus the co-ordination of longitudinal and circular muscle contraction takes place through the intact nerve cord. On the other hand, if the nerve cord is cut, and the body walls of the two halves sewn together by threads, waves will pass the cut in a co-ordinated manner. In other words, as other experiments confirm, waves can be propagated by purely mechanical pull on the muscles of the body wall. There are local reflexes between the segments (Fig. 11.10) so that if longitudinal muscles are stretched, sensory cells are stimulated, and the circular muscles on the next segment are stimulated into contraction producing stretching of the longitudinal muscles of that segment: thus the waves of muscular activity pass down the cord without reference to higher centres.

Giant fibres

There are present in the nervous systems of many different invertebrates, large fibres which pass long distances through the cord. Annelids, arthropods and molluscs are only a few of the groups which possess them (Fig. 11.8). These giant fibres are usually constructed with their cell bodies to one side of a much-thickened fibrous part (Fig. 11.11) containing tubules characteristic of nervous structures. Sometimes these cells extend through many segments, or they may be restricted to a single segment. Generally they abut on to the neighbouring cell so that the cell membranes lie close together without necessarily showing structures characteristic of synapses in other groups (Plate 5). They may be very big reaching as much as 700μ in diameter in squids. These, incidentally, supply very good material for research into the physiological bases of nervous conduction.

Conduction is fast along these paths and therefore they are often implicated in escape reactions. Thus the end-to-end contractions of earthworms which draw them away from danger are mediated in this way. If the tail end is in the burrow and the head end is touched, the worm pulls back upon its tail anchored by its setae in the ground. This response is mediated through

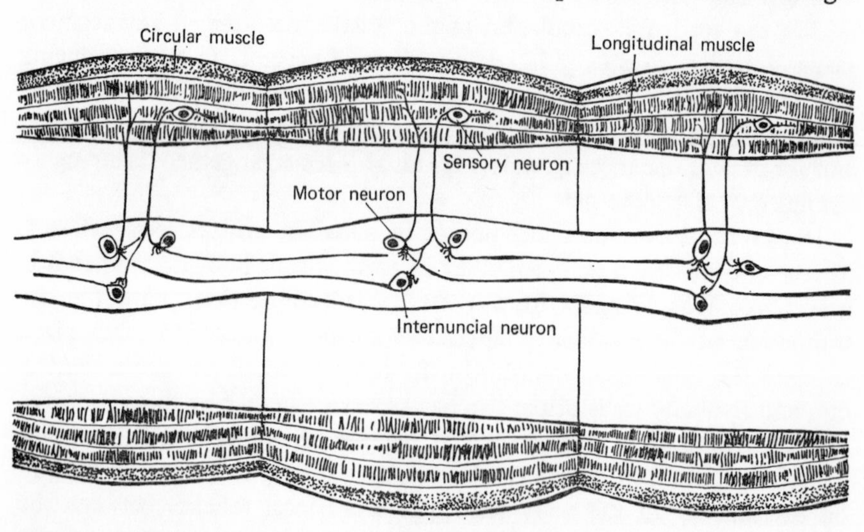

Fig. 11.10 The local reflex arcs responsible for passage of the wave of contraction along an earthworm

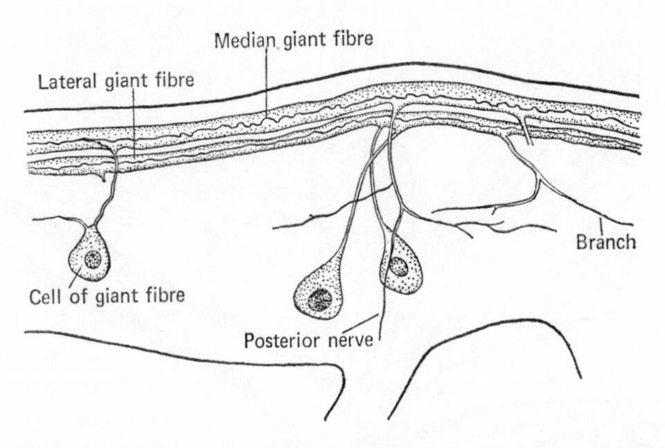

Fig. 11.11 The connections of giant fibres and their cells in a typical segment of *Pheretima*

the middle one of the three giant fibres which can be seen in an earthworm's nerve cord. If, on the other hand, the worm is out of its burrow, a touch on the posterior end causes it to contract up to its head end. This withdrawal is mediated by the two lateral giant fibres. Under experimental conditions these fibres can be shown to conduct in both directions for their synapses are not polarized. But because the median fibre is connected to receptors at the head end, impulses usually pass down it from head to tail, and the connections of the lateral fibres equally decide the direction in which the impulses shall pass. Whereas conduction along ordinary nerve fibres is at about 0·025 metres per second, the lateral giant fibres conduct at 7–12 metres per second, and the median at 17–25 metres per second. The greatly enhanced speed is due to the reduction of the number of synapses and to the greater diameter of the giant fibres in comparison with any other nerve.

The giant fibres of squids arise from giant cells in the pedal ganglion of the brain (p. 000) and make synaptic connections with other giant cells whose fibre passes out along the mantle nerves to the stellate ganglia. Within the ganglia, these fibres make connection with yet a third set of giant cells, in long synapses near the axon base (Fig. 11.12). These last fibres innervate the muscles of the mantle cavity whose contractions cause the expulsion of water through the siphon and the rapid movement backwards of the animal. When a crayfish escapes it does so by a violent flexion of its abdomen, a response which like that of the squid and the earthworm depends on giant fibres. In this case, paired median fibres lead from cells in the brain with an additional system of shorter lateral fibres arising from cells in the ventral ganglion and extending through a few segments only.

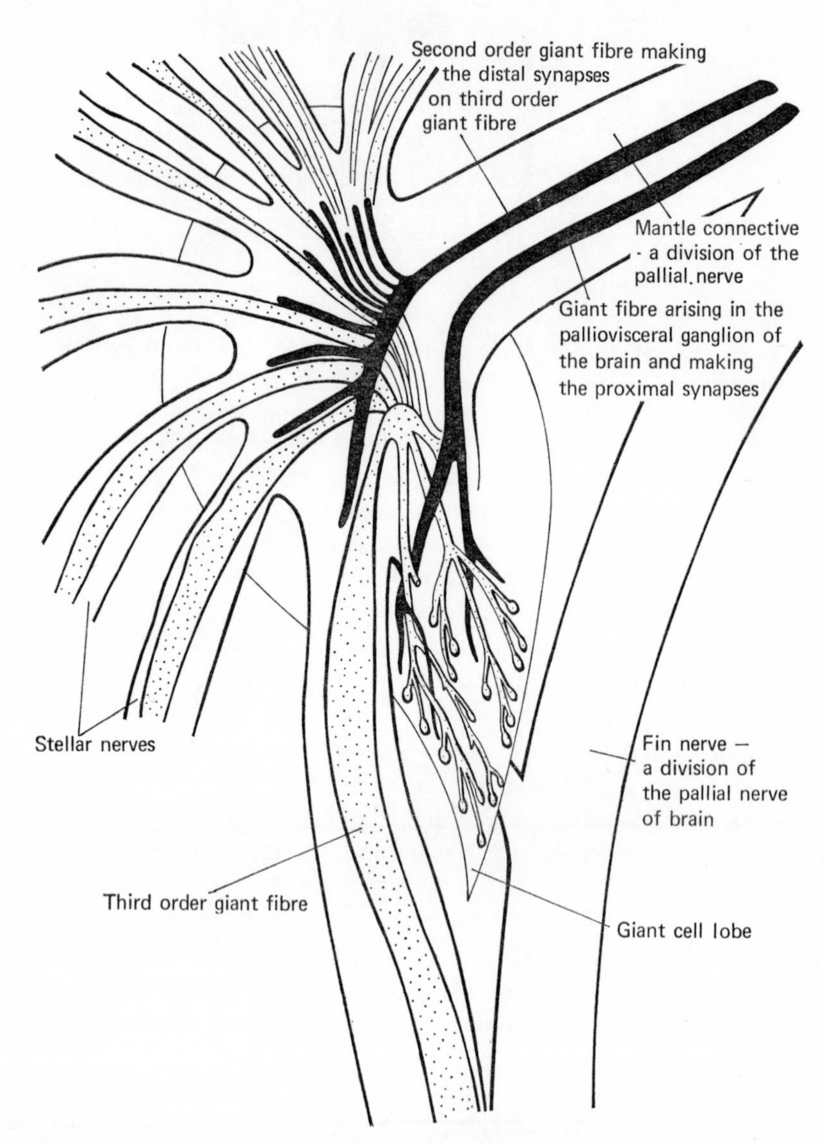

Fig. 11.12 Giant fibres and their synapses in the stellate ganglion of *Loligo pealii*

Spontaneous activity of the central nervous system

Very frequently the nerve cord of an invertebrate shows electrical activity even when it is removed from the body of the animal. This is spontaneous for it cannot be maintained by input from sensory receptors. The activity is, in addition, often rhythmic, sharing a frequency which may coincide with that of some aspect of the animal's behaviour. The rhythm of beat of the swimmerets on the abdomen of a crayfish reflects the rhythm of impulses arising from an isolated piece of the nerve cord.

On the other hand, though an earthworm's nerve cord is rhythmically

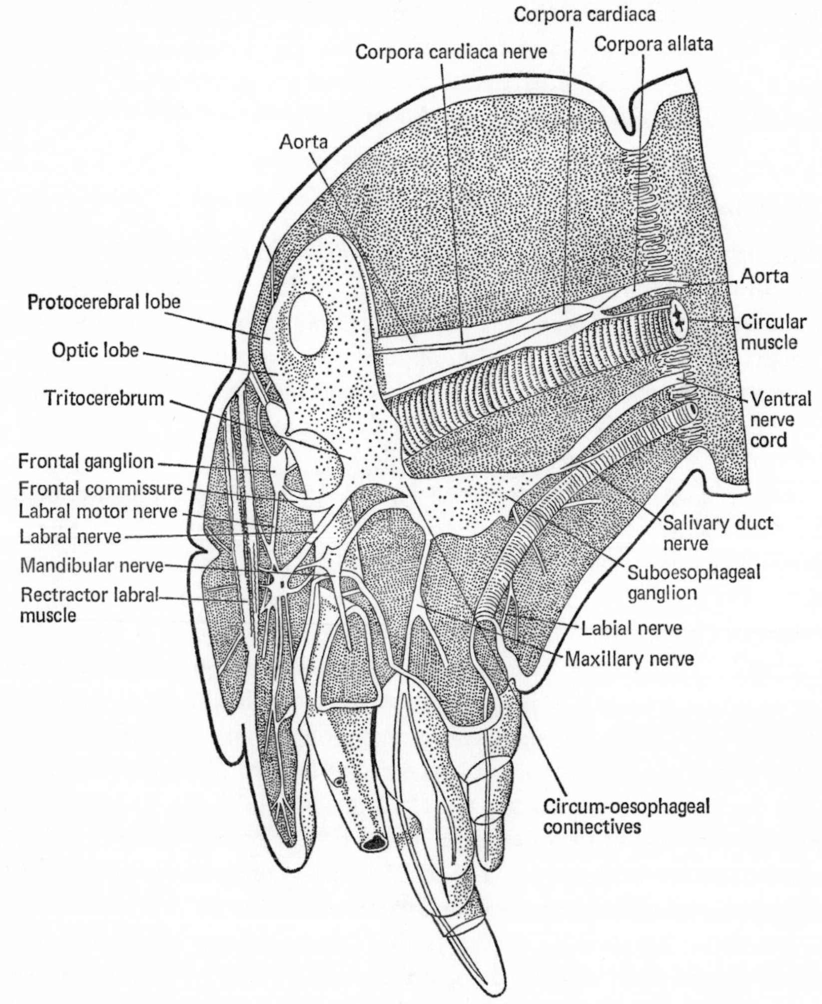

Fig. 11.13 A side view of the brain and its connections in the head of a larva of *Panorpa*

active in bursts at the same frequency as the peristaltic waves, if the cord is removed from the body, the rhythm is not then related to the frequency when connected to the body.

This throws light on the whole question of whether rhythmic locomotory movements (the stepping of insects and the swimming of worms are examples) are determined by the pattern of sensory input, or whether there is central nervous activity which perhaps, by interaction with the input, causes the rhythmic behaviour. It appears, for example, that the rhythm of wing beat in a locust is not determined by sensory input. But insects are capable of altering the sequence of movements of their six legs after one or more has been removed, so that they maintain their balance during their strides; thus the altered sensory input from the reduced number of legs must be having its influence.

The role of the brain

If the brain (the supra-oesophageal ganglion) of an earthworm is removed, the animal becomes restless and unusually active. Though it can burrow and crawl, its activities are less organized than they normally appear to be. This is but one example of what appears to be a general function of the supra-oesophageal ganglia as an inhibitor and moderator of the potentially independent activity of the ganglia in the nerve cord; thus, autonomy is controlled and integrated by this ganglion.

In the insect brain, the cerebral part is the result of the fusion of the nervous elements of three embryonic segments (Fig. 11.13). As in most invertebrate nervous systems it consists mainly of association neurones, and much of it is made up of nerve fibres, the cell bodies being few relative to the number of vertebrate brains, and confined to a rind over the brain. The most important association area is that of the protocerebral lobes which contain a central body towards which fibres converge from many parts of the brain, and a pair of mushroom bodies (corpora pedunculata) each in the form of a cell-covered cup formed of nerve fibres which travel down to make a stalk penetrating into the brain and ultimately joining with that of the other side. These bodies are important association centres and are best developed in social insects, perhaps in correlation with the great variety of their behavioural patterns.

Below the protocerebral part is the deutocerebrum consisting of the fused ganglia from which the antennae are innervated. The tritocerebrum lies below as two separated lobes, which give rise to the connectives to the sub-oesophageal ganglion. This latter part of the nervous system in the

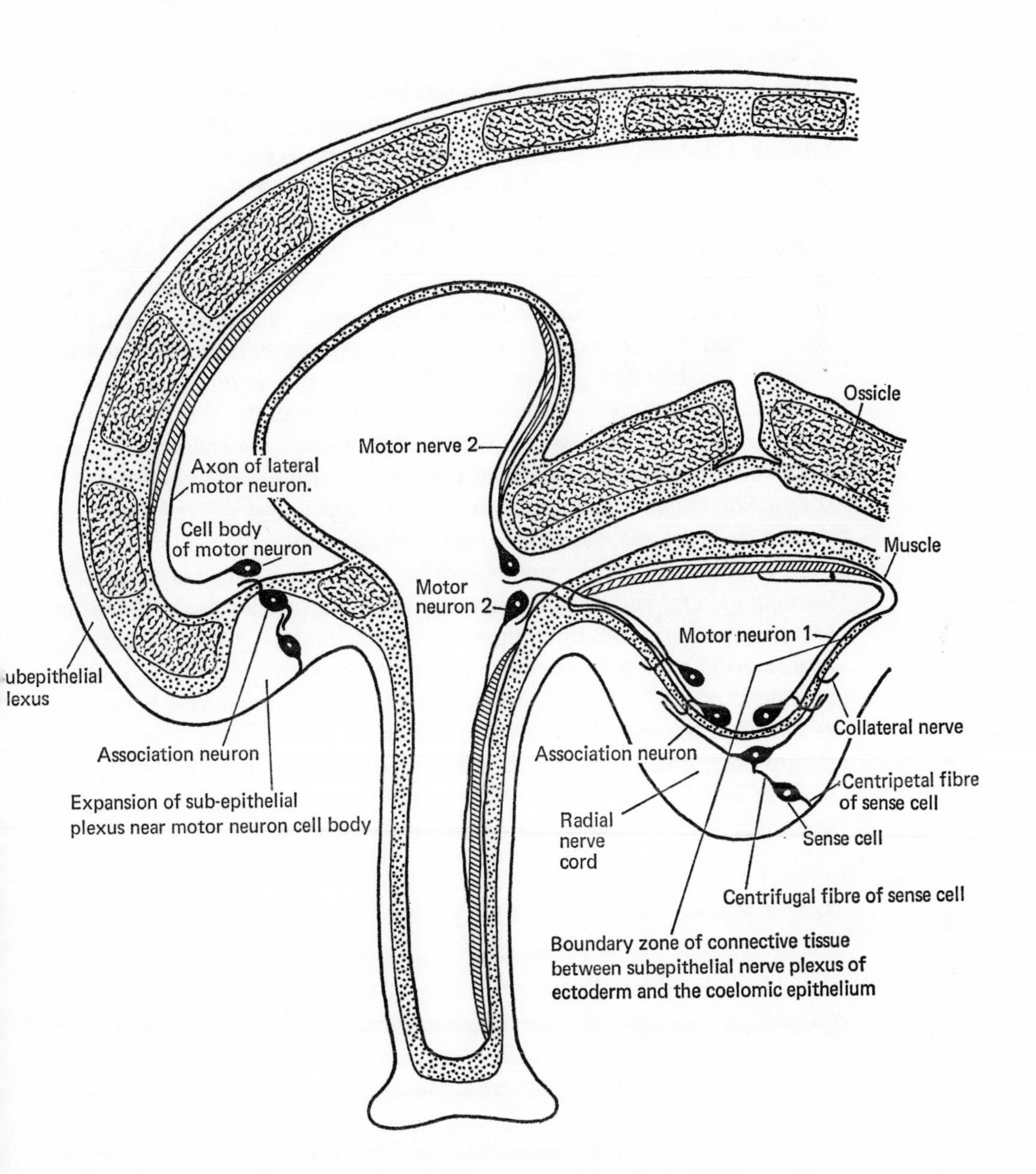

Fig. 11.14 Diagram of the nerves and their tracts in the arm of a starfish

head is the result of the fusion of the ganglia of mandibular, maxillary and labial segments, and thus innervates these mouth-parts.

The inhibitory role of the supra-oesophageal part is dramatically shown when the head of a male mantis is removed during mating. The headless body then makes incessant copulatory movements, though perfectly well organized and directed. It also walks spontaneously, continuously but aimlessly. But removal of the sub-oesophageal ganglia as well produces a quiescent creature, which can right itself but does not move. Thus the stimulation to walk seems to come from the sub-oesophageal ganglion, but is controlled by the supra-oesophageal part. Rather similar relationships seem to exist in the nervous system of the cricket, though here the central body seems to act together with the mushroom bodies. The second thoracic ganglion has all the mechanisms for the co-ordinated muscle movements involved in making sound by rubbing the two fore wings together. But the pulse pattern, which distinguishes the call made in one situation from that in another, is initiated in the mushroom bodies. Electrical stimulation of one of these areas in specific places evokes calls of one of the three main types according to the area of stimulation.

In annelids at least the supra-oesophageal ganglion is not necessary for learning (p. 273) and therefore does not form a memory store as it does in *Octopus* (p. 277). But throughout the invertebrates (possibly excepting cephalopods) the supra-oesophageal ganglion contains cells which, though nervous in origin, have taken on a neurosecretory role, producing hormones which affect behaviour or growth processes, or both (p. 293). The proportion of such cells in the supra-oesophageal ganglion of nereid worms is high. It may be that the main role as integrator of sensory input has, in these animals, passed to the sub-oesophageal ganglion.

Fast and slow innervation: inhibition

The first discovery of a system of innervation which could either bring about fast, short duration contraction or slow, steady contraction through different nerves was made in crustaceans; though such systems have since been demonstrated in other arthropods, coelenterates, annelids, molluscs, and vertebrates. However, the basic mechanisms may not be identical in all of them.

The muscle fibres of crustaceans are innervated by numerous nerve endings. The effect of activation of each ending spreads for only a short distance down the fibre, so that if many sections are activated the contraction will be large, if only a few it will be small. The endings derive from at least two nerves in every case, a fast or a slow fibre, and an inhibitory fibre. One muscle fibre may have more than forty endings upon it. Slow axons require a number of impulses to pass down them before depolarization, and a consequent muscle contraction, takes place at their endings; such facilitation is unnecessary when the fast axons are stimulated. Similar specializations into fast and slow muscle systems occur in bivalves where the shell can be snapped shut by the fast system but held shut over long periods by the slow system.

Inhibition is essential to the control of many movements; it adds new possibilities over movements which can only be brought about by the presence or absence of stimulation. In vertebrates, this is largely a central process brought about by action of inhibitory fibres on the motor nerve cells, but in crustaceans the site of inhibition is peripheral. These inhibitory endings set up a repolarization in the muscle fibre which they supply, this opposes the depolarization brought about by the motor fibres and prevents muscular contraction from taking place.

The radial plan

Though mostly no longer sessile, the echinoderms retain many characteristics of their sessile ancestors. Their bodies are usually built on a radial plan, and their nervous systems are no exception. Much of the system is diffuse and net-like, again reminiscent of what one finds in lower sessile animals like the coelenterates.

Instead of being buried in the body, much of the echinoderms nervous system lies intra-epithelially in the ectoderm, from which it arises. Therefore it is more superficial than similar systems in invertebrates other than coelenterates. A nerve plexus made up of multipolar and bipolar nerve cells covers most of the body. Generally it is thin, but along the underside of each arm of a starfish it thickens considerably to form a radial nerve cord (Fig. 11.14). Each one joins to a circum-oral nerve ring also derived from the ectodermal plexus.

This superficial system is a sensory one, the plexus of multipolar cells acting as an association net for the incoming sensory impulses from the periphery. The motor system, however, is much more variable in its position and pattern, for it is made up of groups of motor cells which collect in the vicinity of the muscles they innervate. Generally this system is associated with the coelomic epithelium (Fig. 11.14). But in addition, in starfish, there are paired cords of tissue, lying above the radial nerve cord in the wall of the radial perihaemal canal, in which the cells grouped in discrete series run across the cords. Immediately above the ectoneural system are the lateral motor nerves which run laterally and outwards to innervate the rest of the musculature. The two systems, motor and sensory, of course, make contact at various points, so that the sensory input can affect the motor behaviour (p. 198).

The molluscan nervous system

Within this phylum a range of nervous system structure can be found stretching from a diffuse type of system reminiscent of the flatworm

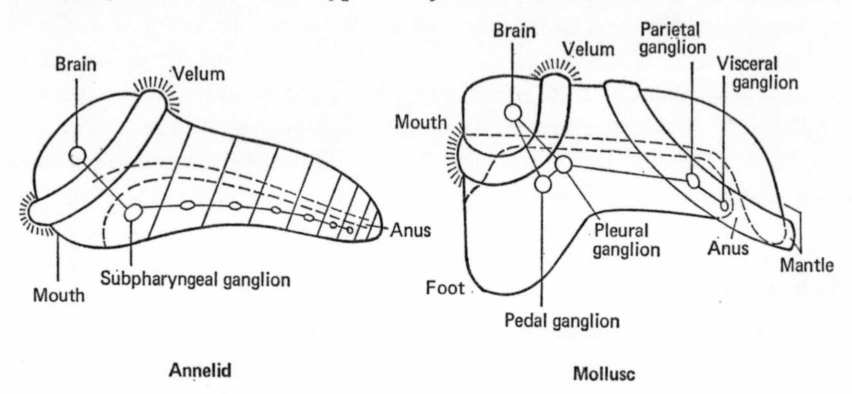

Fig. 11.15 A comparison between the nervous organization of an annelid and a mollusc as shown in their larval forms

through to one which in its complexity recalls the vertebrates. The molluscs do not show any trace of segmentation in their nervous systems (Fig. 11.15); even *Neopilina* with its apparently segmented anatomy nevertheless does not have a ganglionic chain such as there is in an annelid or arthropod. Indeed, the amphineuran type of system is marked by a uniform distribution of cell bodies through the longitudinal cords, so that there are no concentrations of cells forming ganglia except at the head end.

The typical plan of this basic molluscan nervous system (Fig. 11.16) is a ring of thickened nerve cord around the gut forming a cerebral commissure above and a sub-cerebral commissure below the alimentary canal. From the latter arise the connectives to a pair of buccal ganglia supplying the muscles of the pharynx, the radula sac and probably a large part of the rest of the alimentary canal. A sensory pocket in the floor of the buccal cavity is supplied by a pair of sub-radular ganglia connected to the underside of the sub-cerebral commissure. But from either side of the ring of tissue arises a lateral cord which, among other things, supplies the ctenidia. The two cords are joined posteriorly to form a ring, by the supra-rectal commissure. In addition a pair of ventral longitudinal cords arise from the cerebral commissural ring. Many nerves join ventral cords together and to the lateral cords.

From this wide-spread arrangement, processes of condensation can be traced through the Mollusca though among the bivalves this has not

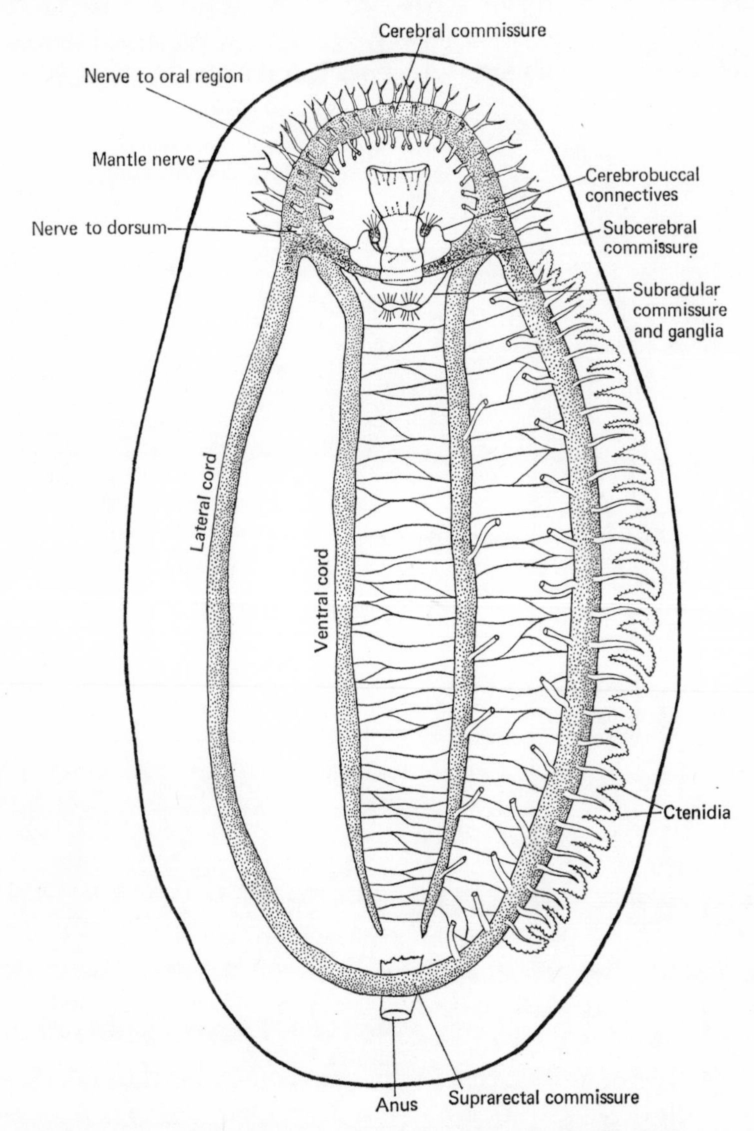

Fig. 11.16 The plan of the nervous system of a polyplacophoran mollusc

proceeded further than the loss of the numerous interconnecting fibres. Most bivalves lead a relatively inactive life; cephalization has not proceeded far. Without the concentration of sense organs which seems to call into being the concentration of nerve cells which is a brain, it is not perhaps remarkable that the relatively simple distribution of ganglia should remain. In the majority, there are three pairs of ganglia (Fig. 11.17), the cerebro-pleural (with which, almost always, the buccal ganglia are combined), to which are connected a pair of pedal ganglia by cerebro-pedal connectives, and the viscero-parietals innervating much of the mantle, the siphons, the

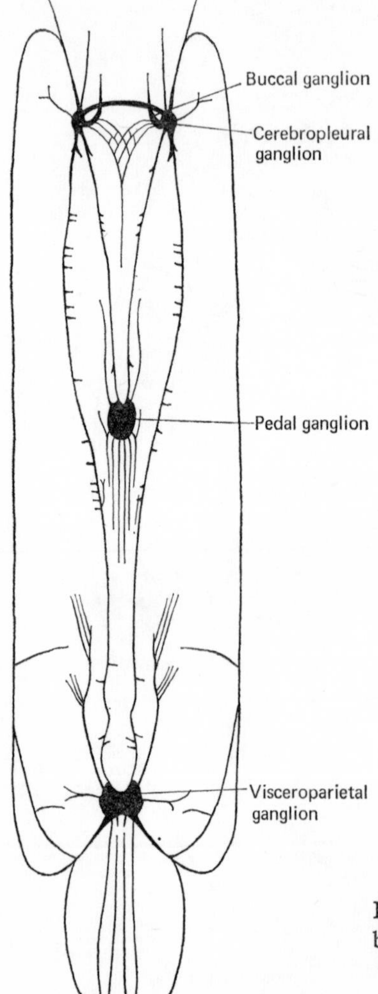

Fig. 11.17 The nervous system of a bivalve

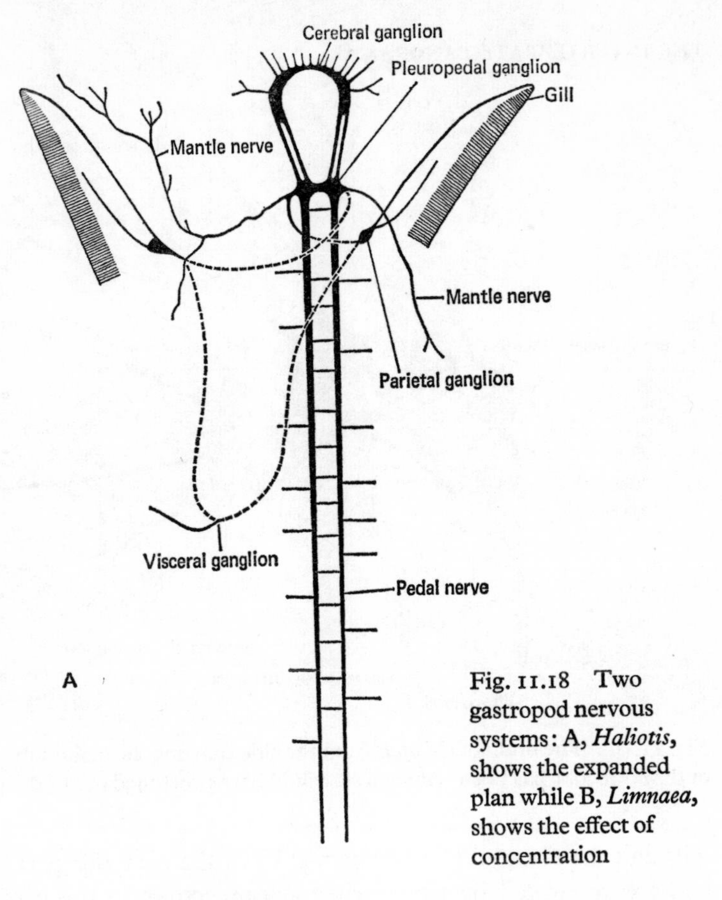

Fig. 11.18 Two gastropod nervous systems: A, *Haliotis*, shows the expanded plan while B, *Limnaea*, shows the effect of concentration

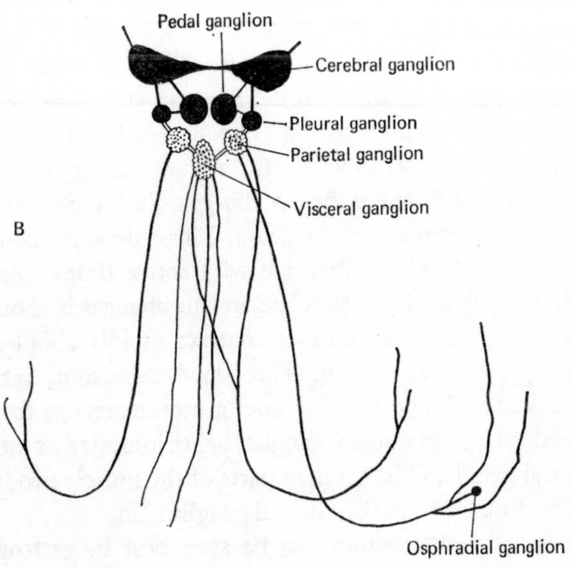

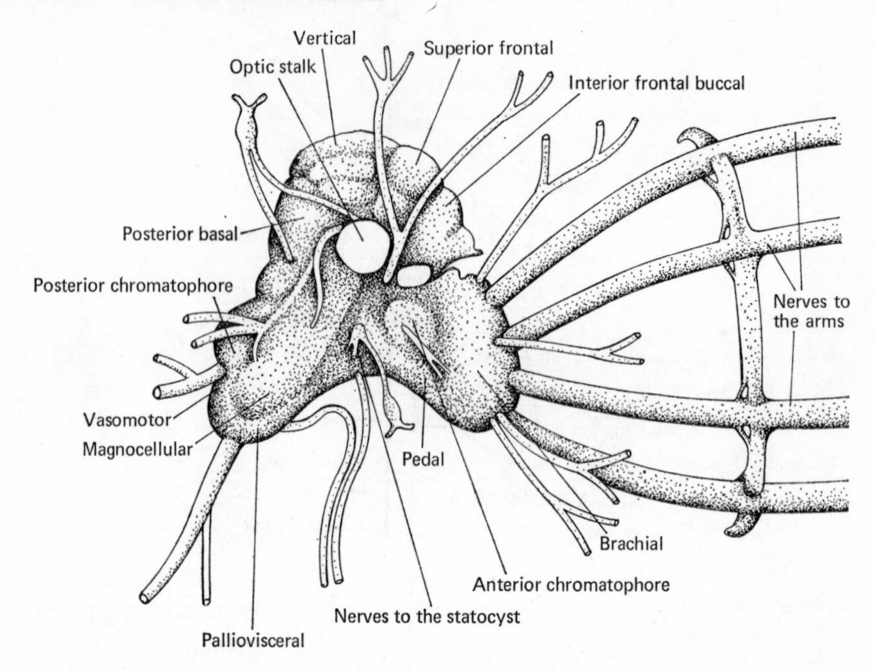

Fig. 11.19 The brain of *Octopus* from the side showing the major lobes; the whole of the optic lobe has been removed; it would have protruded out of the page

gills and the viscera. The last named lie in the posterior part of the body and are connected by pleuro-visceral connectives to the cerebro-pleural ganglia.

These ganglia show independent activity. Thus, after the removal of the cerebro-pleural ganglia of *Mytilus*, the foot can still move the animal and spin the byssus threads by which it is anchored. Removal of the visceral ganglia, on the other hand, abolishes the animal's ability to open and close the valves of its shells. Nevertheless the ganglia interact to produce the typical long-term behaviour of *Anodonta*. This animal remains for periods with its valves closed which alternate with active times when the valves open. The frequency of these 'slow' behaviour changes is about 3–30 times per week. But the adductor muscles contract rapidly about 20 times per hour, exhibiting a 'fast' rhythm. The 'slow' behaviour, taking place in unstriated muscle in the posterior adductor muscle, results from tonic contraction produced by the visceral ganglia, being inhibited at intervals by the cerebro-pleural ganglia. The striated parts of the muscle produce the 'fast' rhythm under the control of the visceral ganglia alone.

The process of condensation can be seen best in gastropods. In the

Diotocardia, for example, the circum-oesophageal ring is composed of dorsal cerebral ganglia and a ventral pair of pleuropedal ganglia (Fig. 11.18). From the latter arise the pedal nerves as long cords containing uniformly distributed nerve cells. Also from the pleural ganglia arises a visceral loop bearing the visceral ganglion in its posterior curve, and a parietal ganglion on each side near the pleural ganglia. This visceral loop may be twisted or untwisted in other orders, until in the pulmonates (Fig. 11.18) the loop has so shortened that all the ganglia are concentrated in a ring around the oesophagus.

The brain of cephalopods is a massive affair (Fig. 11.19) though in *Nautilus* the full fusion and increase in complexity has not taken place. Within the brain of *Octopus*, for example, thirty morphologically distinct lobes can be seen. The portion below the gut is responsible for the control

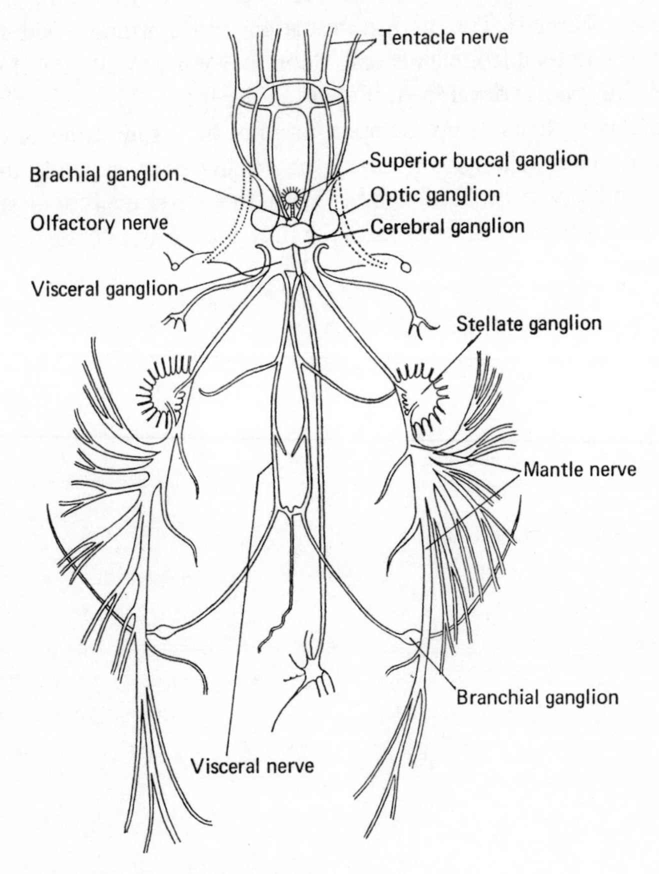

Fig. 11.20 The nervous system of *Sepia*

of movement, thus, the visceral ganglion gives rise to a pair of large nerves running back (and containing giant fibres in squids) to the stellate ganglia which supply the mantle musculature (p. 237), controlling the movements of pumping water in and out of the mantle cavity, which, though mainly used for breathing, can also be used in locomotion. The visceral loop supplies ganglia to the gills and other nerves to the viscera (Fig. 11.20). The pedal ganglion supplies the brachial ganglion which in turn sends a nerve into each arm. It is this ganglion which controls the actions of the suckers.

Into the dorsal side, in the cerebral part, the sensory input flows. The optic lobes are very large in *Octopus* as well as being the site of analysis of the information from the eyes they contribute to the maintenance of general muscular tone. With the superior frontal and vertical lobes they constitute the mechanisms whereby objects are recognized by sight and their appearance learned. Touch discrimination and learning resides in the inferior frontal-subfrontal-vertical lobe systems, well developed in octopods but poorly developed, if at all, in decapods. There are, therefore, two learning systems in the octopod brain. The organization of complex movements like walking, swimming or seizing prey depends upon the anterior and posterior basal lobes. Thus there has evolved a structure unique among aquatic invertebrates for the superiority of behaviour which it endows.

Chapter 12

Aspects of behaviour in invertebrates

MUCH OF an animal's behaviour consists of its reactions to aspects of its environment. These are of two kinds, the physical and chemical characteristics of the surroundings such as light, temperature and water currents, and those stimuli which arise from other living things. Historically the study of the behaviour of the great majority of invertebrates has centred upon the individual's responses to the major stimuli present in the environment. Indeed, the distribution of many of these animals is a record of their behavioural responses, which are part of the whole spectrum of adaptiveness by which the animal's survival is ensured. Such behaviour with that of feeding, burrowing and so forth, is summed as maintenance behaviour, while reproductive behaviour comprises courtship, mating and brood care. The study of those aspects which can, then, be called social behaviour, in that one individual reacts to another of the same species, has only recently begun to attain the level of sophistication reached by research into vertebrate behaviour in the immediate pre-war period.

The role of behaviour in the distribution of animals can be exemplified by that of various species of woodlice in atmospheres of different relative humidity. If a group of *Porcellio scaber* are observed at different humidities a greater proportion of its members will be active at low humidities than at high (Fig. 12.1). The activity is random as might be expected when the stimulus has no source towards which the animal can orientate itself. This behaviour, consisting of changing rates of movement with different levels of stimulation, is known as an orthokinesis. The behavioural reactions may also include a higher rate of change of direction (klinokinesis) at the higher humidities. They are most obvious in *Oniscus asellus* and least in *Armadillidium*. Thus the behaviours reflect the extent of morphological and often physiological adaptation to land life found in these animals, *Armadillidium* being the least likely to lose water in a drying atmosphere.

Responses to light have been studied most and can often be correlated with behaviour in natural surroundings. The movements evoked are frequently directed with respect to the light source even in cases where a morphologically distinguishable eye cannot be found. The sea-urchin, *Diadema setosum*, shows reactions to daylight which are probably mediated in this way (p. 202). The return of the spider *Agalena* to the corner of its web which it uses as a lair is directed with relation to the light source. If a web is rotated through 180° (Fig. 12.2) so that the spider now rests in the diagonally opposite position to its original one, the animal after feeding will endeavour to return to the position in space to which it formerly went after taking prey from the centre of the web.

Although experiments are designed to investigate the effects of one source of stimulation alone, under natural conditions an animal is responding to a number of different stimuli. The interplay of simple responses will serve to explain the distribution of the mollusc, *Lepidochiton cinerea*, on

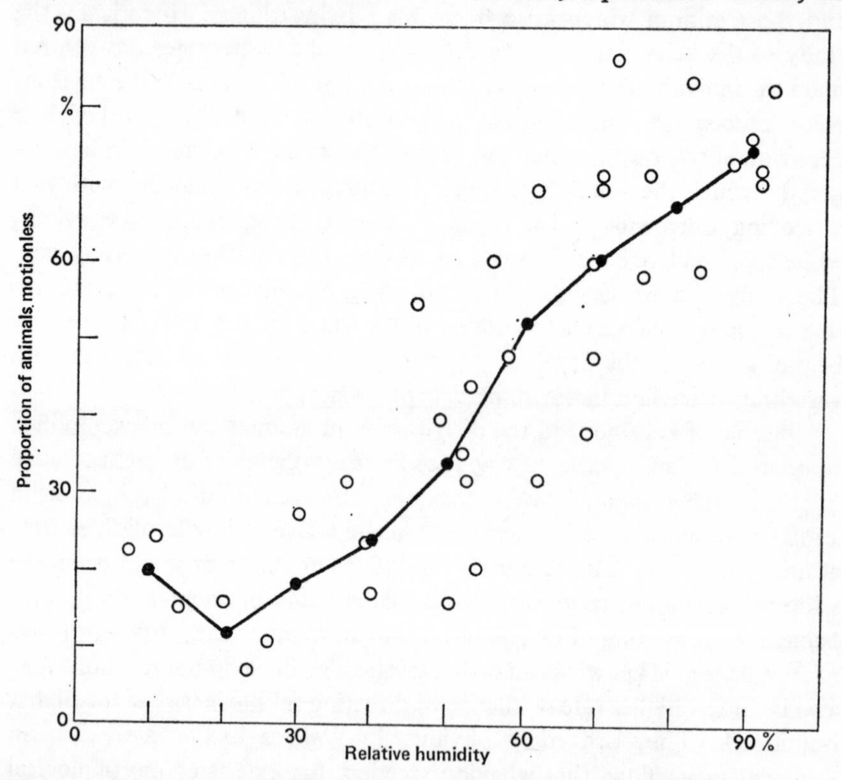

Fig. 12.1 The proportion of woodlice immobile at various humidities. Each point records the proportion which are quite motionless during a set of observations each lasting 30 sec. carried out at 15-minute intervals at a given humidity. The line joins the average points

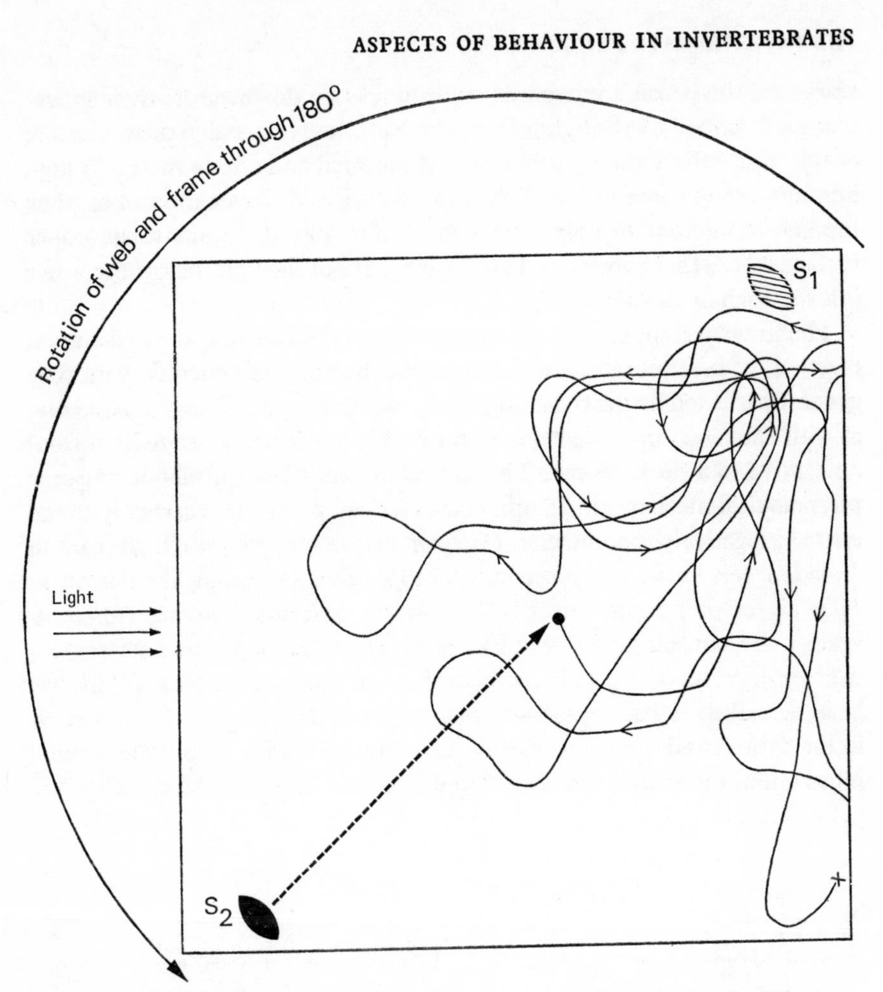

Fig. 12.2 The effect of rotating a web of *Agalena* through 180° with the light from one side. S_1, original position of place where spider rested; S_2, position after rotation. A fly was put down at ●, picked up by the spider which moved towards S_1 then wandered for 22 minutes until it sucked the fly's juices at X

stones on the shore. When the tide is out, the animals are found on the underside of the stones. There they are protected from the effects of sun and wind which dry the upper surfaces of the stones. When the tide is up, they emerge to browse on the algae coating the stones. Put on a vertical glass plate in the laboratory, the chitons move in all directions so long as the plate is immersed, but when the moistened sheet stands in air they move downwards reacting positively to gravity (Fig. 12.3). In addition they show orthokinesis since they move faster in strong light than in dim; further, they move faster when exposed than if immersed in water. Thus, when the stones on the beach are exposed, the animals go downwards, their movement accelerated by the light of the sun. But when they reach the underside of the stone where the light intensity is minimal and where there is a high humidity, they come to rest. When the tide rises to cover the stones, they become indifferent to gravity and move out over the stone to its upper surface where they move slowly – the intensity of the light below the water is lower than in air – and feed.

The gravity responses of the chitons stress the fact that the orientation reactions of invertebrates are not fixed, the direction in which they move is generally affected by the conditions under which they are living. Frequently, as with *Diadema*, an animal kept in the dark alters its response to light from a negative to a positive one. This is undoubtedly brought about by some physiological mechanism. In other cases we can point more certainly to the cause. Negatively phototactic *Daphnia magna* exposed to high carbon dioxide levels in the water become photopositive. Or, again, the flatworm, *Planaria alpina*, reacts positively to water currents, moving upstream against the current when about to lay eggs. After depositing its eggs, a worm will cease to be positively rheotactic and move downstream. This has been described as having the advantages that firstly the eggs are deposited in the better conditions of cooler water upstream, and secondly the parents move out of the area so that food supplies do not become limiting when the young are hatched.

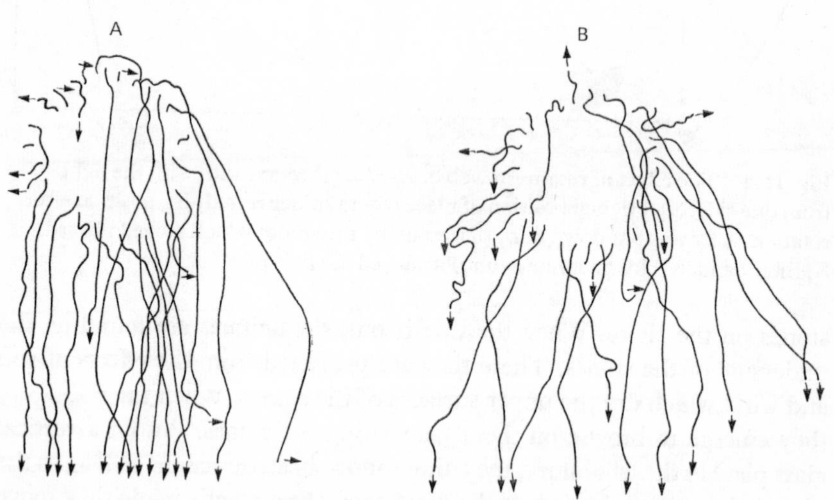

Fig. 12.3 The movement of chitons placed on a surface held vertically in the air. A, on plain glass; B, on roughened perspex. The results show that the downwards movement is a truly directed one and not merely due to the animals slipping down the moistened surface

Migration

A number of invertebrates perform movements which, if they are considered in terms of the equivalent number of body lengths the animal traverses, are long-distance ones. One of the most striking is found in the daily movements of plankton in the upper layers of the sea. Though many different organisms are involved, most of the animal species involved are adult or larval crustaceans. There are two main ways in which plankton moves, one, a daily vertical movement and the second, a horizontal movement over longer distances taking very much greater lengths of time.

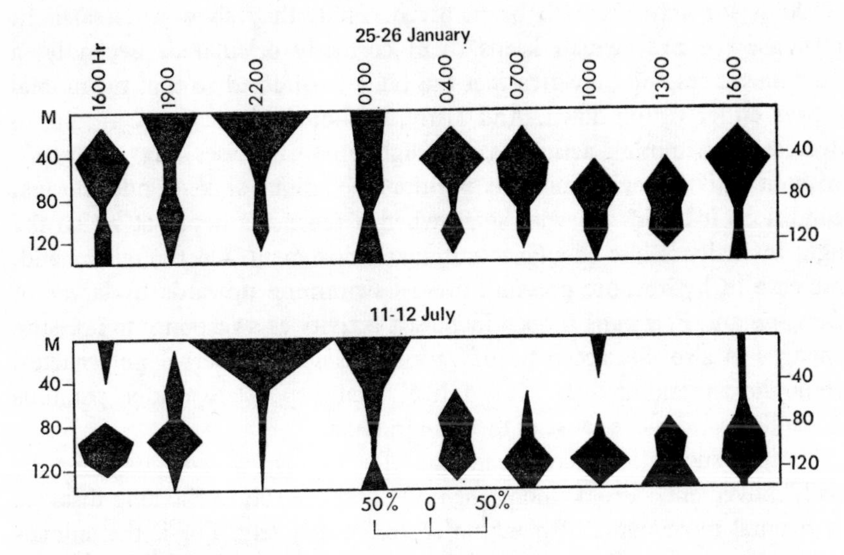

Fig. 12.4 Diagram of the vertical distribution of female *Calanus finmarchicus* at 3-hr intervals on 25–26 January (sunset 1627 hrs; sunrise 0837 hrs) and on 11–12 July (sunset 2007 hrs; sunrise 0358 hrs)

Calanus helgolandicus (Plate 6) is a copepod common in the plankton and important as the food of herring. It provides a good example of the pattern of movement (Fig. 12.4). Essentially the animals become randomly dispersed at midnight, sinking gradually thereafter to their lowest position at midday. Movements of 50–150 metres are common, but even great movements, of 300–400 metres, are made by some bathypelagic crustaceans.

The movements of *Calanus* are due to active directed swimming. The observation that they are synchronized with the diurnal changes in light intensity immediately suggests that light is the major directing factor.

However, the force of gravity and changes in pressure with depth also have their effect though they are more difficult to demonstrate. Groups of *Calanus* in a glass tube taken beneath the surface divide into individuals swimming upwards and others swimming downwards. As the tubes are carried deeper, more animals swim upwards. When the direction of illumination is reversed, the direction of gravity remaining unchanged, the animals react mainly to the light.

Some evidence on the mechanisms producing the vertical migration can be obtained from experiments on *Daphnia*. Though not a marine animal, nevertheless in ponds these water fleas carry out similar movements. Three reactions to light seem to be involved. First, they show a dorsal light response (p. 214) which keeps them correctly orientated. Secondly, a phototactic response ensures that the body is directed so that the animal moves either up or down. And lastly, a phototactic response results in decreased swimming activity as the light intensity rises (e.g. at dawn); without actively maintaining its position the animal sinks. Under normal conditions it begins to swim again when it reaches a depth at which the light intensity is low enough to stimulate movement. On the other hand, increase in hydrostatic pressure evokes swimming upwards by larvae of *Carcinus* and *Portunus*. Indeed increased activity as a response to pressure changes is also characteristic of various pelagic and larval polychaetes, ctenophorans and medusae. It is unlikely that response to a single stimulus is implicated as a sole cause of these migrations.

Various suggestions have been made for an adaptive advantage for the daily movements. From among them the possible role in the long-distance horizontal movements of plankton seems most likely. For if the animals move from one water current near the surface into another level and another current, probably moving faster and possibly in a different direction, the planktonic animals will be dispersed passively over considerable distances.

Some of the larger planktonic animals seem to move horizontally in fairly straight lines; *Palaemon northropi* has been followed for distances of thirty metres or so while maintaining a straight course. Another crustacean, *Mysidium gracile*, has been shown to orientate itself by the plane of polarization of polarized light falling on the water surface. The animal is not analysing the plane directly (p. 208) but reacting to the light pattern produced in the water by the scattering of the light when it strikes suspended particles. Higher light intensities are produced at right angles to the plane of polarization in this way and it is to these that the mysids orient themselves.

But an ability to analyse the pattern of polarized light in the sky is basic to the guidance of honey-bees on their flights from the hive to the food and back, as well as to the movements of many other arthropods. The sand-hopper, *Talitrus saltator*, is an example. This amphipod crustacean lives in the moist sand of some sea-shores. If it is put on dry sand high up the beach it will move directly towards the sea. In some of the original experiments animals were moved from the west coast of Italy to the Adriatic shore. On release they now continued to move westwards despite the fact that the sea now lay to the east.

This fixed direction of movement is related to the sun, for animals shaded from the direct view of the sun, but illuminated by sunlight reflected from a mirror, will alter their headings so that they maintain the same angular direction with respect to the new, false position of the sun as they did when they were able to see it in its correct place. However, they can also keep their direction when the sun is obscured. A sight of the sun itself is not necessary. Like honey-bees their orientation is disturbed by placing a piece of polaroid between them and the sky. This can have the effect of altering the sky pattern of polarized light into one which is not to be seen naturally, thus causing the animals to be confused. But if the position of the polaroid is such that it only moves the patterns into unexpected parts of the sky, they alter their direction to restore their relationship to the patterns in their new positions.

Since the direction taken remains unchanged throughout the day, allowance must be made for the changes in the sun's position between dawn and dusk. In common with other animals which utilize the sky polarized light pattern for orientation, these talitrids have an internal clock mechanism which enables them to do this. Like many rhythms of activity (p. 270) the 'clock' is set by the day/night cycle of the environment. In this way, the clocks can be reset by keeping the animals in an artificial light/dark regime different from that of the normal diurnal cycle. Thus, ones kept under reversed light conditions (i.e. 12 hours dark/12 hours light opposite to the natural day) on release went in a direction which was almost 180° from that of control animals kept under normal lighting conditions (Fig. 12.5). The clock, like those of other animals such as honey-bees and spiders, is very resistant to other environmental changes. Thus, sand-hoppers from Italy taken by air to South America orientated at an angle to the sun there which would have been correct had they remained in Europe. There is some evidence which makes it appear that *Talitrus saltator* has also a clock timed to fit the lunar day for it can orientate by the moon.

Talitrids under moist conditions show another set of responses for they tend to move up the shore towards low points in the silhouette of the upper parts of the beach. In the laboratory animals will move towards the base of a V cut in a sheet or to a hole (Fig. 12.6). This is similar behaviour to that of moving bands of crickets (*Dociostaurus maroccanus*) which will make for openings and holes in any obstacle in their path.

Some of the spiders (*Arctosa* species) skating on the water surface from which they take their prey also use polarized light to orientate their return to the bank of the river they usually inhabit (Plate 7). Species living within the Arctic Circle in Northern Finland show correct orientation

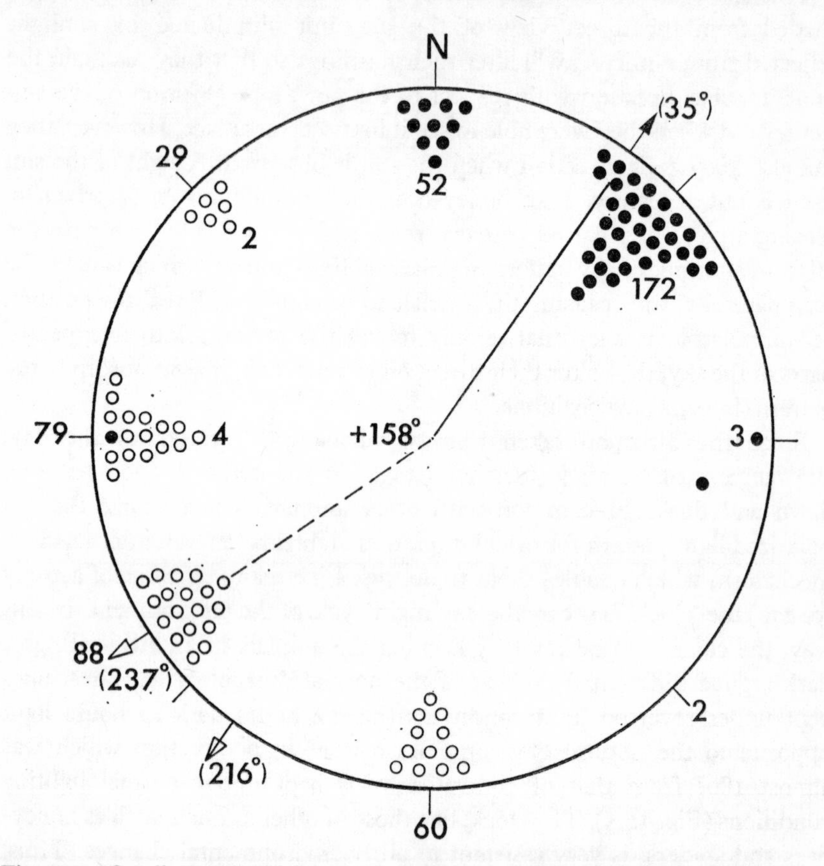

Fig. 12.5 Sandhoppers which had been kept in artificial day/night conditions twelve hours out of phase with the actual conditions (•) moved in an average direction almost 180° away from that taken by control animals (o). The actual direction of the shortest way to the sea was 216°. The figures within the circle show the numbers of experimental animals and, those outside, the controls, each solid or open circle is equivalent to five observations

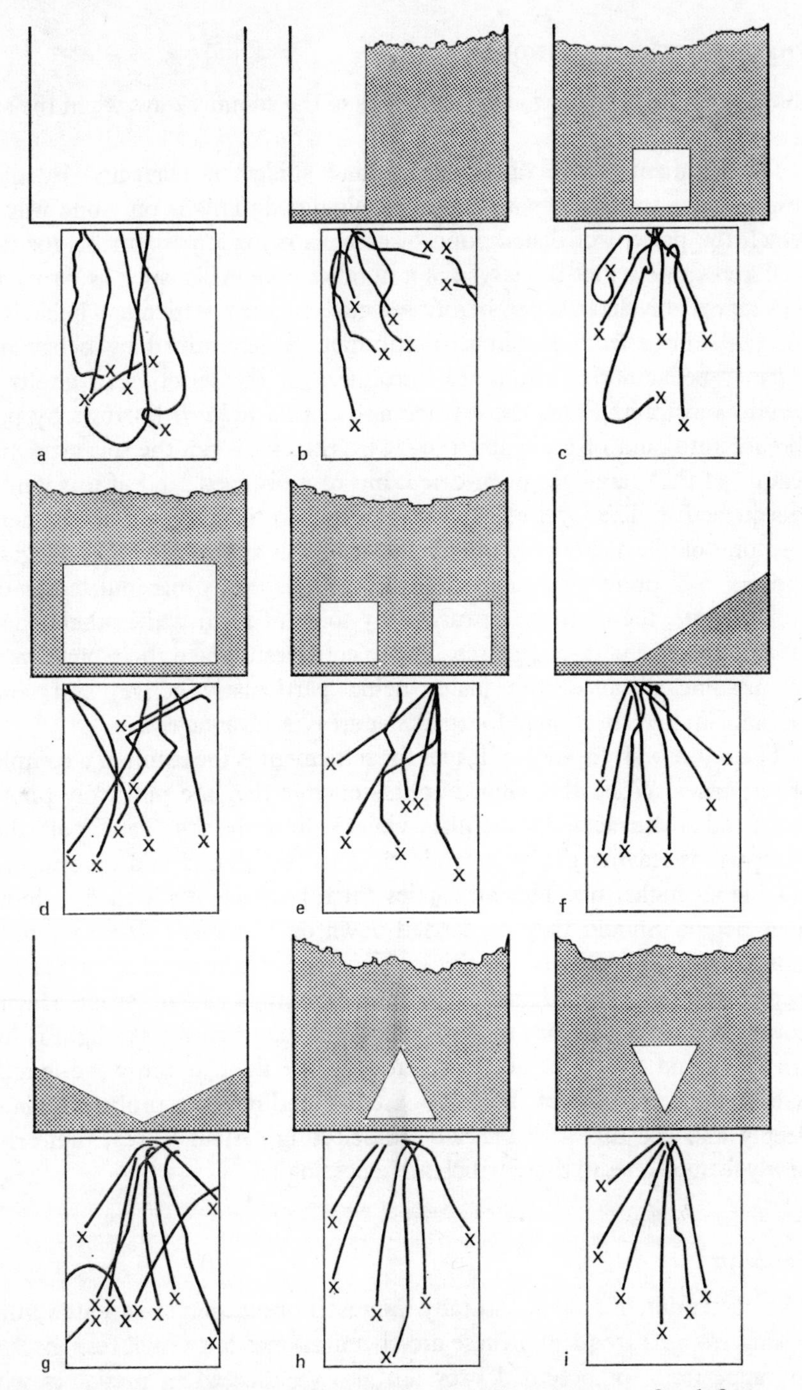

Fig. 12.6 Visual orientation in *Talitrus saltator*. The upper part of each figure shows the shape projected on a vertical screen; the lower the tracks of hoppers released at X on the horizontal board in front of the screen

throughout the whole twenty-four hours of the summer days when the sun is continuously visible.

Displacement of the sand-hoppers and spiders is corrected by their responses so that their positions are maintained. This is only one way in which the pattern of distribution of animals is made permanent, for it is well known now that the larvae of a number of animals, such as barnacles and many polychaetes, are highly selective in their settlement behaviour (p. 322). Thus from the first, distribution is determined by behaviour. Further behavioural factors are introduced in the selection of adults of certain substrata. Thus, the crustacean *Cumella vulgaris* burrows by preference into sand of grain size 149–295μ. But size is not the sole attractive feature of the grains, for oven-dried, but re-moistened sand of this kind is less attractive. The layer of micro-organisms on the grain surface is therefore one of the factors by which the choice is made. The size of grain appears to be preferred because it is the one most easily manipulated by the mouth-parts, the grain being turned by some of them while others shave off the film of micro-organisms. These cumaceans leave their burrows to swim some distance, the males being particularly active, and some mechanism to restrict them to particular areas is advantageous.

The selection by aphids of leaves for settlement is the result of a complex of responses. When the winged aphids emerge they are positively phototactic and are attracted by the ultra-violet light of the sky. Taking off, they fly upwards leaving the quieter air close to the ground and entering the moving air higher up. This air carries them passively while they maintain their height. Should they be carried down on to a plant they may settle momentarily and even probe the leaf with their mouth-parts. But at first locomotion takes precedence over feeding and they take off again. Having flown for a longer period, they pass into a phase when the sky light is less attractive and they are more attracted by the light of the wave-lengths reflected by green leaves. In this phase, on landing, they probe the plants deeply and use this as a method for selecting certain leaves, preferring newly formed ones to those which are senescing.

Feeding

The perception of food is naturally the result of reaction to various stimuli arising from it; frequently these are chemical ones but visual responses to the appearance of potential prey are also widespread in predators with well-developed eyes.

When the penetrant nematocysts borne on the tentacles of many coelenterates puncture the surface of the prey animal, they release tissue fluids which contain glutathione. This substance causes the muscular responses by which the paralysed prey is swallowed; sensitivity to it is so great that the threshold may be as low as 10^{-4} to 10^{-8} M. The mouth of the coelenterate opens and applies itself to the food, stretching wide as it does so. Similar responses are made to pads soaked in glutathione solution. In crustaceans, such as the spiny lobster *Panulirus*, the important chemicals evoking feeding are glutamic acid, trimethyl amine, its oxide and betaine, all substances likely to be found in the flesh upon which this animal feeds. Electrophysiological studies show that the receptors must be on both the antennules and the dactylopodites of this lobster.

The habit of earthworms of drawing leaves into their burrows was made use of to determine what chemical substances were most attractive to them. Pine needles were coated with pure gelatine and with gelatine to which various substances were added. Quinine and salt-coated leaves were rejected. But, under natural conditions, it is strikingly obvious that leaves are selected. Those of particular age and dryness are chosen with a special choice of rotten ones. This may well be determined by the chemicals released in the breakdown of the leaf tissues.

Touch is not unexpectedly another important stimulus arising from the food, one which has the advantage of localizing the position of the food. In the jelly-fish, *Tiaropsis indicans*, a touch on the bell edge causes the manubrium to contract at first but then to bend towards the stimulated area with great precision. If other places are then touched the manubrium will move to each in turn. This jelly-fish feeds by catching food organisms on the marginal tentacles and picking them off with the manubrium.

Aiming, in order to grasp prey effectively, is a common feature in insects. A dragonfly larva catches its food by rapidly extending its specialized labium, the mask, bearing two jaws at its tip (Plate 10). The prey caught by these jaws is then drawn to the mandibles of the insects which chew it. The distance to which the mask extends, forced out by blood pumped quickly into it, is relatively fixed. The anisopteran larvae find the range visually, the mask being shot out when the image of the prey falls upon certain ommatidia on each eye (Fig. 12.7). The optical axes of these visual elements intersect at the point which the extended mask will reach. On the other hand, zygopteran larvae detect the position of their prey by means of their short antennae, judging the range by touch.

Jumping spiders locate their prey visually and the effective visual stimuli have been studied in *Salticus scenicus*. These spiders were kept in a cage in which black figures on white backgrounds could be displayed to them; it was usually essential that the test figures were moved (Fig. 12.8). Plasticine balls evoked more attacks than flat shapes, so that a three-dimensional solidarity was important. But objects less than 0·6–0·8 cm across, or very large ones, were not attacked. A variety of shapes were attacked but on the whole ones with long outlines were preferred to shorter so that the whole silhouette of a fly was attacked more frequently than a shape without wings, head and legs.

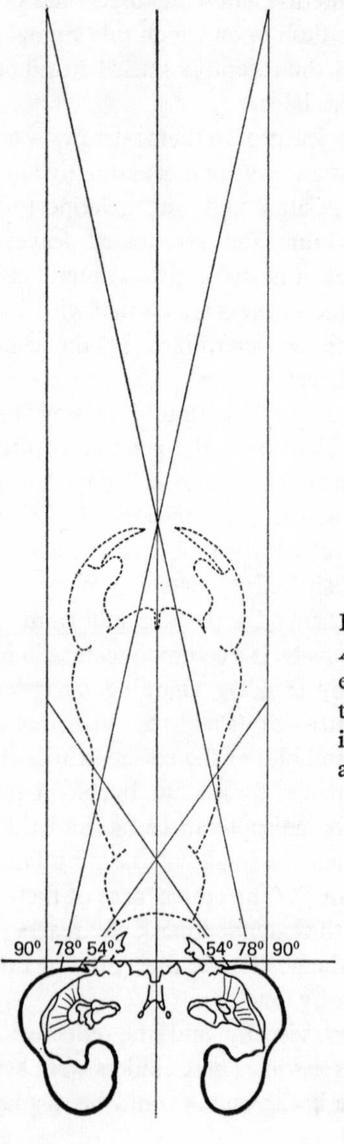

Fig. 12.7 Visual field of the larva of *Aeschna* showing the position of the extended labial mask. The lines indicate the visual axes of various ommatidia, the intersections determining how far away an object is

It is usual for animals to attack a great variety of objects when they first hatch; only later do they begin to reserve their attacks for a restricted class of objects. The reverse seems to be true of the young of *Sepia*. Immediately after leaving the egg, with yolk still in the yolk sac, they attack only those objects which bear a similar appearance to mysid crustaceans. Later their attacks become generalized not apparently through a learning process but rather as a function of the maturation of the nervous system. Learning follows when they specialize upon certain prey.

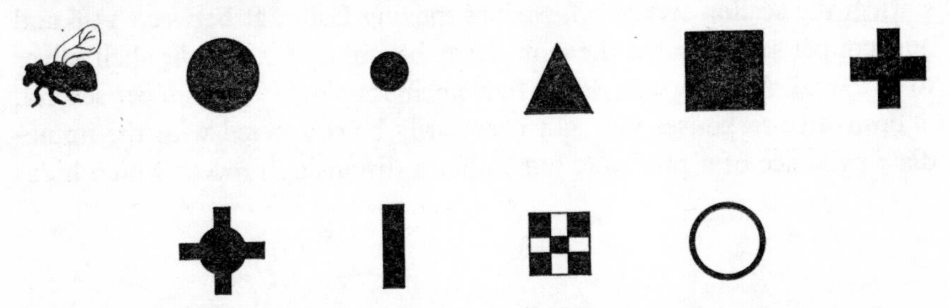

Fig. 12.8 Shapes releasing attack from spiders, *Salticus scenicus*, when moved in front of them

Prey-predator responses

The reverse of the coin of feeding is the behaviour by which animals avoid becoming prey to others. Starfish tissue fluid is sufficient to send queen scallops (*Chlamys opercularis*) swimming rapidly away. The touch of a starfish on the hind-end of the foot of *Nassarius reticulatus* causes this gastropod to rotate its shell, then throw itself off the substratum. This violent response can also be evoked by a tube-foot torn from the starfish and applied to the snail's foot (Fig. 12.9).

Shell twisting is also part of the escape responses of *Physa fontinalis*, for this fresh-water snail may be attacked by the leech, *Glossiphonia complanata*, if it cannot obtain its usual diet of *Chironomus* larvae or *Tubifex*. The *Physa* flexes its shell through 180° back and forth. *Bithynia tentaculata* will even nip a leech's head between its operculum and the shell rim.

But a response to moving shadows is commonly found among sedentary invertebrates. In fact it was the withdrawal response of the siphon of the clam, *Mya*, which Hecht used in his classic work which laid the foundation for our understanding of the way in which visual pigments function. The decrease in light intensity necessary to produce such a response in the tubicolous polychaete, *Branchiomma vesiculosum*, is remarkably small for in weak light it will withdraw its crown of tentacles at a reduction of 0·3 metre candles. These responses are no doubt protective and they are well fitted for this function as they show rapid adaptation, an important feature if reactions are not to be made to every decrease in intensity made by water movements and so forth unaccompanied by the dangers of a predator.

The rate of movement of the shadow is important in determining the reaction of the scallop *Pecten jacobeus* (p. 207). This can be tested by moving black and white stripes in front of the animal. Slow movement causes it to put out the mantle tentacles; if these detect chemicals typical of starfish the scallop swims off. Stripes moving faster, at between 11·6 and 29·4 cm per second cause the tentacles to be withdrawn and the shell valves to be closed. A scallop can detect movements as slow as 1·7 mm per second.

Protective responses may not necessarily be connected with the immediate presence of a predator, but rather a disguising process which hides

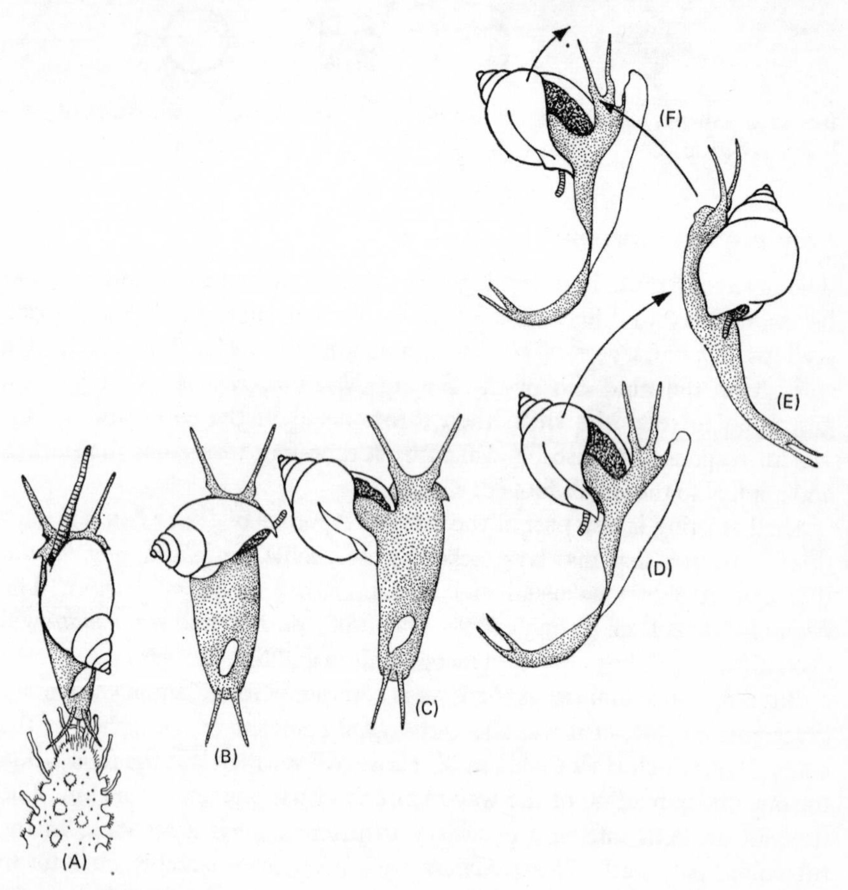

Fig. 12.9 The escape of *Nassarius*

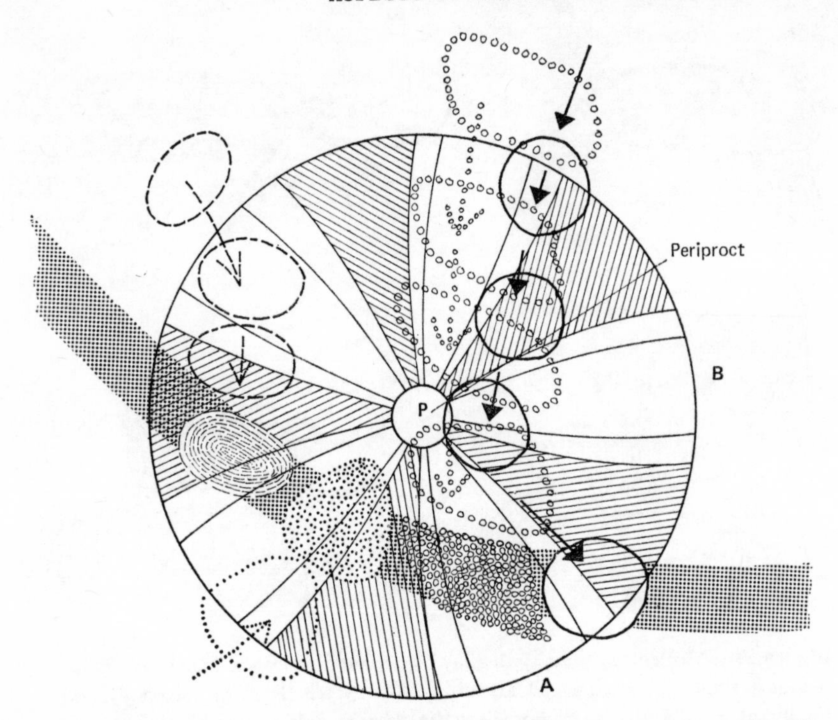

Fig. 12.10 The placing of cover over localized brightly lit areas of the surface of *Lytechinus*. Four stones are moved into a narrow band of sunlight (stippled) by the routes shown

the animal from danger. The urchin *Lytechinus* will pick up and move stones over its surface until they come to lie over areas which are illuminated strongly (Fig. 12.10). Even a clear glass coverslip can be selected and lifted into place, being held there although the tube-feet grasping it are not shaded when it is in place.

Courtship and reproduction

Apart from the maintenance activities described so far, the reproductive behaviour of invertebrates show a wide variety of patterns and forms. The main features are those common to courtship wherever it is found, that is, a system of signalling by which the sexes are brought to receptiveness and by which species can recognize each other. Courtship patterns are generally highly specific ensuring that interbreeding between species does not take place.

The essential signals may be given in any sensory mode to which the

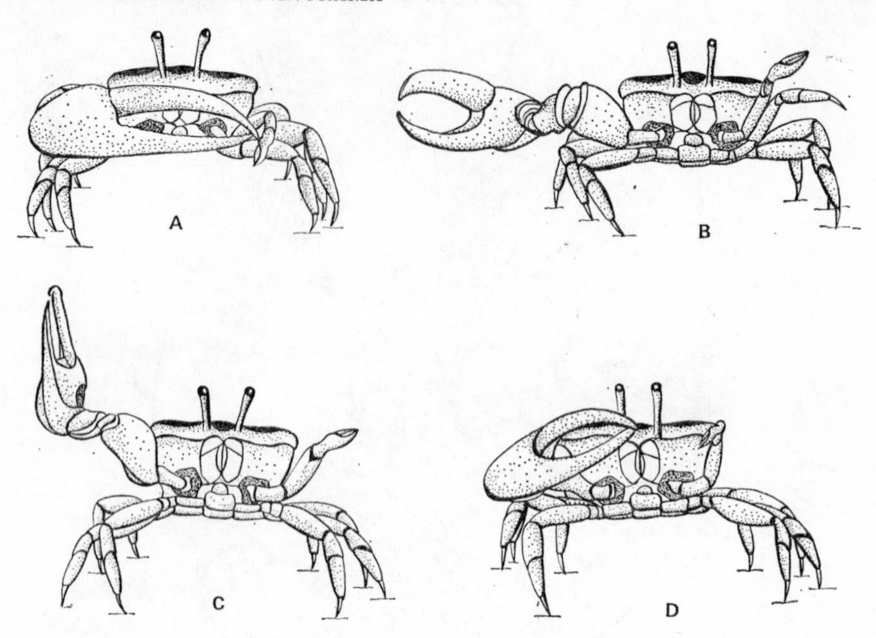

Fig. 12.11 Moderate intensity display by a male *Uca lactea*. The large chela begins in the flexed position (A), is unflexed outwards (B), then raised (C) and finally returned to the start. Note how the carapace is also raised

receiver is sensitive. Each mode has particular advantages and disadvantages; visual signalling, for example, may be effective over greater distances and clearly establishes the position of the signaller, but in so doing exposes the animal to the possibility of predation. Sound signalling can be carried out in environments where many obstacles lie between the mates, such as in the grass layer of a field, but yet the location of the two mates may be more difficult than by vision. Whatever may be the means of signalling used there is ample evidence that the range of signals which are distinguishably different from each other has by no means been exhausted. This principle has the few exceptions one expects from a generalization. Certain moths, for example, share essentially the same chemical signalling substances which can be shown in experiments to be attractive between species but other aspects of their behaviour such as the times at which they fly serve to keep the species separated. The signal common to all species is, however, only responded to by the one flying at that time.

The courtship display of fiddler crabs (*Uca* species) is a good example of visual display. The males make a burrow and take up position beside its entrance (Plate 8). There they signal by movements of the large chela moving it in a manner characteristic of the species. As the chela is strikingly coloured the display is obvious (Fig. 12.11). The male of *Uca pugnax*, for example, raises its body until it is clear of the ground. At the beginning of the display movements the chela is flexed in front of its mouth. Then it raises the major chela unflexing it so that it moves obliquely upwards. At the top of its stroke the chela may open and the crab may kick some of its

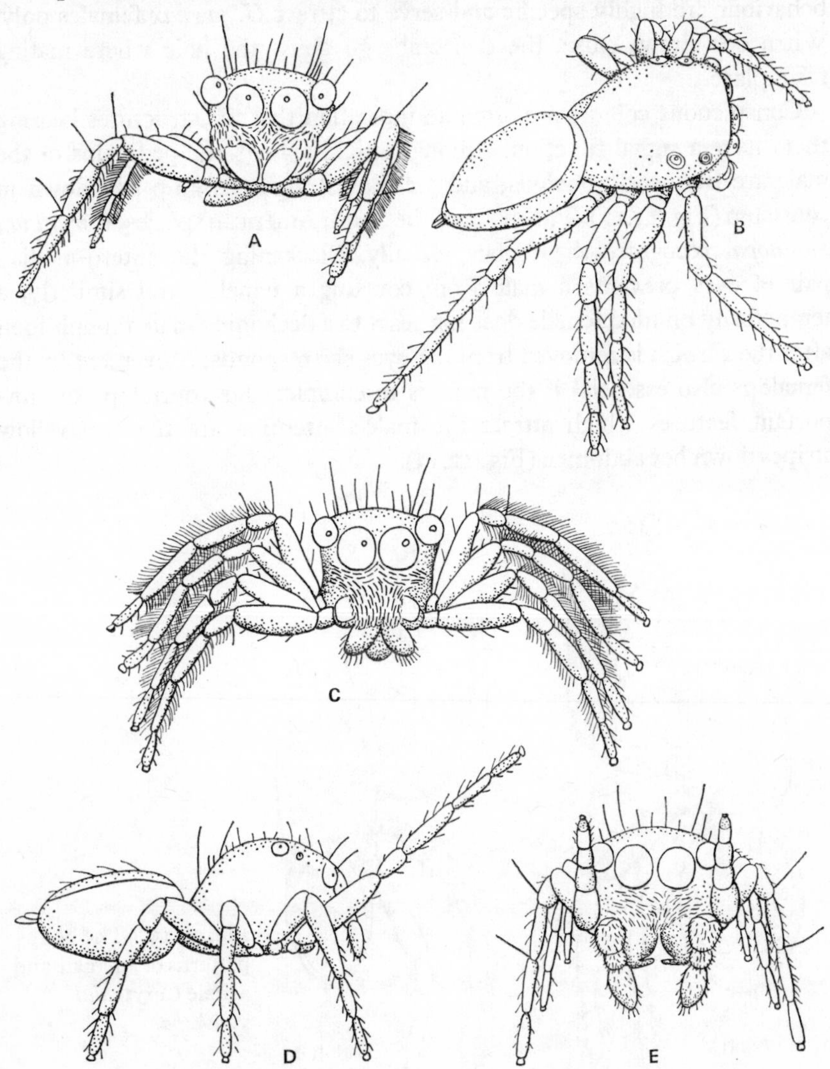

Fig. 12.12 Display by a male *Corythalia xanthops*. A, frontal view of rocking preface to threat and courtship; B, dorsal view of the same; C, threat display; D, beginning of courtship from the side, and E, from the front

other legs outwards. The major chela is then brought down again in a series of jerks 'as if let down in worn notches so that it slides down with the least hint of "braking" ' as Crane has described it. After a series of such waves the male makes a number of 'curtsies' which are quick light movements, lowering his body with chelae stationary in front of his face as he drums on the ground with his other legs. The details of the elements of this behaviour are highly specific and serve to attract *U. pugnax* females only. When one draws close, the two crabs go down the hole where mating takes place.

Conspicuous colours are often an indication that the structures bearing them have a signal function. Among salticid spiders the pedipalps of the male are frequently obvious, and are moved vigorously up and down in courtship (Fig. 12.12). The males of the South American species, *Corythalia xanthopa*, recognize the female visually. Blackening the antero-median pair of eyes prevents a male from courting a female. And similarly, a temporarily blinded female does not react to a beckoning male though soon after the varnish is removed from her eyes she responds. Movement by the female is also essential if the male is to complete his courtship; the important features which attract the male's attention are the two yellow stripes down her abdomen (Fig. 12.13).

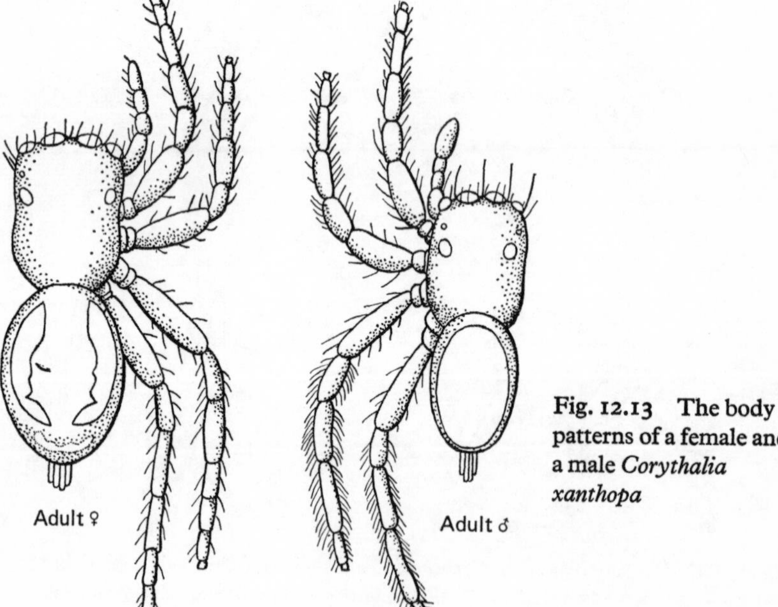

Fig. 12.13 The body patterns of a female and a male *Corythalia xanthopa*

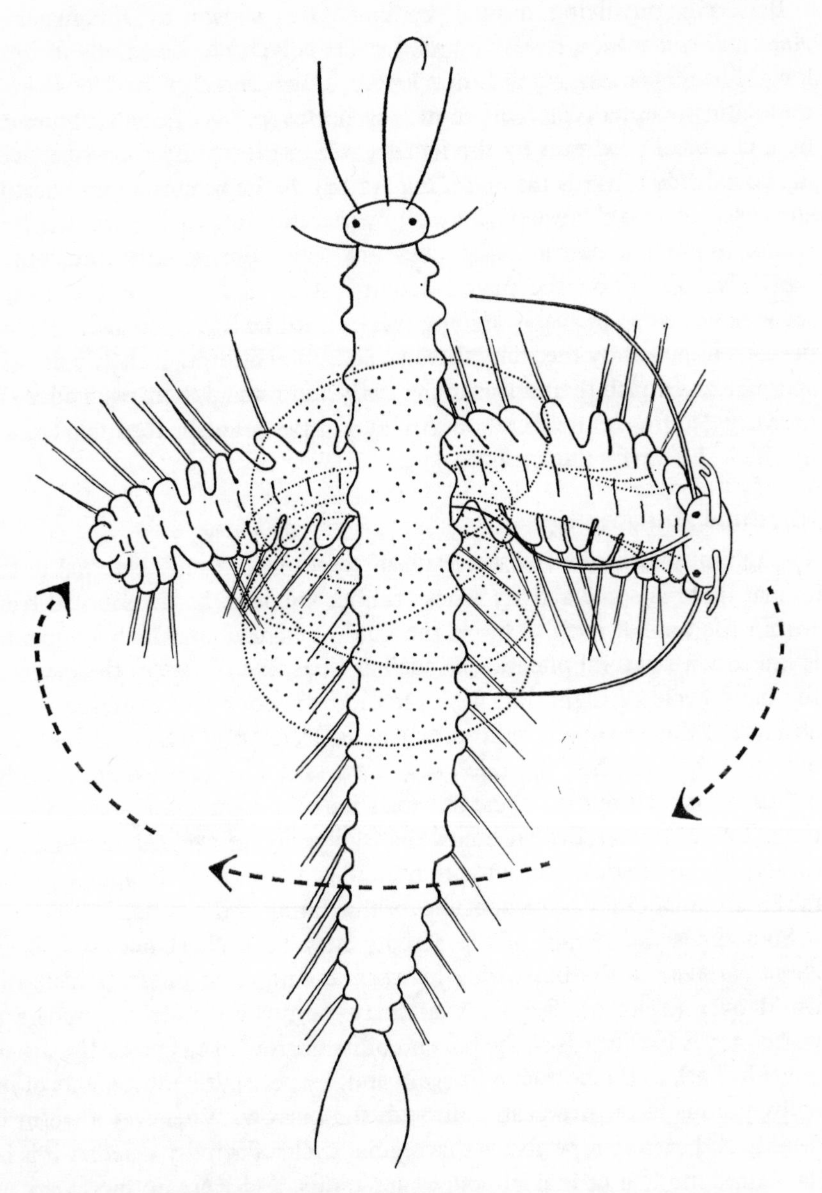

Fig. 12.14 Mating of *Autolytus edwardsi*. The male is swimming around the female in the direction of the arrows, ejecting sperm and secretion partly as threads. The long appendages are extended but bent by the resistance of the water and at least the median antenna is in contact with the female. The semen is indicated by dots. Dorsal cirri are omitted

Behaviour involving mutual movements is shown by the males of *Autolytus edwardsi*, a free-swimming syllid polychaete worm about 3 mm long. The female has less well-developed parapodia and swims slowly in an undulating manner typical of errant polychaetes (p. 185). A male, influenced by a chemical produced by the female, will swim rapidly round her with his dorsal side towards his mate (Fig. 12.14). As he rotates around her, he sheds semen into the water plus a slimy secretion the threads of which he winds round his partner. She does not swim during this time but is passively rotated by the male's activity. After a matter of twenty-five seconds he swims in ever-widening spirals until he leaves the female, who, perhaps impeded by the slimy threads, sinks down through the water. The spermatozoa attach themselves to her body and remain there even after she has wriggled free of the secretion. An hour or two later she forms an egg sac in which she carries the fertilized eggs.

Rhythms and activity

Though most behaviour is the result of reaction to stimuli external to the animal there is some activity whose timing seems to be determined from within the animal itself. This is the kind of behaviour which frequently is linked with natural phenomena such as tides, the phases of the moon or the daily cycle of night and day, yet the behaviour will continue in the absence of these external indicators. Anemones expand when the tide is in only to contract when the tide goes down. Yet in a tank where no tidal influences can be felt the cycle of expansion and contraction goes on for a time. There is much evidence now that the period of cyclical behaviour of this type is not made by the alterations in the environment though factors in the environment may be responsible for the timing of the cycle.

Sometimes the period of the activity may be a short one as it is in *Arenicola marina*, the burrowing lugworm common on intertidal flats the world over (Plate 9). Regularly at forty-minute intervals an immersed worm moves backwards to the tail end of its burrow to eject a casting, then it creeps back to the bottom of it again and, by peristaltic movements of its body, pumps water headwards through the burrow. Whenever a worm is feeding and casting it performs this regular cycle of activity whether it is in its natural mudflat or in a laboratory apparatus. Tides are not necessary as timers of this behaviour. The oesophagus of the worm removed from the body shows the same cyclic activity and seems to be the pacemaker for the behaviour, even pieces will show the same rhythm as the whole tube.

The approximately twelve-hour period of the tides is commonly found in the cycles of shore animals' activities (Fig. 12.15). *Uca pugnax* has two peaks of activity each day correlated with dawn and dusk. Kept in perpetual dim light the times of activity show a shift of some fifty minutes each day. Despite the absence of tides this change in timing correlates with the change of the time of low tide on the beach from which the crabs come.

Cockroaches and other insects have daily activity rhythms, their greatest numbers of movements rising to a peak once a day. In a normal day/night cycle, cockroaches are most active just after dusk, but if they are kept in continuous light, the peak of this activity comes later each day. The true

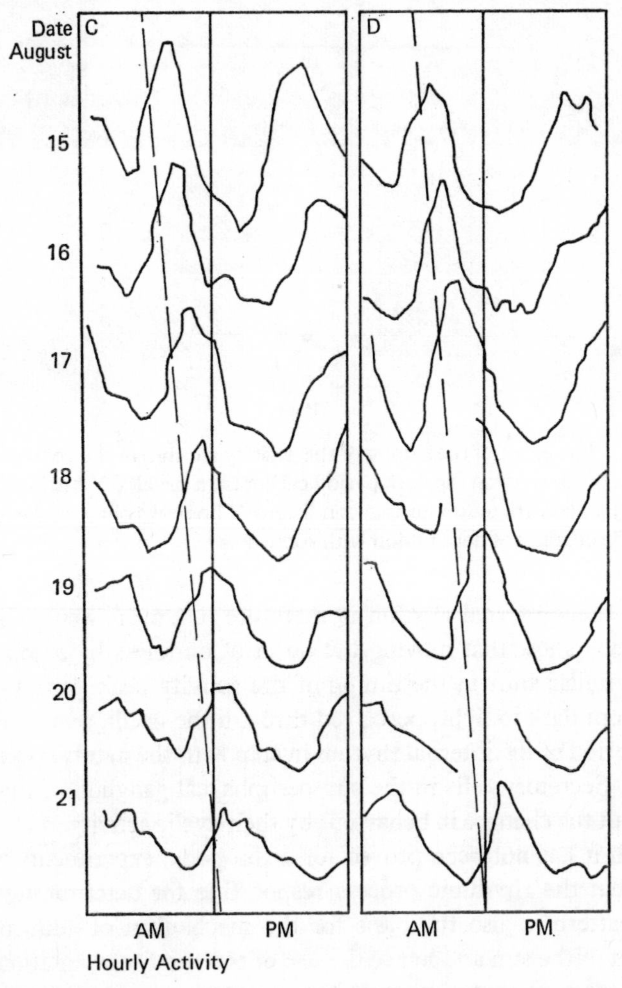

Fig. 12.15 The locomotor activity of two independent groups (C and D) of *Uca pugnax* during one week in constant conditions. The peaks of activity can be seen to move each day so that they follow a tidal, or lunar, periodicity

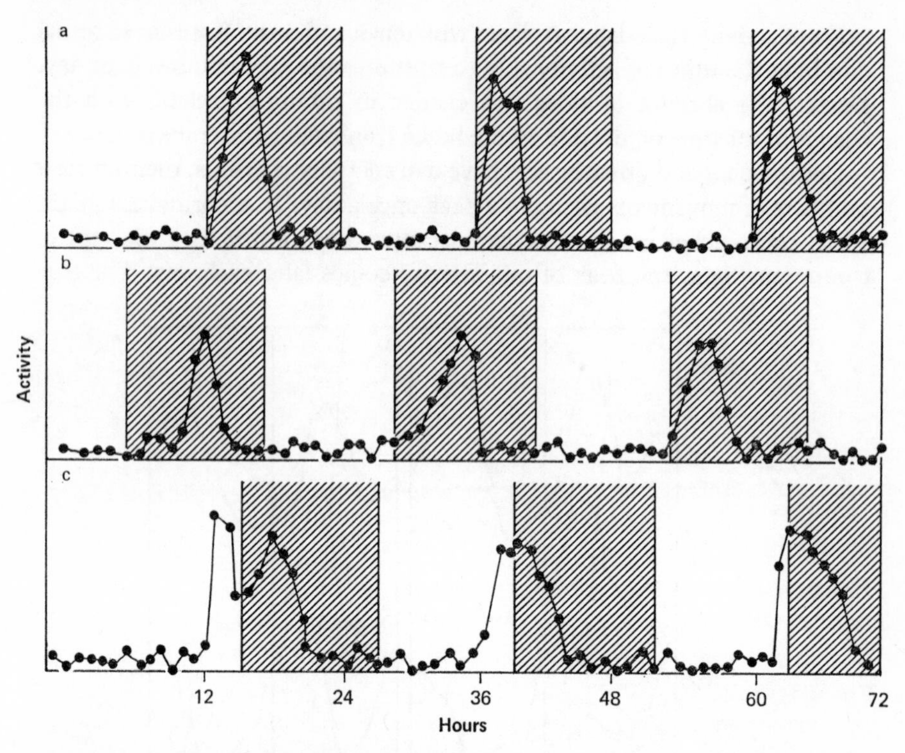

Fig. 12.16 The effect of the timing of the activity rhythm of the cockroach *Periplaneta* of (b) starting the dark period earlier than usual, (c) starting it later than usual; (a) control under normal conditions. The peaks of activity can be seen to follow the changes of dark period with some delay

period of their internal rhythm is therefore just over twenty-four hours. Experiments show that moving the onset of darkness by a small amount causes a similar shift in the timing of the activity peak (Fig. 12.16). The change from dark to light, perceived through the ocelli, serves to keep the natural period of the internal rhythm in step with the natural rhythm of the solar day. Secretory cells in the sub-oesophageal ganglion of these insects bring about the changes in behaviour by their cyclic activity.

Though it has not been proved for arthropods, experiments with birds suggest that the rhythmic process responsible for determining the daily activity pattern is also the basis for the mechanism of allowing for the movement of the sun and hence the use of the sun for orientation (p. 257). The resistance of these internal rhythms to changes in the environment such as temperature fluctuations are an essential part of the necessary constancy of the timing device used for navigation.

The analysis of the longer period rhythms is more difficult and nothing is known of the mechanism of the control of, for example, lunar period activity. There are many examples of such periodicity of which the palolo worm is perhaps one of the most famous. This polychaete worm (*Eunice viridis*) spawns once a year at the third quarter of the October or November moon. The hind-ends of male and female worms, laden with sexual products, break off and swim to the surface where eggs and spermatozoa are released into the water. *Odontosyllis enopla* is another polychaete, found in West Indian waters, which shows a lunar periodicity swarming luminously at the third quarter of the moon. Christopher Columbus' crew observed a mysterious light in the sea on 11 October 1492 when they were on the western side of the Atlantic – and the moon was in her third quarter!

Learning

Much of invertebrate behaviour may be innate, but by no means all. Behaviour can be acquired by learning though it is arguable whether true learning has been demonstrated in protozoa where a nervous system of the kind found in metazoa is absent. However, it is clear that worms, for instance, can learn. *Nereis pelagica* shows habituation to mechanical shock, moving shadows and a change in light intensity (either increase or decrease). The contraction of the whole body which follows the first of such stimuli rapidly wanes. This indicates that the worms have learned not to respond (Fig. 12.17). *N. virens* will learn to select the correct arm of a T-maze if it is rewarded for a correct choice by a period of five minutes in a darkened chamber.

The role of the supra-oesophageal ganglion (p. 240) in learning is uncertain for removal of it makes little effect on the habituation of *N. pelagica* (Fig. 12.18) though decerebrate *N. virens* are thereafter unable to make a correct choice. But this latter effect may be due to the destruction of the sensory input from the tentacular cirri, for trained animals upon which the operation has been done with great care to avoid cutting the cirri nerves sometimes are able to discriminate in the way that they had learned. It appears that the supra-oesophageal ganglion, much of which consists of neuro-secretory cells, does not act as a memory store for learning but rather that the learning process takes place in the segmental ganglia.

Representatives of most of the major phyla have been tested in mazes or puzzle situations for evidence of learning and it would be fair to say that none of them has been found to lack this ability at least in some measure.

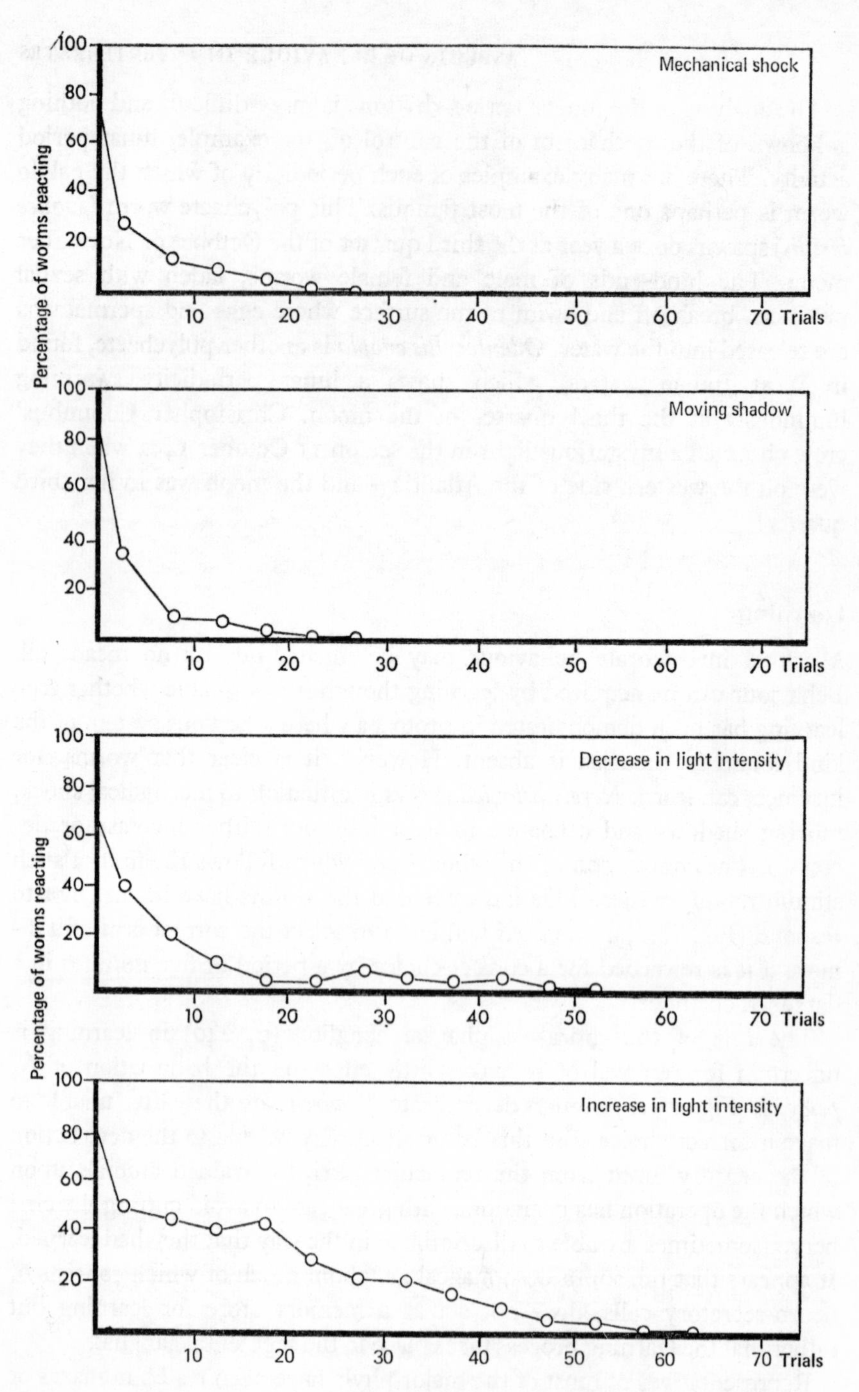

Fig. 12.17 Habituation of the withdrawal reflex of *Nereis pelagica* to stimulation by mechanical shock, shadows, decrease and increase in light intensity. Except for the first trial, the average percentage of worms reacting of five successive trials is shown

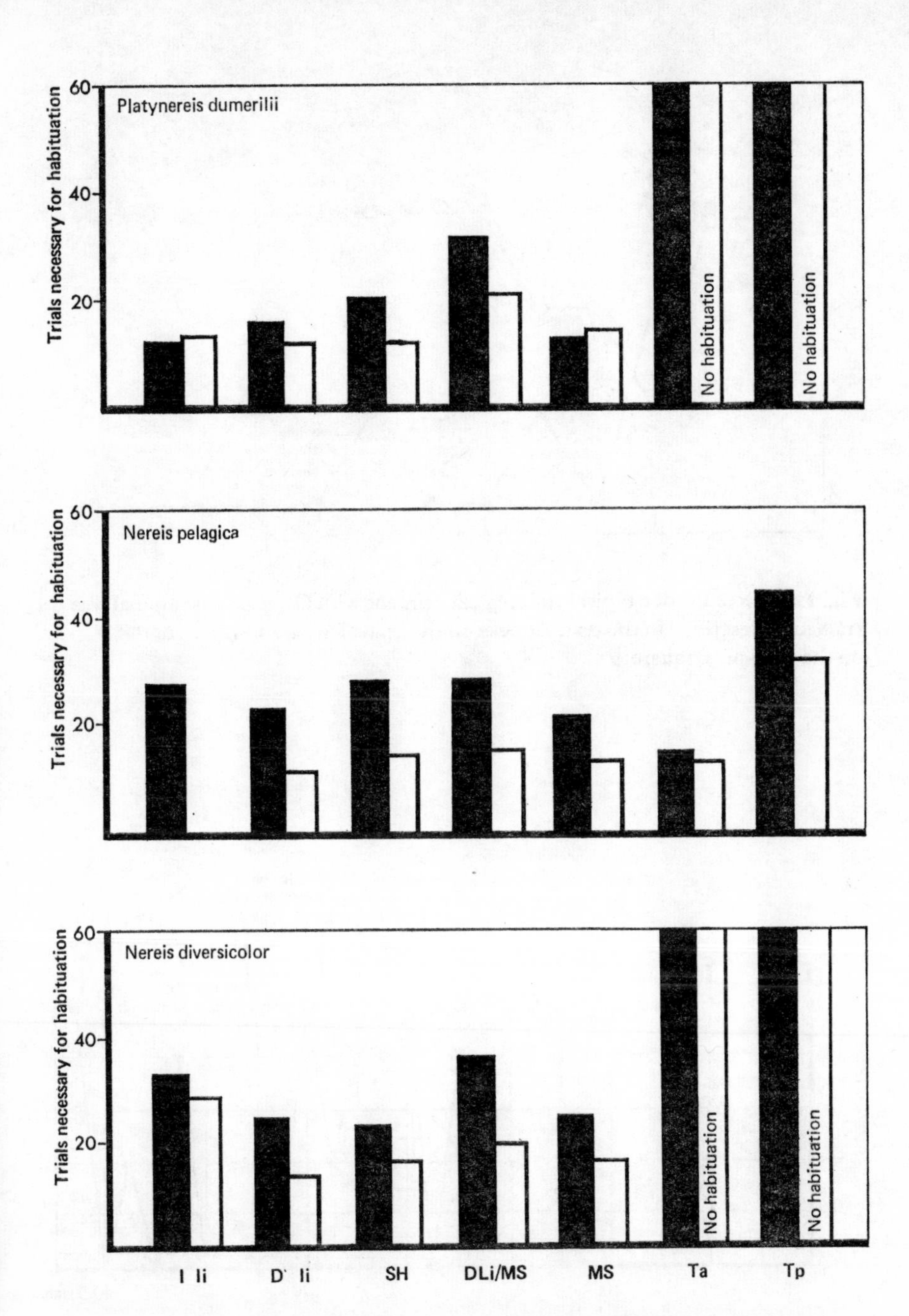

Fig. 12.18 Effect of removal of the supra-oesophageal ganglion of various polychaete worms on their habituation rate. From left to right, to increase in light intensity (I li), to decrease in light intensity (D li), shadows (SH), light decrease coupled with mechanical shock (D li/MS), mechanical shock alone (MS), tactile stimulation at anterior end (Ta), tactile stimulation at posterior end (Tp). Filled blocks, intact animals; open blocks, decerebrate worms

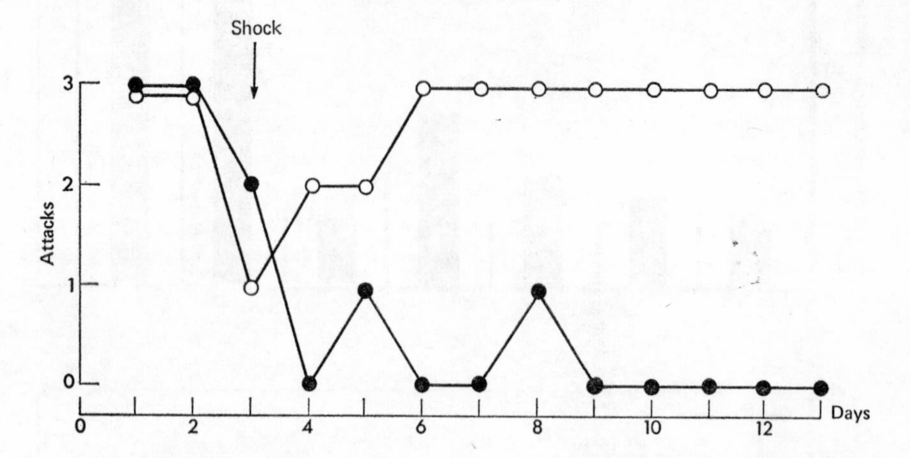

Fig. 12.9 Results of a typical training experiment with *Octopus*. The animal was trained as described in the text. Attacks on the crab alone are shown as o, those on the crab-plus-square ●

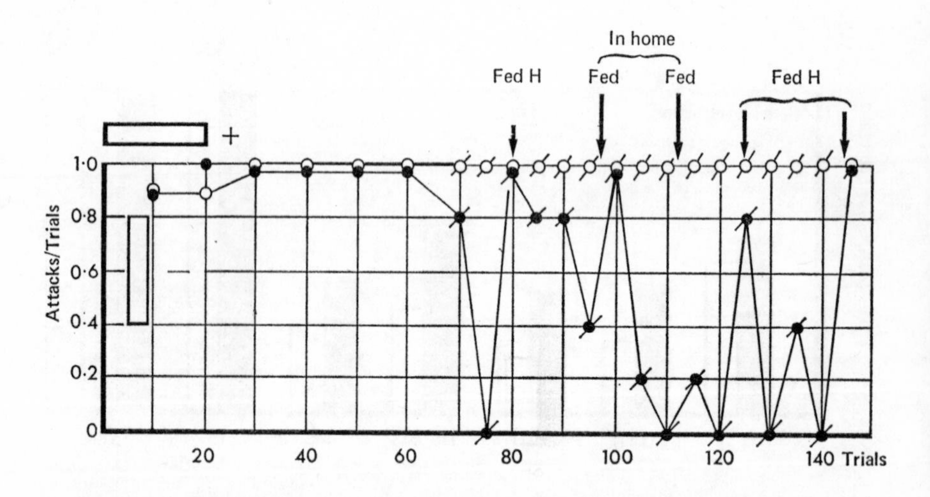

Fig. 12.20 The responses of an octopus with 85% of its vertical lobe removed. Trained to attack a horizontal but to avoid a vertical rectangle (●) it seemed to learn nothing in 60 trials, attacking nearly every time. In a series of tests without rewards (ø) and ● performance was correct but again deteriorated when food was given (arrows) either in the home (Fed H) or with the rectangle (Fed)

However, the detail of the work and the correlation of behaviour results with the morphology of the nervous system make it worthwhile singling out the octopus for further consideration.

This animal can be trained to leave its 'home' at one end of a tank and to come to seize a crab when this food is shown alone but to remain in its home when the crab is shown with a white square (Fig. 12.19). After the first two days, on which the octopus attacks on all occasions, it is given a mild electric shock if it grasps the crab-with-square. After two more days of three trials a day, that is, after six trials only, the animal ceased to attack the crab-with-square.

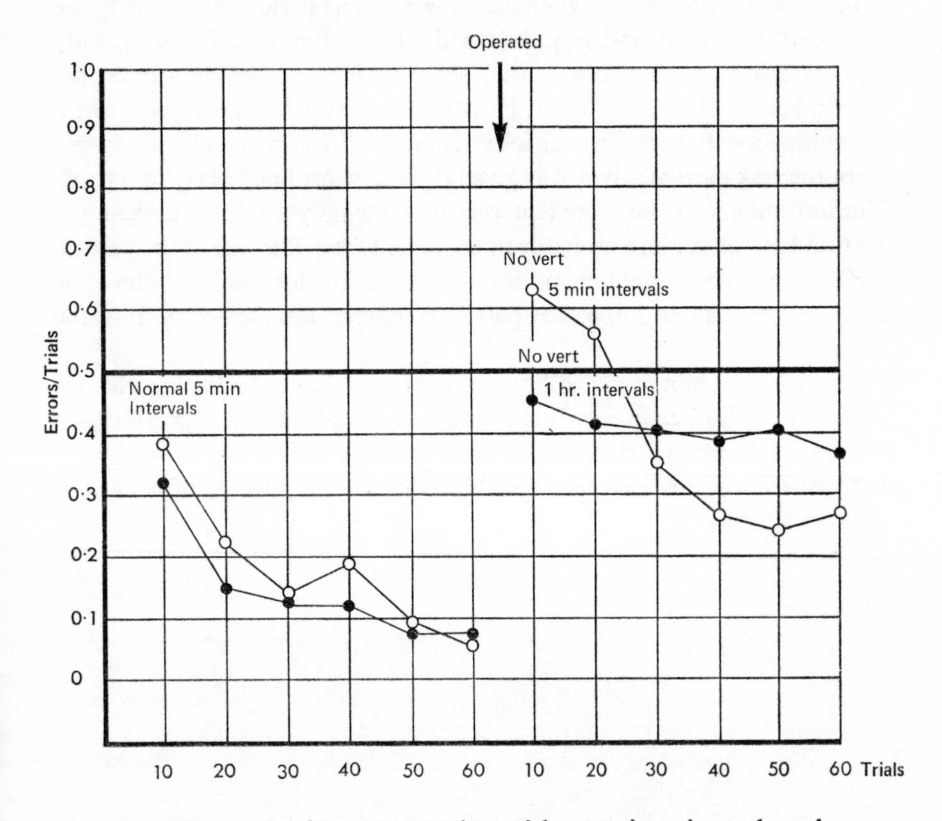

Fig. 12.21 Responses of 10 octopuses given trials at 5-minute intervals, and 10 at 60-minute intervals, before and after removal of the vertical lobe (vert.). Food was given for attack at a horizontal rectangle (o), shock for attack at a vertical one (●) Before the operation the 5-minute group was slightly more accurate. After the operation it was initially less accurate, but finally more so than the 60-minute group. Thus, the vertical lobe appears to hold short-term memory before storage in the optic lobe

The same situation can be used to test the ability of the octopus to discriminate between shapes. They can be trained to distinguish between a number of different shapes, though two of a pair may be equal in area or have the same length of outline. They can also tell the difference between the same shape orientated in one way and another. Thus, they easily learn to distinguish a vertical from a horizontal rectangle, but they only obtain chance scores when the rectangle is offered slanting obliquely to the left and slanting to the right. This could be explained if the analysis of shape were done on a basis of rows and columns of receptors (p. 206).

Removal of the vertical lobe of the octopus brain (p. 248) produces animals which learn to discriminate for very short periods only if they are rewarded with food and punished with shock. But tested subsequently without reward or punishment they show evidence of having learned the discrimination (Fig. 12.20). Or, if an animal which has already learned a discrimination has its vertical lobe removed and is re-taught it can re-learn the task particularly if it is given trials at short intervals (Fig. 12.21). The indications are therefore that long-term memory does not reside in the vertical lobes, but more probably in the optic lobes. The role of the vertical lobes seems to be to hold memories for a short time for storage in the optic lobes and to play an important part in retrieving the memories from the store.

There is no doubt of the ability of many invertebrates to display what we can call complex behaviour. It is possible, as Dethier has suggested, that it has been a desire to eschew anthropomorphism which has prevented us in the past from seeing anything vertebrate-like in the patterns of behaviour of lower animals.

Chapter 13

Reproduction and development

REPRODUCTION is much more than simply replacing one pair of animals by a pair of offspring that survive long enough to breed in their turn. It provides an opportunity for preserving a degree of variability in the population, which may be important in a short-term evolutionary sense, is often the occasion for members of a species to explore and colonize new environments, and it is also the occasion when animal numbers may be adjusted to exploit the resources of the environment to the full. The pattern of reproduction in any animal is therefore closely related to the circumstances in which it lives and generally represents some sort of compromise, balancing advantages and disadvantages of a particular breeding biology to produce the most beneficial outcome for the species as a whole.

A very common method of reproduction is sexual and probably all animals employ this method at some time, though not necessarily in every generation. The importance of sexual reproduction lies in the fact that the offspring receive a genetic inheritance from two parents and so are not exact replicas of the female, with the result that the population remains varied and heterogeneous. The benefit of this heterogeneity is felt when the environmental conditions change. Then variants, which were previously at a disadvantage and did not survive in such great numbers as other better adapted variants, may be more favoured and the balance of the population changes. In this way the adaptability of the population extends beyond that of any individual in it.

The ability of the population to produce new individuals (i.e. its reproductive potential) depends ultimately upon the number of females in it, but so long as sexual reproduction is practised, the proportion of females in the population is limited by the need for sufficient males to ensure that they are all fertilized. In situations in which it is advantageous to have a very high reproductive potential, sexual reproduction may be abandoned so that even though variability may be sacrificed, all members of the population can produce offspring. Asexual reproduction is achieved in two ways, either by parthenogenesis, in which eggs are laid and develop in the normal way but without requiring to be fertilized by spermatozoa, or by vegetative reproduction in which new individuals are produced by division or are budded from an older animal by a special growth process.

Vegetative reproduction and the alternation of generations

The whole hazardous process of mating and embryonic development can be avoided if the animal is able to reproduce vegetatively: to grow, bud and proliferate new individuals that are replicas of the parent. A surprisingly large number of animals multiply in this way even though they also engage in sexual reproduction from time to time, and only the Mollusca, Arthropoda, Nematoda and Echinodermata, among the major invertebrate phyla, never resort to asexual reproduction.

It is not always easy to make a clear distinction between normal growth and asexual or vegetative reproduction. Daughter organisms are often produced by a process which is simply an extension of normal growth without any interruption of the parent's activities, and if the new individuals remain attached to the parent body, the result is a colonial or compound animal in which there may be a degree of mutual co-ordination between members of the colony, and sometimes a division of labour, so that the colony behaves very much as a single, complex organism.

Asexual reproduction coupled with a high degree of mutual interdependence is characteristic of coelenterates and appears to be the coelenterates' answer to the problem of developing complex structure in animals that have no mesoderm and lack the capacity of forming systems of internal organs. After the larva of a hydroid has settled on a suitable substratum it begins to form a small polyp with a stolon containing both body layers growing from its base. The stolon spreads on the surface to which the polyp is attached, sprouting more polyps as it grows. More often, instead of forming a tangled stolon or hydrorhiza from which individual polyps arise, the stolon grows vertically as a main stem, or hydrocaulus, with lateral branches bearing polyps, giving the hydroid a delicate plant-like form. In these, the growth zone is immediately behind a polyp so that the stalk on which it is formed elongates and eventually buds another polyp behind it. In all of these the coelenteron extends throughout the polyps, stalks and stolons, so that the entire colony is organically continuous.

In many hydroids there is more than one kind of polyp in the colony and some division of labour between them, but the greatest degree of morphological differentiation and mutual interdependence is found in the Siphonophora, an order of pelagic Hydrozoa. The colonies include individuals of both medusoid and polypoid types, which are modified to form gas-filled floats, pulsating swimming bells, feeding polyps, tentacular polyps, and others armed with batteries of stinging cells (nematocysts), and reproductive individuals that produce eggs. These may be arranged in a linear series, growing from a long stolon, or suspended in a compact mass beneath a large gas-filled float, as in the Portuguese man-of-war, *Physalia*. Co-ordination between the individuals is so complete that they might almost be regarded as the organs of a single complex organism.

Multiplication of individuals by growth, leading to the production of colonies, though generally without differential modifications of the individuals and the division of labour between them that is so highly developed in coelenterates, occurs also in the Entoprocta, Ectoprocta, Urochordata and Phoronida. Like most colonial hydroids, they form encrusting or erect branching structures and often have a strong superficial resemblance to one another. The branching forms particularly recall delicate fronds of ferns and at one time ectoprocts and hydroids were collectively known as 'zoophytes', or 'plant-animals', on this account.

The vegetative process might reasonably be regarded as growth rather than reproduction, but the distinction breaks down completely if the individuals that are formed break free and lead an independent life. This is a common method of reproduction in the hydrozoan medusa *Sarsia* which buds small medusae at the edge of the bell near the ends of the radial canals, and several anemones, including *Metridium*, produce small anemones simply by putting out lobes of the pedal disc which are then separated off and grow.

The great capacity for growth displayed by these animals in asexual reproduction is also shown in many animals by their considerable power to regenerate lost or damaged parts of the body, and this, too, may be exploited to produce new individuals. The detached arm of a starfish, provided the fragment includes part of the central disc of the body, will grow four additional arms and the larger fragment a new arm, and in this way the reparative process of the animal leads to asexual reproduction as the result of a fortunate accident. This occurs more frequently in fragile animals like the synaptid holothurians, which fragment very easily on disturbance, with a subsequent regeneration of a whole individual from each fragment. The same is true of many of the longer nemertean worms. Fragmentation and regeneration as a means of asexual reproduction is spontaneous and has become a regular feature of the life-cycle of the cirratulid polychaetes *Dodecaceria* and *Ctenodrilus*.

A more orderly and more complicated type of growth is found in small aquatic oligochaetes and in some turbellarians. Oligochaetes grow by adding new segments at a growth zone immediately in front of the terminal, non-segmental piece, known as the pygidium. When fresh-water oligochaetes such as *Nais* or *Aeolosoma* have grown a certain number of segments – the critical number of segments varies from species to species – a zone of fission develops at some position along the body and a new pygidium and a new prostomium appear there, so that there are now two worms arranged in tandem. They may separate at this stage, each adding further segments and undergoing fission in its turn, but often formation of new fission zones outruns the rate of separation of the individuals so that chains of small worms are formed. Very much the same process, though without the formation of segments, takes place in some rhabdocoel and planarian turbellarians, and in the rhabdocoel *Stichostemma* chains of as many as six or seven individuals may be formed before they begin to separate.

Although vegetative reproduction has the advantage of permitting new animals to be produced with a complete adult structure without interrupting the life of the parent, all the offspring are genetically identical. Periodic rejuvenation of the population by sexual reproduction is normal in all animals that rely heavily upon vegetative reproduction (although some small oligochaetes have never been observed to become sexually mature). Often, sexual reproduction is seasonal or is confined to periods when environmental conditions become arduous, but in some urochordates and coelenterates sexual and asexual generations alternate regularly.

Alternation of sexual and asexual generations proceeds with the greatest regularity in the Thaliacea, a class of pelagic urochordates. They include the salps, in one generation of which the individuals are solitary and contain gonads. These reproduce sexually, and the fertilized eggs develop via a larva into the asexual generation, similar anatomically to the sexual animals except that they have no gonads. They proliferate a chain of individuals from a stolon and these eventually break free to reproduce sexually. In the salps the two kinds of individuals are known as blastozooids, which are produced vegetatively from a stolon, and oozooids which develop from eggs. The balance between the two generations is lost in the Pyrosomidea, another sub-class of the Thaliacea, and the numerous small individuals formed vegetatively and so corresponding with the salp blastozooids, are fused together to form a floating colony. They produce very large yolky eggs, and the oozooid is a transitory embryonic stage attached to the yolk sac, from which the colony is budded. Reduction or even complete suppression of one or other of the generations is common also in the Cnidaria.

Here the sexually reproducing form is typically a medusoid and the vegetative generation a polyp. In the scyphozoan jelly-fishes, the medusoid phase is dominant, but a small scyphistoma polyp is formed and proliferates numerous little medusoids by strobilation so that by the time the strobilae are released the polyp looks like a miniature stack of dinner plates. In hydroids it is the medusoid which is often reduced, in some remaining attached to the polyp and in such forms as *Hydra* never being produced at all.

Particularly in the Scyphozoa, it is obvious that the alternation of generations is merely a simple means of producing large numbers of sexually reproducing individuals from a small number of eggs by interposing a phase of vegetative multiplication in the life-cycle. Fundamentally the same phenomenon occurs in tapeworms although in them the process can hardly be described as an alternation of generations. The anterior end of the adult is armed with hooks, suckers, or commonly, both, and attaches the worm to the wall of the host's intestine. In the majority of tapeworms the posterior end of the body then begins to proliferate a large number of identical segments or proglottides. These form the long ribbon-like chain of the more familiar tapeworms and each proglottis undergoes a sexual evolution in which it is first a functional male and, later, a functional female. Ultimately it constitutes no more than a sac filled with fertilized eggs which have already begun to develop (self-fertilization is common) and in this form it is shed and leaves the body of the host. In consequence of this association between growth and egg production a very high reproductive rate can be maintained over a long period. It has been estimated that a single sheep tapeworm, *Moniezia expansa*, produces some 4,000 proglottides each containing 10,000–20,000 developing embryos during its life in the host, which is a little less than one year. Such continuous high productivity could not be achieved without this sequential development of maturing eggs and the continuous proliferation of new reproductive units.

A number of syllid polychaetes produce chains of sexually reproducing individuals in a manner not unlike that of cestodes. Segment proliferation is accelerated as the breeding season begins and a number of individuals is produced, each composed of a relatively small number of segments and often with a newly formed prostomium with eyes and other sense organs. In these worms, while stolonization may increase the total egg production, the process improves breeding success in another way. Like a number of other polychaetes, these syllid stolons swarm at the surface of the sea and pair for spawning. The chances of surviving the hazardous process of swarming and of males meeting females are increased by multiplying the number of reproductive individuals. Furthermore, since most polychaetes die after swarming and spawning, the reproductive potential of the species is increased by the fact that the main part of the worm does not swarm but lives to form additional stolons.

It is impossible to separate the different reproductive phenomena that occur in the Metazoa; the growth processes that are involved are often basically similar and the biological advantage is always much the same. Much was made of the alternation of generations in coelenterates when it was first discovered, and it was regarded by some as a very primitive and basic reproductive pattern in the Metazoa. In fact, it is impossible to make a meaningful distinction between it and the proliferation of proglottides in the Cestoda or of reproductive stolons in the Syllidae. Viewing the great variety of growth patterns associated with reproduction in this family of polychaetes, we can see that it is a development from the metamorphosis of the whole animal such as we find in nereid polychaetes and a few syllids. In these there is no growth of new segments, merely a modification of the animal to adapt it to its brief life in the plankton. It is only a small step from this to separating a certain number of segments in the form of a reproductive stolon and thence to the production of numerous stolons.

Embryological development

The alternative to vegetative reproduction is development from an egg, whether or not this process involves the fertilization of the egg by a spermatozoon. The egg is a single cell and the whole animal is developed from it by repeated cell division in the course of embryological development. The development of metazoan animals shows interesting and important variations which are often valuable in indicating relationships between otherwise dissimilar groups, but underlying this variation there is a basic pattern of development which is followed by almost all animals.

The fertilized egg divides into two cells which divide in their turn, and so on, so that the embryo passes through 2-cell, 4-cell, 8-cell, etc., stages. By the time it has reached the 64-cell or 128-cell stage, the embryo consists of a hollow ball of cells known as the blastula and the cavity within is the blastocoel. By the process of gastrulation, cells of one half of the blastula come to lie within the other half and the embryo, now at the gastrula stage, consists of a hollow ball with walls composed of two cell-layers. The blastocoel between the two walls may persist for a time as a cavity but generally is almost entirely obliterated. The inner wall of the gastrula constitutes the endoderm, the outer wall the ectoderm and at about this stage a third cell layer, the mesoderm, is formed between the two.

Up to this stage of development, the embryo relies upon yolk stored in the egg and generally concentrated in the endoderm cells, or upon material in solution in the medium surrounding the embryo, which is absorbed across the cell walls. Shortly after gastrulation and with only slight modification, the embryo can begin to feed on particulate matter for the first time. The central cavity in the gastrula, bounded by endoderm, forms the digestive cavity, a mouth and anus are formed by intucking of the ectoderm and the embryo is now a larva. Quite often there is a rather more elaborate modification of the gastrula than this and the larva develops a number of temporary, specialized structures as adaptations to swimming and feeding. Subsequently, at the larval metamorphosis, the larva is converted into a small adult by a more or less radical modification of its structures.

Variation occurs in the pattern in which cell division occurs in the early stages of development and in the source of the cells that constitute the mesoderm. In both respects, the variation can be used as an aid to deciding the relationships between different animal groups. But much more conspicuous variation is introduced by the quantity of yolk that is included in the egg, and animals in any phylum may have much or little yolk, with corresponding modifications of their embryology, depending upon their ecological adaptations. Yolk tends to accumulate in one half of the egg and ultimately in the cells that constitute the endoderm. Large concentrations of yolk impede cell division so that in those parts of the embryo where there is little yolk numerous small cells are formed while in other parts of the embryo where there is more yolk, cell division is slower and fewer but larger cells are produced. The fact that the endoderm cells contain more yolk and therefore tend to be bulkier than ectodermal cells introduces complications at the time of gastrulation. If there is very little yolk and all the cells are much the same size the embryo gastrulates by a simple invagination. If the endoderm cells are too bulky for this, individual cells may migrate inwards, or the ectodermal cells may multiply rapidly and grow around the yolky cells, completely surrounding them. In those animals which have eggs containing an enormous quantity of yolk cell division is confined to a small area of the egg surface and the embryo is formed floating on top of a great yolky mass.

If there are substantial food reserves in the egg, the stage at which the embryo must find the raw material for further growth and development may be postponed. Indeed, in many animals with very yolky eggs, the whole of embryological development may be completed without recourse to an external food supply so that the young hatch as miniature adults. Animals with yolky eggs do not invariably avoid a larval phase, however.

Larvae serve two purposes: they can be an early feeding stage or if they drift in the plankton for a period, they can distribute the species over a wide area. Generally they do both, but animals with yolky eggs may have non-feeding pelagic larvae when dispersion of the young is important, just as animals with small eggs may have non-pelagic larvae which feed but remain near the parents.

Parthenogenesis and sex

Although most, if not all, animals reproduce sexually at least occasionally, whether or not the egg is fertilized is in many ways of minor importance so far as its subsequent embryological development is concerned. Parthenogenesis, or development without fertilization, occurs naturally in some members of almost every group of animals and can be induced artificially in many more. Penetration of the egg by a spermatozoon does two things: it stimulates the egg to begin dividing and it contributes genetic material to the egg. Both can be achieved in different ways. Pricking eggs with a needle, heating them, or subjecting them to chemical stimulation may all induce the initial stages of development in a considerable variety of animals, and although such artificial parthenogenesis may not proceed very far, it is obvious that sperm penetration is not essential for development to begin.

Making good the genetic contribution of the spermatozoon is more difficult. All cells in the body of an animal, with rare exceptions, contain a double set of chromosomes. During the formation of the eggs and spermatozoa, the number of chromosomes is halved and the fusion of the spermatozoon with an egg restores the double set of chromosomes, with one set derived from each parent. In almost all animals it is important that there should be a double set of chromosomes, otherwise cell division does not proceed normally and in most parthenogenetic species, the full number of chromosomes is restored even though there is no contribution of chromosomes from a spermatozoon. Either there is a failure in the reduction of the chromosome number so that by the time the egg is fully formed it already contains a double set, or alternatively, a normal egg is produced with only a single set of chromosomes and at an early stage in development, cells fuse together in pairs and restore the chromosome number in that way.

Chromosomes are the carriers of inherited genetic material, and eggs with their single set of chromosomes have a selection of genes of the parent, although, unless there has been a mutation (a rare event), the eggs can have no genes that were not present in the mother. Each spermatozoon similarly contains a selection of genetic material from the male. Sexually produced offspring with a genetical inheritance from both parents therefore show a good deal of variation while parthenogenetic young eventually develop into replicas of their mothers.

From the point of view of the population or the species (but not of the individual), variety is valuable in a varied and varying environment because it helps ensure that a proportion of animals will find congenial surroundings and will survive to breed in their turn. But although sexual reproduction provides this variability in the population, it has the disadvantage that the population must contain a significant proportion of males who use up the resources of the environment but produce only inessential spermatozoa. A parthenogenetic population can consist entirely of females, all producing offspring, and in this way the population can multiply at a very high rate indeed, but because of the uniformity that parthenogenesis entails these animals are very vulnerable to changes in the environment and the population may be almost completely wiped out at an even greater rate than it grew.

Parthenogenetic species always undergo wide fluctuations in numbers, but many environments can support only a small fraction of the summer population during the winter months, so fluctuations are not unknown in sexual animals. Under some circumstances, the very high population growth rate resulting from parthenogenesis allows a species to exploit fully a sudden favourable change in the environment and this can more than offset the disadvantage of relative uniformity, particularly if sexual reproduction can occasionally be interposed to prevent the population from becoming genetically completely stagnant and inflexible.

Parthenogenesis is particularly characteristic of some insects (which we shall not consider here), and the fresh-water rotifers and cladoceran water fleas. Superficially the life history of rotifer and *Daphnia* populations is similar, though in detail they differ substantially. Both produce resting, or overwintering eggs from which females hatch in spring. These females reproduce parthenogenetically and there is a succession of parthenogenetic generations during which the population increases dramatically. What causes them to do so is unknown, but eventually a generation of females produces two kinds of eggs which develop parthenogenetically into male- and female-producing females respectively. The final generation contains both males and females from which are produced fertilized eggs which remain dormant until the following spring. Water fleas are less co-ordinated than rotifers and in them fertilizable eggs may be produced at any time, whether or not males are present to fertilize them. If these eggs are not fertilized, the female may resume laying parthenogenetic eggs for a time.

It is very likely that this pattern of reproduction evolved originally as an adaptation to life in small temporary pools of water, but the problems of life under these peculiar circumstances are not essentially different from those in other environments in which some animals have adopted much the same reproductive system. A resistant, dormant stage is necessary for some, at least, of the population to survive periods when normal life is impossible, and parthenogenesis allows the rapid replacement of the population as soon as normal conditions of life are restored.

Hermaphroditism

Animals are generally sexually ambivalent and have the potentiality of developing in either a male or a female direction. Which direction development takes depends upon genetic factors, hormones and, sometimes, environmental influences, but the distinction between the sexes is not nearly as complete as the structure of the mature, adult animal might suggest. Hermaphroditism, when both sexual elements are present and functional in the same individual, is quite widespread, either as protandrous or protogynous hermaphroditism in which the animal first functions as a male or female, respectively, and then changes its sex, or as simultaneous hermaphroditism. The latter is the more conspicuous and familiar and is normal in platyhelminths, oligochaetes and leeches, the Ectoprocta, the Urochordata and in all but the most primitive gastropods, though not in other molluscs. Simultaneous hermaphroditism also occurs sporadically in most other major groups of animals.

Hermaphroditism presents a number of interesting biological problems. It is possible for an animal to fertilize its own eggs, as is often the case in tapeworms, but to do this is to lose much of the genetic variability that the contribution of genetic material from two reproductive partners gives, which seems to be the chief advantage of sexual reproduction. Most hermaphrodites are, in fact, at some pains to prevent self-fertilization. In the urochordates, for example, there is a neat anatomical arrangement whereby the passage of eggs along the oviduct automatically closes the sperm duct, and vice versa, so that the simultaneous release of eggs and spermatozoa is impossible. Many hermaphrodites practise copulation and exchange spermatozoa, and in these, too, the anatomical arrangement of the genital openings is such as to prevent self-fertilization. When the common earthworm *Lumbricus* copulates, the two partners lie head to tail, bound together by a mucous secretion, spermatozoa are released from the male openings and travel along grooves in the body wall to the openings of a pair of chambers, the spermathecae, of the partner, where the spermatozoa are stored. The spermatozoa do not reach the female openings of the worm that produced them, and, in any case, the eggs are laid and inseminated from the contents of the spermathecae at a later time, after the worms have separated. In leeches, many tropical earthworms, turbellarians, and in most hermaphrodite gastropods, a penis is developed and the spermatozoa are deposited directly into the spermathecae or equivalent organs so that that chance of accidental self-fertilization is further reduced.

Self-fertilization is uncommon, although it forms the normal pattern of reproduction in some animals. One polychaete, *Nereis limnicola*, living in fresh-water lagoons or in brackish estuaries on the Californian coast is probably a self-fertilizing, viviparous hermaphrodite. Eggs and a few spermatozoa are released into the coelom, the embryos develop there and the young leave the parent's body when they are some segments long. Reproduction presents special problems for these worms because the eggs and very young worms are unable to control the uptake of water from their environment. For this reason it is necessary for them to remain in the controlled environment of the parental coelom until the young worm has developed an adequate osmoregulatory apparatus and can survive in the brackish or fresh-water conditions outside. Cestodes, too, which appear to resort to self-fertilization regularly, live under rather special conditions for although very dense tapeworm populations have been found in some host animals, it is unusual to find multiple infestations of tapeworms. Individual tapeworms often have an inhibitory effect on one another and the presence of an adult tapeworm in a host prevents further infestations by the same species. While this may confer some advantage on the tapeworm, it obviously raises reproductive difficulties and these seem to have been met by self-fertilization. Either the penis is inserted into the vagina of the same proglottid, or the tapeworm loops round and younger proglottids inseminate older ones further along the chain, in which the female reproductive system is more advanced.

Examples of habitual self-fertilization are sufficiently uncommon to be remarkable and since most hermaphrodites practise cross-fertilization the advantages of hermaphroditism are not obvious, particularly as such a heterogeneous collection of animals is bisexual. The chief advantage to animals such as some internal parasites (trematodes) or in animals which are not very mobile (earthworms, snails) is that every encounter between two individuals may potentially lead to the fertilization of eggs, and fruitless encounters between members of the same sex are avoided. This enormously increases the reproductive potential of these animals which might otherwise be limited by the difficulty of congregating for reproduction. This can hardly be the whole explanation of hermaphroditism, however; we do not know, for example, why the Ectoprocta, which are densely crowded together, should be self-fertilizing, as a rule, and the ovum already fertilized before it leaves the parental coelom. Nor is it clear why some species are hermaphrodite when the great majority of their near relatives are not, although apparently living under much the same circumstances.

Breeding seasons

The times at which animals breed and the lengths of their breeding seasons vary enormously. Some have a short, sharp breeding season lasting only a few days of the year. A famous and extreme example is the palolo worm, *Eunice viridis*, a polychaete of the south-west Pacific. At Samoa and other islands in the region, the entire population of this species swarms at the surface of the sea for breeding in only a single night in most years. At the other extreme, some animals show no synchronized breeding and are apparently reproductively active at all times. This is commonly a feature of the very small animals that make up the interstitial fauna of sandy deposits in the sea.

The reason for this wide variation in the pattern of breeding is very largely, though not entirely, the varied nature of the periodic changes in different environments and the animals' responses to them. The environment of an interstitial fauna is a very constant one, it is not affected by seasonal climatic changes and at no stage do the animals or their young leave the substratum; there is little to be gained by seasonal breeding. But most environments, particularly on land or in fresh waters, show pronounced seasonal fluctuations. Temperature, light, rainfall and food-availability all show recurring annual patterns and even the depths of the ocean, which are relatively isolated from the climatic changes at the surface, are not immune. Much of the food resource of this environment is derived from the bodies of dead plants and animals that drift down from the surface waters, and if the productivity of the surface waters shows a seasonal fluctuation, its effects will ultimately be felt in the depths.

Seasonal changes may affect reproduction in two ways. In a negative sense, adverse conditions at certain times of the year may prohibit reproduction because the adults are incapable of growing and maturing eggs or even of sustaining normal life. Alternatively, some seasons may favour reproduction because they present optimal conditions for the developing and growing young. The latter is probably by far the more influential.

Environmental factors which prevent reproduction do not act uniformly on all species. The Indian earthworm *Eutyphoeus* burrows deeply into the ground and aestivates during the hot dry summer and, in the south of France, *Allolobophora icterica* behaves in the same way. Low temperatures in temperate and sub-polar regions have a similar effect: the snail *Helix* seals the mouth of its shell with a secretion and becomes inactive during the winter and will do so at any time if the temperature or humidity falls. Fresh-water organisms are particularly vulnerable and in small bodies of water must contend with both freezing and drying out of their environment. To counter this, many have a resistant dormant stage and the production of protected eggs is one of the easiest ways in which a population may survive such unfavourable circumstances. As we have seen, rotifers and water fleas commonly overwinter as eggs, the adults dying in the autumn and being replaced by a new generation in the spring. The fresh-water sponge *Spongilla* produces gemmules, or small capsules containing a few cells, protected by a thick coat, from which a new sponge grows as soon as favourable conditions are restored. Fresh-water ectoprocts produce statoblasts which are exactly comparable to sponge gemmules.

In the sea, physical conditions are much more constant than in fresh water or on land. The extreme seasonal fluctuation in sea temperature is nowhere more than about 15°C and often much less. Since animals are generally adapted to the conditions in which they live, a direct prohibition of maturation and breeding by low temperatures is uncommon. We can detect signs of it, however, in animals living near the polar extremes of their geographical range. The polychaete *Clymenella*, an American species which has been introduced to the British Isles and is here at the northern fringe of its range, breeds in May. The eggs develop in its coelom over a long period and it is interesting to see that their growth is almost completely halted during the coldest months of the year. Presumably the low winter temperatures depress the metabolism of the animal to the point at which laying down food reserves in the maturing eggs ceases.

For probably the majority of animals, extreme conditions in their environment do not prevent reproduction or any other activity, and a more potent factor influencing the time of breeding is the availability of food for the young animal. Food chains can almost always be traced back to plants and while the annual cycle of the vegetation on land is obvious, it is even more pronounced in fresh waters and the seas in temperate and sub-polar regions. Most plants, including unicellular ones, require light and a source of inorganic salts for photosynthesis and growth. Salts may be available during the winter, but the short dark days provide too little light for rapid photosynthesis and multiplication. In spring, the day length increases rapidly and a combination of factors may increase the supply of nutrient salts – snow melt and melting ice releasing salts from the land, the churning up of waters by storms bringing salts from deep waters. The result is a dramatic outburst of plant growth especially in the phytoplankton of the lakes and seas, and many animals breed at this time. Often larvae are specialized in their feeding habits and they appear in the plankton shortly after the increase in numbers of their food organism, dying down again as the food disappears, to be replaced by other larvae feeding on other plants. A succession of this sort involving several species of diatoms and several species of copepods has been detected in the North Sea and is certainly not an isolated example.

Many animals show synchronized spawning even though food is available for much of the year, so that it is most unlikely that the presence of a specific food organism is the sole reason for co-ordinated breeding, important though that may be. It certainly cannot be the explanation of periodic synchronized spawning such as occurs in many marine animals during a prolonged breeding season of several months. The intervals between bouts of spawning are often two weeks or a month and are related to the tidal cycles or the phases of the moon. In one species of polychaete, *Platynereis dumerilii*, which shows a well marked lunar periodicity of breeding in the Mediterranean, we know something of the synchronizing mechanisms. The worms are sensitive to light intensities that prevail at full moon, but not to those at or near new moon. The worm's 'day' is thus some hours longer around the time of full moon than in other parts of the month. This increased period of illumination sets in train maturation processes which take about two weeks to complete and the worms engage in massed spawning at about new moon. Professor Hauenschild, working at Naples, has been able to induce swarming in the laboratory at any time in the natural month by suitable manipulation of the daily period of illumination. We have not got such a precise analysis of the controlling mechanisms of other marine animals that show a lunar periodicity, but it is quite possible that they use periodic fluctuations in the pattern of environmental stimulation as an alarm clock in much the same way as *Platynereis*.

Synchronized spawning appears to be advantageous for its own sake. The chief benefit is the improved chance of successful fertilization of the eggs it brings. If only one or two members of a population spawn at the same time the likelihood of spermatozoa encountering eggs is very much reduced. It is this factor, as much as anything, which accounts for the very precise co-ordination of breeding in some tropical animals like the palolo worm, that live in an environment in which food resources do not show such striking seasonal variation as they do in temperate waters.

Internal control of reproductive processes

Reproduction involves very many changes in an animal. There may be a growth of new structures that are used only when the animal breeds, and there is always the production of eggs and spermatozoa which entails a fundamental change in the metabolism. These developments must proceed in harmony with one another and are generally co-ordinated with environmental events which mark the onset of the breeding season. It is safe to predict that the control and co-ordination of these changes in the animal are through the agency of hormones, although except for insects and the higher vertebrates, the only animals in which these control mechanisms are understood in any detail are polychaetes. The polychaete control system appears to be a particularly simple one; other invertebrates may prove to be more complicated, but there is fragmentary evidence that in some, at least, very similar processes take place.

In the nereid polychaetes a hormone is secreted by the supra-oesophageal ganglion, or brain, throughout the immature life of the worm and prevents it undergoing the considerable changes associated with sexual maturation. As soon as the hormone level begins to fall the worm starts to mature; yolk is deposited in the developing oocytes, spermatozoa mature and become mobile, and in those species that form a special reproductive form known as the heteronereis, the body-wall muscles begin to break down, more powerful parapodial muscles are formed, the parapodia enlarge, and swimming chaetae are secreted to replace the old ones which are shed. By the time these processes are nearing completion the brain has ceased to secrete the hormone altogether. By removing the brain (and so the source of the hormone) from young worms precocious maturation can be initiated at any time, and the implantation of the brains of young worms into old worms postpones sexual maturation indefinitely by maintaining a high level of hormone in the body of the older animal.

'Juvenile' hormone is probably produced in neurosecretory cells in the central nervous system. These are cells with many of the characteristics of ordinary nerve cells but are specialized for the production of hormones. The hormone is synthesized mainly in the cell body and, like other secretions, appears in the form of small spherical packets which can be seen in the electron microscope. These secretory droplets travel along the axon of the neurosecretory cell and accumulate at the end of the axon. The secretion is then released into the circulatory system. With the aid of appropriate staining methods it is possible to reveal accumulations of secretion in the brain and other parts of the nervous system, that are visible in the light microscope. Neurosecretory cells have been discovered in the central nervous system of almost all metazoan animals and it is likely that in all of them these cells are the sources of hormones although we do not yet always know what the functions of the hormones may be.

In the nereid worms there is a seasonal fluctuation in the quantity of stainable secretion in the brain and presumably this reflects the amount of juvenile hormone synthesized and released into circulation at various time of the year. Similar seasonal cycles have been observed in the neurosecretory systems of other invertebrates, including sipunculids, earthworms and leeches, gastropods and spiders, and it is tempting to suppose that in these also sexual development is controlled by a juvenile hormone. This may well be so, but it is safest to defer judgement on this point until the supposition has been confirmed experimentally, for we know of entirely different control systems which produce very much the same result as a juvenile hormone.

In octopuses the onset of maturation is determined by the stimulatory effect of a hormone secreted by the optic glands, small spherical bodies lying on the stalks of the optic lobes, and the activity of these glands is regulated by an inhibitory nerve from the posterior part of the supraoesophageal mass of the central nervous system. If this nerve is severed, the optic glands become active and the animals become precociously mature. In malacostracan crustaceans there is a still more complicated system and gonadial development is controlled by two hormones, an inhibitory hormone secreted by the X-organ in the eyestalk and a stimulating hormone secreted by the Y-organ in the antennary or maxillary segment. The former of these two endocrine organs is neurosecretory, but the Y-organ, like the octopus optic gland, is not.

Even in polychaetes, sexual maturation is not invariably controlled by a juvenile hormone. In the lugworm, *Arenicola*, there appears to be only a spawning hormone. In a number of polychaetes, the eggs are shed and are fertilizable before they are fully mature, but in *Arenicola* this is not so. The eggs cannot even be spawned until all their maturation processes have been completed and the final stage of maturation only takes place after the release of the spawning hormone from the brain. Within a short time of the release of the hormone, the worm spawns. In male worms, the spermatozoa become active under the influence of the hormone and, like the eggs, are spawned almost immediately. Possibly a similar system is effective in lamellibranch molluscs. In the common mussel, *Mytilus edulis*, and the scallop, *Chlamys varia*, the removal of the cerebral ganglia, which contain numerous neurosecretory cells, is followed by discharge of the eggs and spermatozoa although, as in *Arenicola*, no hormone is known to control the earlier stages of sexual development.

Despite the important variations in the manner in which maturation is controlled in different animals, two features are common to them all: sexual maturation is delayed and does not begin until it is allowed or stimulated to do so, and ultimate control rests with the central nervous system either by direct nervous control of the endocrine glands, or what amounts to the same thing, by the hormones being secreted by neurosecretory cells in the central nervous system or extensions of it.

A delay in sexual maturation is universal. At first sight it is strange that one organ system – the reproductive system – should remain inactive for most of the life of the animal when all the other systems become active as soon as embryological development is completed, and sometimes even before that. The most likely explanation of this is that in most populations, animals live to the limit of the nutritional resources of their environment. Such food excess as there may be over the immediate requirements to sustain life are directed either to growth or to the development of eggs and spermatozoa. They cannot do both at the same time if they are to do either efficiently. There is thus an incompatibility between growing and breeding although special reproductive adaptations, such as the production of reproductive stolons, may override this in some animals. At least in polychaetes and oligochaetes the brain is a source of growth hormones which are necessary for both normal and regenerative growth. In nereid polychaetes it is even likely that the growth hormone and the juvenile hormone are identical. The disappearance of the hormone inhibits further growth and at the same time permits sexual maturation to proceed. Since the ultimate nature of hormone action is likely to be upon the metabolism of cells, the most probable function of hormones controlling growth and reproduction is to switch the metabolism of the animal from one controlling the laying down of new body tissue, to another which relates to the development of the gonads, eggs and spermatozoa.

It is also evident that the central nervous system is the ultimate controller of maturation. In almost all animals breeding is related in some way to events taking place outside the animal, to seasonal changes in the environment, to the local abundance of a food organism or, more immediately, to the presence of a mate. These external events are detected by the sensory system and information is conveyed to the central nervous system via the sensory nerves. When an appropriate pattern of stimulation is received, this information must be converted in some way into appropriate action in the body. Nervous impulses are an inappropriate way of controlling such long-term activities as growth, and, instead, control is exercised by chemical means, i.e. by hormones. We are beginning to understand something of the working of these control systems in a few animals. For example, in the octopus, cutting the optic gland nerve, the optic nerves or removing the sub-pedunculate dorsal lobe area of the brain, all cause precocious enlargement of the optic glands and sexual maturation. Evidently the most significant sensory input is to the eyes, the information is passed via the optic lobes and nerves to the sub-pedunculate dorsal lobe area of the brain which is the central control centre, and instructions are passed to the optic glands via the sub-pedunculate nerve.

In all invertebrates that have been examined so far, the hormones involved in such fundamental processes as growth and sexual maturation are secreted either by neurosecretory cells which form an integral part of the central nervous system or, at one stage removed, by endocrine glands which are under nervous control. In this way the brain acts as a direct link between the outside world and the reproductive processes of the animal.

Chapter 14

Larvae and larval behaviour

RELATIVELY few invertebrates develop directly from the egg. Rather than hatching as a small version of the adult, they leave the egg as a larva, often bearing little resemblance to the adult. The change to the adult may therefore be a profound one necessitating a complete metamorphosis into the final form of the animal. The larva is thus a stage in the development of the animal and is often free-living, hence it is not usually regarded as an embryonic stage, though some authorities consider that the larvae of insects can be considered as free-living embryos. Gurney has defined the larva of Crustacea as 'a free-swimming phase in the life-cycle of the individual which differs in form and habit from the adult and is commonly transformed into it by a sudden and radical change which constitutes metamorphosis'. This can be applied equally to most other invertebrate groups, though the larvae are not necessarily free-living, for in some species of marine invertebrates these stages are passed in the egg. It is difficult to explain the advantage of this and it seems to bear no relation either to the evolution of the group or to the habitat of the species. Of closely related species one may swim freely from an early stage and the other develop totally within the egg.

The marked difference from the adult often indicates that the larva lives a life quite apart from that of its parent. Thus the typical trochophore larva of sedentary or burrowing polychaetes is adapted to existence in the plankton feeding on the abundant minute plants of the nannoplankton. Indeed, a great many of invertebrate larvae pass this part of their lives in the plankton and their numbers contribute a very large part of the total biomass of these floating organisms.

The fact that larvae live in different places and in different ways is adduced as evidence for a belief that the adaptive function of free-living larvae is to avoid competition for food between adults and young. It is plainly true that a caterpillar feeding on a leaf is not competing with the nectar-feeding butterfly, or that the zoea larva of a crab in the plankton is not depriving the adult scavenging crab of food. Indeed, by occupying different niches in the environment, adult and young are not brought into competition for space. This would seem a stronger argument if we could show that natural populations of many of these creatures are really taxing the food resources of their environment. However, since they are free-living and active there is no doubt that they do not require food reserves such as occur in a heavily yolked egg in which the embryo passes through a long series of complex changes. During this period development proceeds by the utilization of the stored food. The larva can obtain its own. The eggs of invertebrates which pass through a larval stage are therefore generally relatively small, and thus can be, and are, produced in very large numbers without undue strain on the metabolism of the parent.

In temperate and tropical seas, it has been estimated that some 10 per cent of the species with pelagic larvae have larvae with sufficient yolk to supply their needs, even when they spend a long period in the plankton. These lecithotrophic larvae contrast with the planktotrophic larvae, dependent mainly on obtaining their own food. Larvae may swim actively for 2–4 weeks or longer in the summer and even for as long as three months in the winter. Such larvae occur in 55–65 per cent of the species in northern seas and are even more common in tropical waters where they occur in 80–85 per cent of the life histories. Yet other planktotrophic larvae spend only a few hours or days in the plankton. Such young stages occur in about 5 per cent of the life histories of species in all areas, even in the arctic, where free-living larval forms are relatively rare.

Some general impression of the wastage in the planktonic stages of larval life can be gained by comparing the reproductive capacities of a number of invertebrates. *Solaster endeca* is a starfish whose larvae are entirely lecithotrophic and therefore are not likely to be controlled by the availability of food in the plankton. Yet this animal is estimated to produce some 10,000–20,000 eggs in the main spawning period which suggests that losses to various predators and by other factors than food supply are great. But *Spirorbis borealis*, a serpulid polychaete whose larvae spend at most a few hours in the plankton, produces no more than a few hundred eggs each year. The short time spent as a floating larva reduces the risk and hence the losses.

There is no doubt of a further suggested biological advantage of larval stages, that is, that they act as dispersal stages, carrying the young away from the place where the parents live. This would seem particularly advantageous to the many invertebrates which are either sedentary or move about very little. Yet the distribution of free-living stages among different species does not always fit such a generalization. The three common nemertean worms to be found on the shore might be expected to have similar life histories. However, *Amphiporus lactifloreus* has young which leave the egg in the crawling stage, while the young of *Lineus ruber* has a short free-swimming stage and those of *Cephalothrix* spend two or three weeks in the plankton. Nevertheless, in general, this is true.

The daily vertical movements of the larval forms in the plankton may well result in wider dispersion of these stages (p. 255). The sea is not a homogeneous mass of moving water, but currents at different depths travel in different directions and at different speeds, often changing over a remarkably small depth range. Larvae which are at one time of the day in a current carrying them in one direction move up into another as they migrate during the night. Thus they may be scattered widely.

But the time has to come when the dispersed larvae must change to the adult, settling down in some place where conditions are suitable for adult life. This is not a random process, but one in which the larvae of a number of different invertebrates have been shown to be selective, choosing the substratum on which they will settle (p. 322). This will no doubt lower the mortality rate by reducing the number which die through settling on an unsuitable place, but, as has been pointed out above, there is a mortality rate which is far greater at other periods in the larva's life. In comparison with a viviparous animal, or one in which the eggs are protected by a hard covering or something similar, the losses must be enormous.

Larvae may evolve in directions quite different from those of their parents, occupying as they often do quite separate niches from those occupied by the adults, the characters favoured by natural selection will similarly be different. This has been seen as providing a further source of variation to be utilized in the evolution of the species particularly by the process of neoteny by which larvae, which almost by definition are sexually immature, become capable of reproduction. Possibly a potent source of change in the evolution of the vertebrates, this may have played only a small part in the evolution of the invertebrate phyla.

The specialization of the larvae for their special ways of life has led to some bizarre forms, but before dealing with some examples of them, it would be better to consider the main types of larvae.

Larval form

The simplest larva found among the metazoan invertebrates is the coelenterate planula. The fertilized egg divides so that at each division equal-sized cells are cut off. The ball thus formed enlarges until the cells push apart to form a hollow blastula with a single layer of cells, the ectoderm, bounding the cavity in which are a mass of cells, the endoderm. The ectodermal cells are ciliated and thus the planula can swim or float. These larvae may remain thus for a considerable period before sinking to the bottom and attaching by one end. This forms the base for the polyp, the mouth and tentacles appear at the free end of the larva. The status of the polyp depends upon the species; in the Scyphozoa (e.g. *Aurelia aurita*), it is responsible for forming the ephyra larvae by transverse fission (strobilation). These larvae are eight-armed immature medusae (Fig. 14.1) which become adult by further growth and change, thus the spaces between the arms become filled in and the system of canals becomes fully developed throughout a much-thickened mesogloea layer. Other coelenterates have large yolky eggs which form actinula larvae (Fig. 14.2);

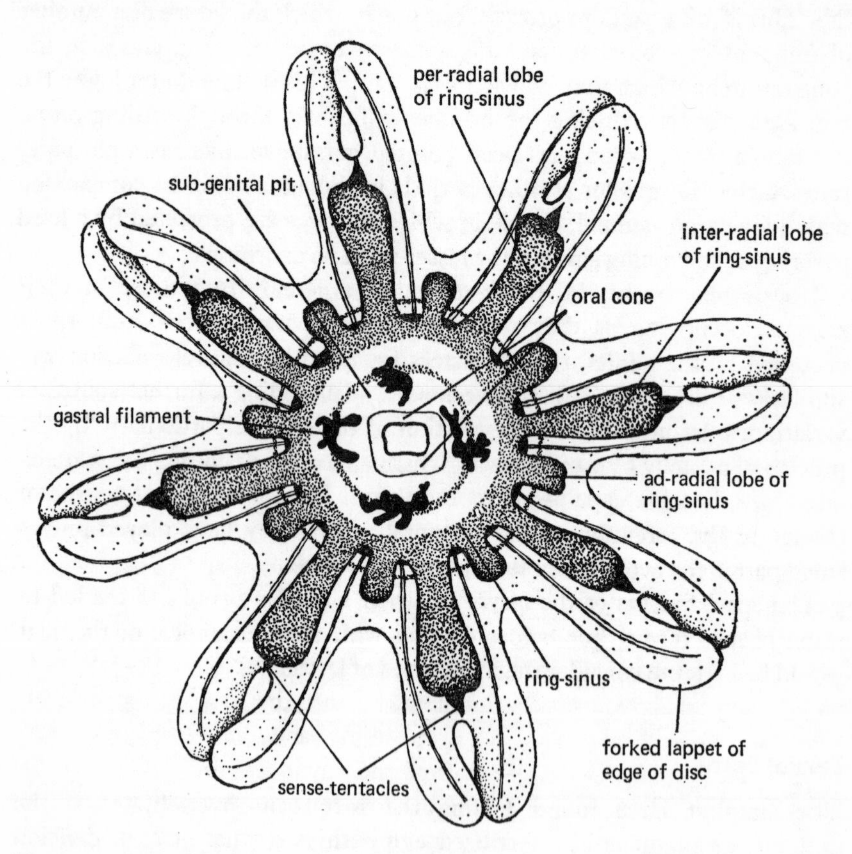

Fig. 14.1 An ephyra larva of *Aurelia aurita*

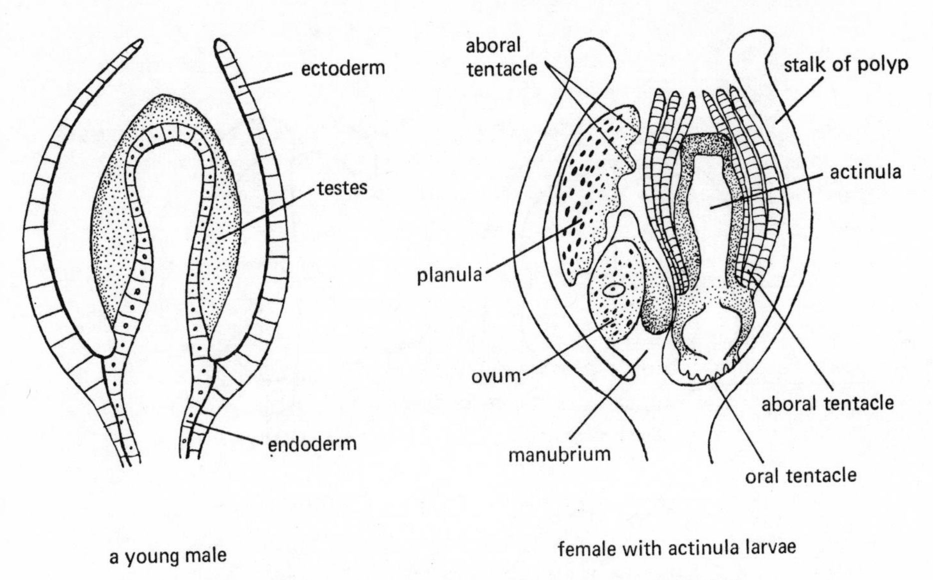

Fig. 14.2 Longitudinal sections through the gonophores of *Tubularia*

these are more developed than the planula for they already have arms and a manubrium. Each can develop directly into an ephyra; the planula stage is passed through in the genital pouch.

The remainder of the metazoan invertebrates can conveniently be grouped according to the way in which their eggs divide. Thus, the Annelida, Arthropoda, Mollusca, Platyhelminthes and Nematoda have spiral cleavage, while the Echinodermata, Pogonophora, Hemichordata, Cephalochordata and Urochordata have radial cleavage (Fig. 14.3). The mitotic spindles in a spirally cleaving egg are orientated vertically but at an angle to the polar axis in the third and subsequent divisions; what is more, the inclination from the vertical is alternately to one side of this axis or to the other (Fig. 14.3). On the other hand, radial cleavage produces a simple pattern of meridional and latitudinal furrows, the cells lying directly above each other (Fig. 14.3).

These differences in cleavage pattern are correlated with differences in larval form and mode of development, so that the first group, showing spiral cleavage, is known as the Protostomia, and the second group the Deuterostomia. The names derive from the way in which the mouth of the larva is formed. In protostomes the blastopore becomes the mouth, but in deuterostomes (*deuteros*, second) the mouth is a new penetration, the blastopore becoming the anus.

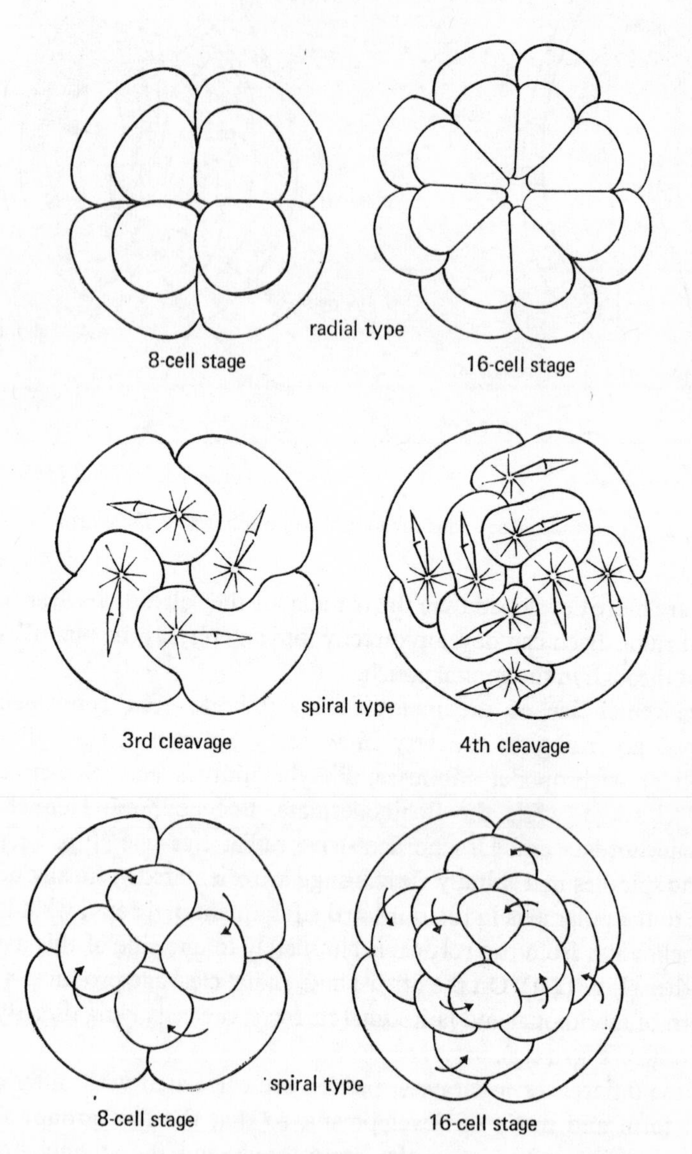

Fig. 14.3 Comparison of radial and spiral cleavage. 8 and 16 cell stages, spiral type; 3rd and 4th cleavages, spiral type; 8 and 16 cell stages, spiral type

The protostome larva

The typical form of these larvae is that of the trochophore. It is a multicellular larva shaped like two cones placed base to base. Around the equator of the larva is a band of cilia, the prototroch, which is the means of locomotion of these larvae. Other ciliary bands may circle the body between the prototroch and the hind end (Fig. 14.4). The mouth opens below the prototroch and the gut loops up inside the body forming a stomach before turning backwards to form the anus near the tip of the lower cone. The cavity in which the gut lies is a blastocoel; within it also lie a pair of excretory organs, the protonephridia, and a pair of bands of cells destined to form the mesoderm of the adult.

Despite its relatively simple appearance the larva has a nervous system of some complexity. A group of sensory cilia form an apical tuft arising from an apical plate of thickened ectoderm. A ganglion lies below this, connecting to nerve rings related to the ciliary bands and the prototroch in particular. Eye spots may also be found. Later we shall see that such development of the nervous system brings with it the possibility of complex settlement behaviour (p. 322).

Such a larva is unsegmented, but gives rise to an adult showing strikingly developed metameric segmentation. This is brought about by the elongation of the part of the larva between mouth and hind-end. As this occurs the mesodermal bands elongate, their cells proliferating. Blocks of this tissue appear, each of which acquires a coelomic cavity by the splitting apart of the cells. The segmentation is made more obvious by the appearance of parapodia and chaetae.

Larvae such as these are termed nectochaetes (Fig. 14.5). They still remain in the plankton, indeed their muscular parapodia aid in movement while the long slender chaetae are an adaptation to floating. Ultimately the apical organ and its nervous connection develops into the prostomium of the adult worm whose body has been formed by the continuing extension of the nectochaete body (Fig. 14.5). At least in *Polygordius* the larval mouth forms that of the adult. Immediately before settling on the bottom the prostomial rudiment and the rest of the body are separated by the remains of the trochophore which shrivels and drops away.

Molluscan larvae. The early stages of the development of the larvae of molluscs are remarkably similar to those of annelids and the trochophore which results bears strong resemblances to that of a worm (Fig. 14.4). However, as development proceeds, structures appear which are unique to molluscs. Dorsally a mantle appears between prototroch and anus; despite the youth of the animal a shell may be secreted by it (Fig. 14.6 and Plate

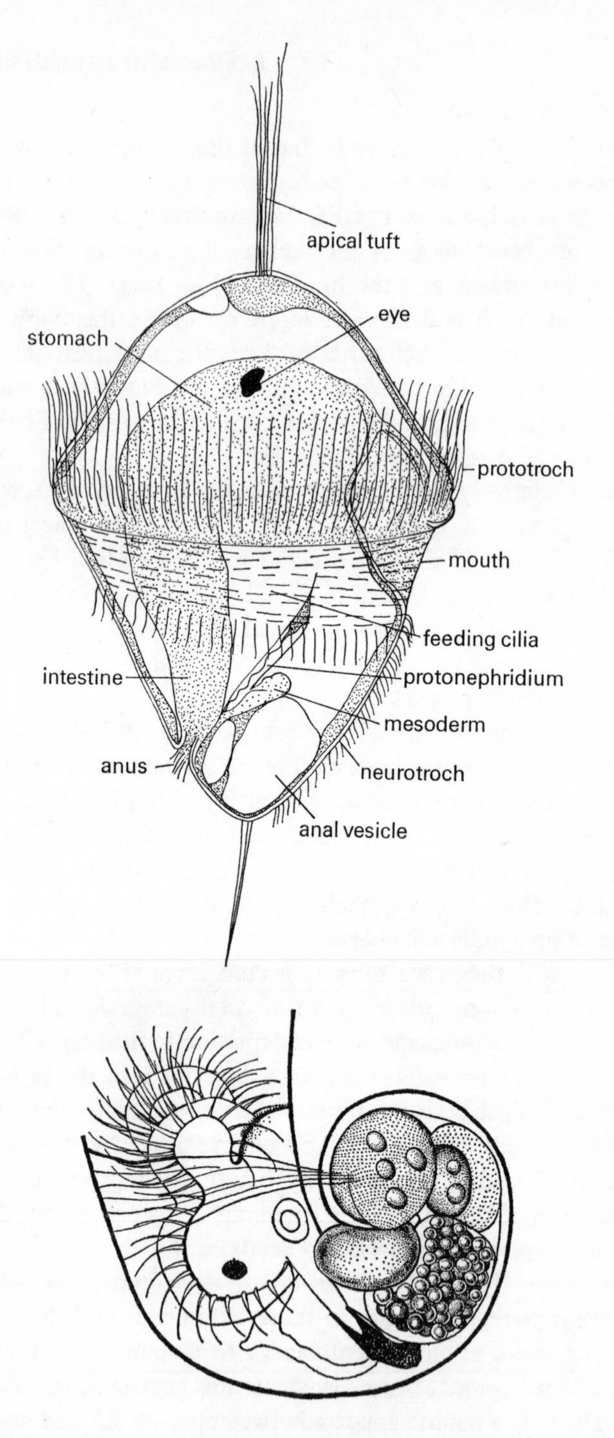

Fig. 14.4 Trochophore larvae of annelid and mollusc. (top) advanced trochophore of *Pomatoceros triqueter*; (bottom) larva of *Nassarius reticulata*

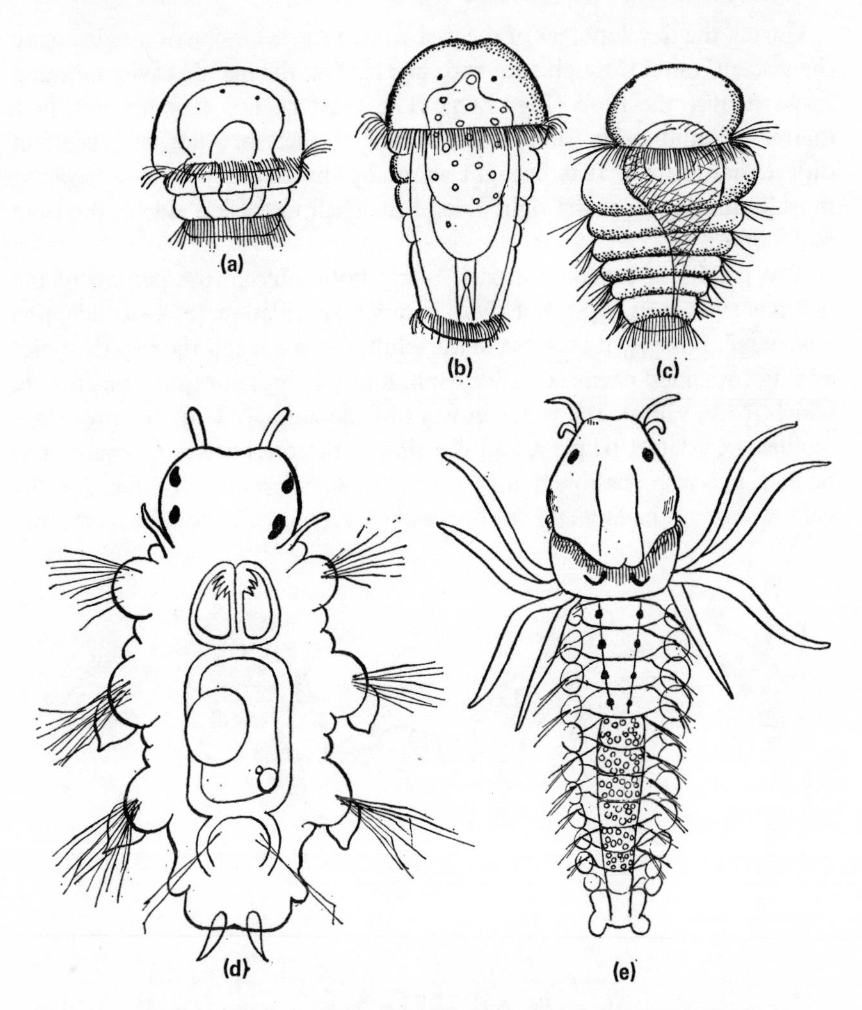

Fig. 14.5 Stages in polychaete life histories: (a) young trochophore (*Nephtys ciliata*); (b) and (c) metetrochophores, older than (a) (*Nephtys ciliata*); (d) nectochaete (*Nereis pelagica*); (e) old larva of *Phyllodoce maculata*

20). Ventrally a foot becomes evident, a projection formed by the fusion of two previous ones. Perhaps most striking of all is the considerable enlargement of the prototroch to form a velum. This ciliated organ serves for locomotion as well as feeding; its greater development may be correlated with the need to keep this larger and heavier larva afloat. Such a stage is found in the life histories of gastropods and bivalve molluscs in particular, and is called a veliger.

During the development of the gastropod veliger torsion occurs bringing the viscera round through 180° and thus leaving the mantle cavity pointing forward over the head (Fig. 14.7). The twist occurs very quickly, in a matter of minutes in some species, and is therefore not the result of differential growth. It is brought about by the contraction of a retractor muscle running from the right side of the shell to the left side of the head and foot.

The results of torsion are particularly noticeable in the pattern of the nervous system (p. 248), but there is much speculation about its adaptive advantage. Possibly it is of use to the adult, for as a result the mantle cavity and its contained chemosensitive osphradium is brought into a position in which it can sample the water in front of the animal. Thus the predatory mollusc can detect its prey, and warning of the presence of an enemy can be had. It is also possible that this is a specialization for larval life. For the velum, vital to the larva for feeding and movement, can be withdrawn into

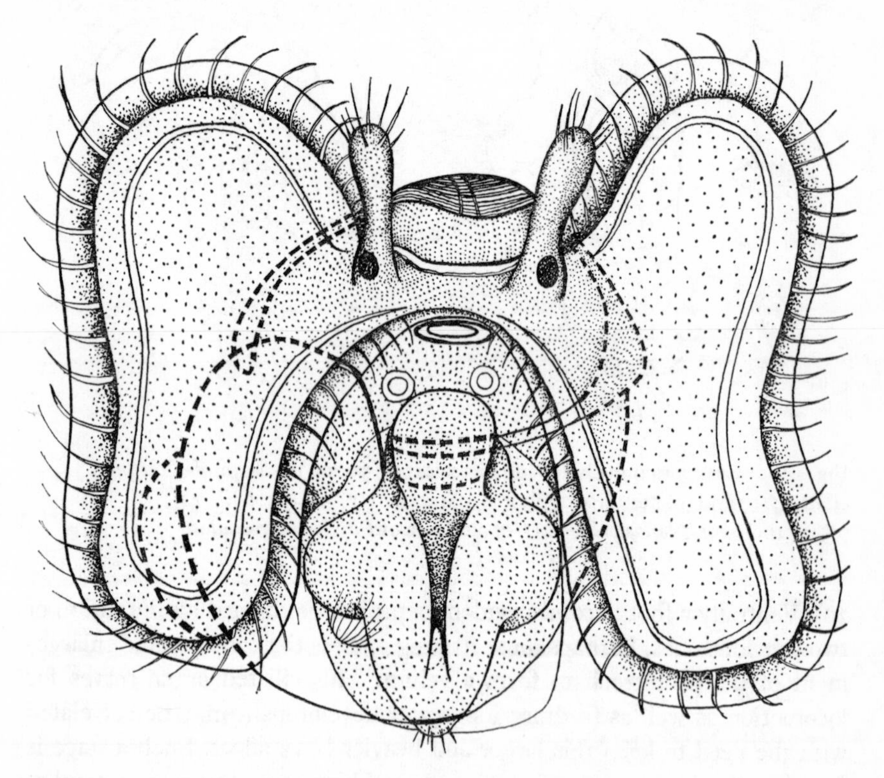

Fig. 14.6 Older larva of *Nassarius reticulata* with shell forming (compare with Fig. 14. 4)

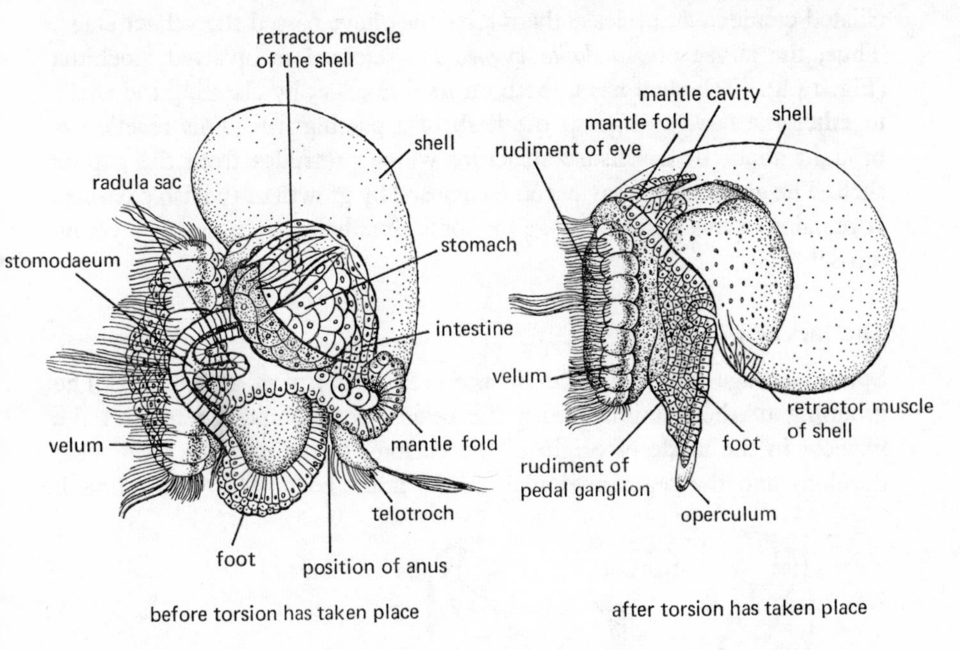

Fig. 14.7 The veliger larva of *Patella coerulea* before (left) and after (right) torsion

the mantle cavity, thus protecting it in a way which was impossible so long as the mantle cavity opened to the posterior. It has been claimed that a predator cannot therefore bite off the velum, though plainly this adaptation has little value as a defence against filter-feeding fish like herrings who will swallow the larvae whole. These are not the only theories which have been produced to account for this remarkable event in the veliger's life; it is one of those corners of biology where speculation is alone possible. We cannot experiment to prove a use nor can we observe the evolutionary process by which it came about.

A point of interest in considering the ancestry of the molluscs is that though mesodermal bands are formed in much the same way as in annelids, they do not become divided off into segments, but the cells wander off to form muscles and so forth. The complete absence of segmentation even at this early stage gives strong evidence for a molluscan origin from a non-segmented ancestor.

Not all molluscs are oviparous, some, like *Littorina rudis*, are viviparous. Fresh-water animals tend not to have free-living larvae, connected no doubt with the possibility of the young forms being swept away by the current. The family of bivalves, Unionidae, incubate their eggs within the ciliated ctenidia, and release them after they have passed the veliger stage. Thus, the larvae of *Anodonta cygnea* are released as bivalved glochidia (Fig. 14.8). Each shell has a tooth on its free edge; by clapping the shells together the larva can grasp the flesh of a passing fish. This reaction is brought about by a sensory structure which protrudes from the gaping shell. The glochidium may become enclosed by growth of the fish's tissues. It remains in this parasitic stage for some time before escaping as a young mussel.

The larvae of crustacea

Spiral cleavage does not occur in the great majority of arthropods. The grouping of these animals with the other Protostomia depends on the likeness in the mode of origin of the coelom, of the way the mesoderm develops and the segmented plan of the adult body. Moreover, though

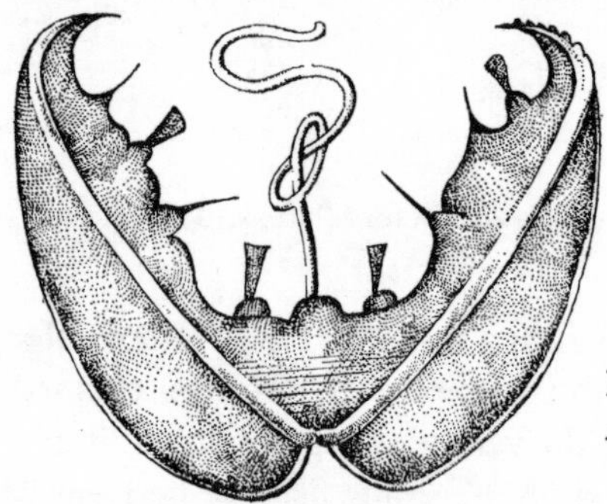

Fig. 14.8 The glochidium larva of *Anodonta cygnea*

there is a great range of forms of larvae among these animals, and in particular among the Crustacea, none have anything which remotely resembles a trochophore.

Crustacean larvae are highly modified for their way of life, usually in the plankton. Throughout larval life, they add segments to the body, gaining appendages at each moult. Typically they hatch as nauplius larvae with a basic set of three pairs of appendages, antennules, antennae and mandibles (Fig. 14.9). These are tiny creatures with rounded unsegmented bodies which feed in the plankton on the smallest of the algae, grinding them by means of gnathobases on the antennae and mandibles. The nauplii of various groups are very similar though those of Cirripedia can be distinguished by their frontal horns (Fig. 14.9b). It is on the evidence of the possession of larvae of this sort that the parasitic cirripede, *Sacculina*, can be placed with its relations, though the adult has none of the features even of a crustacean. In many species the nauplius stage is passed in the egg, so that the animal hatches with a more complete set of appendages.

The subsequent stages in the life history differ widely from group to group. There may be gradual development into the adult by the addition of further segments and appendages. Thus the metanauplius larva which is the first free larva in many groups has a longer body (with segments) but

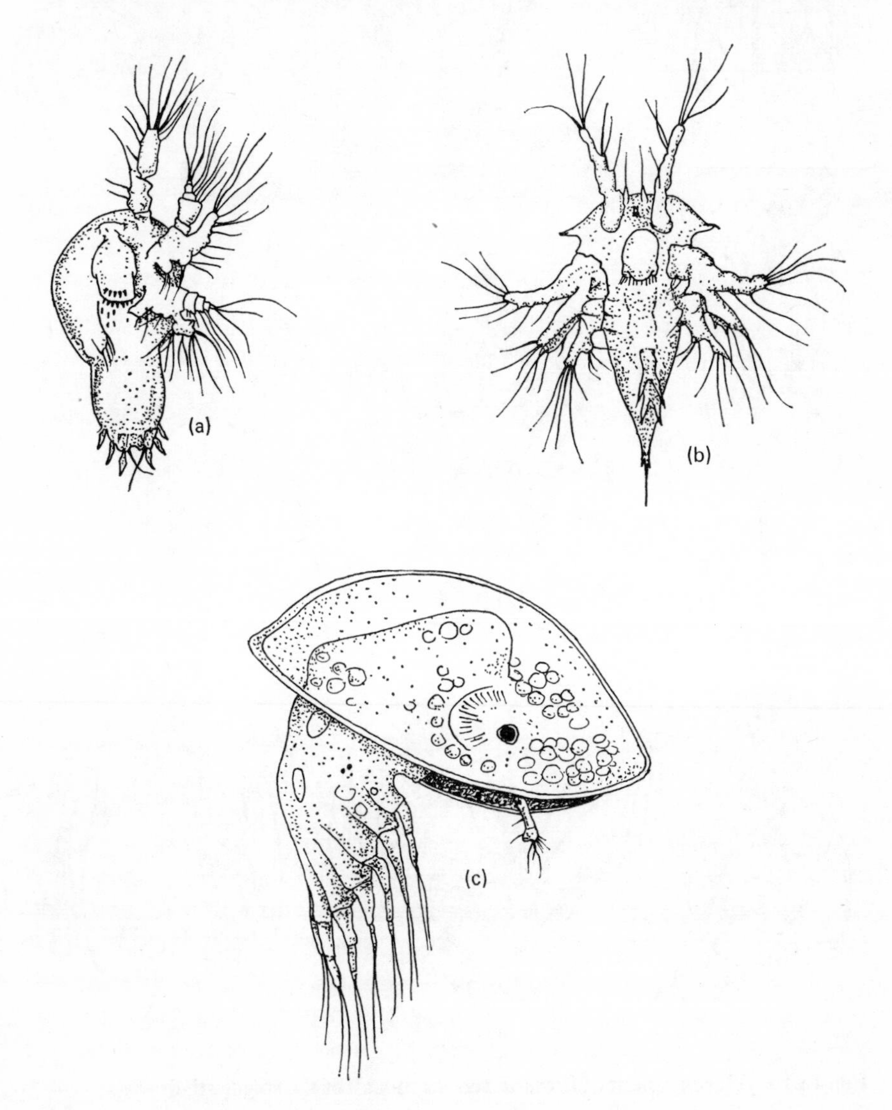

Fig. 14.9 Larvae of crustacea: (a) fourth nauplius of *Calanus*, (b) second nauplius of *Balanus*, and (c) cypris of *Balanus*

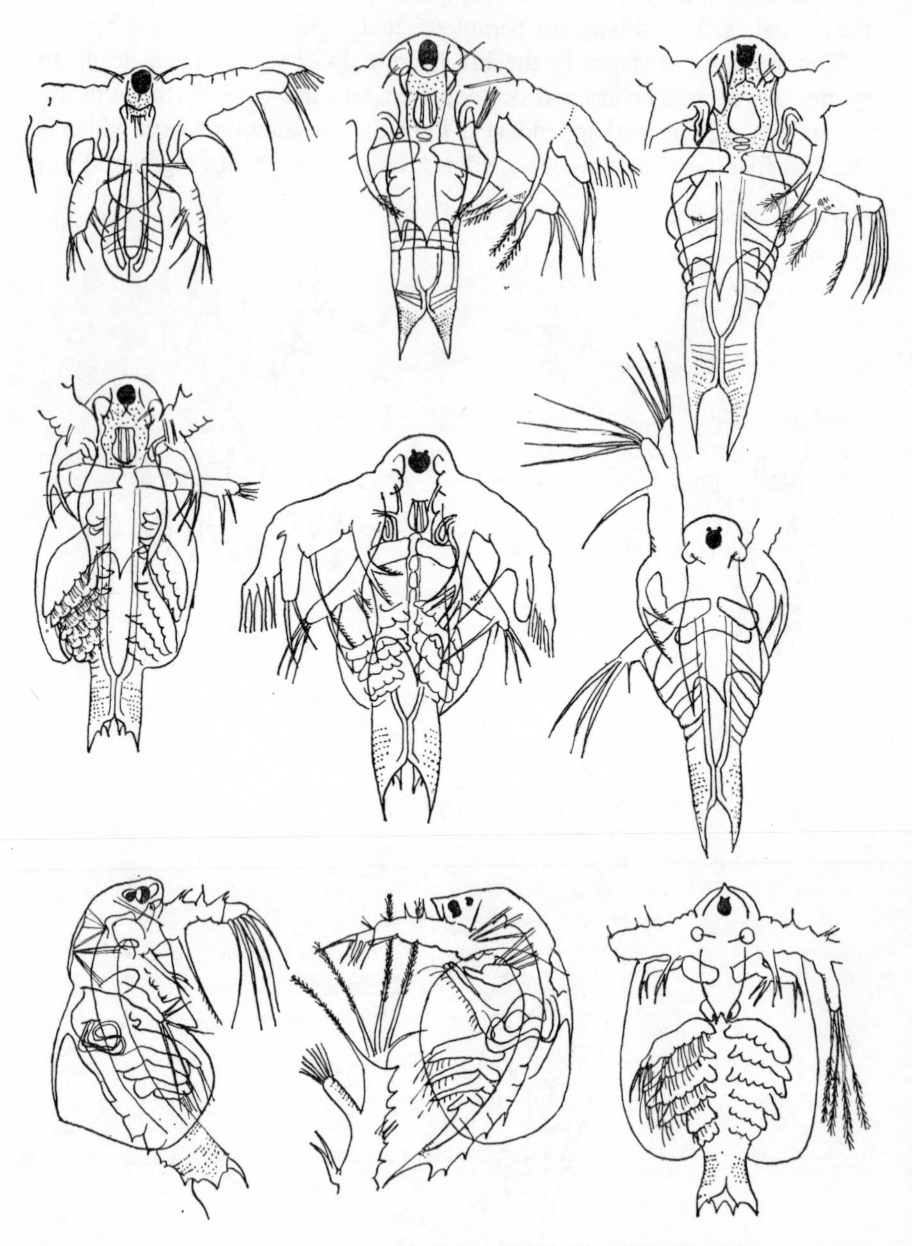

Fig. 14.10 Development of *Estheria syriaca* (not all of the stages are drawn)

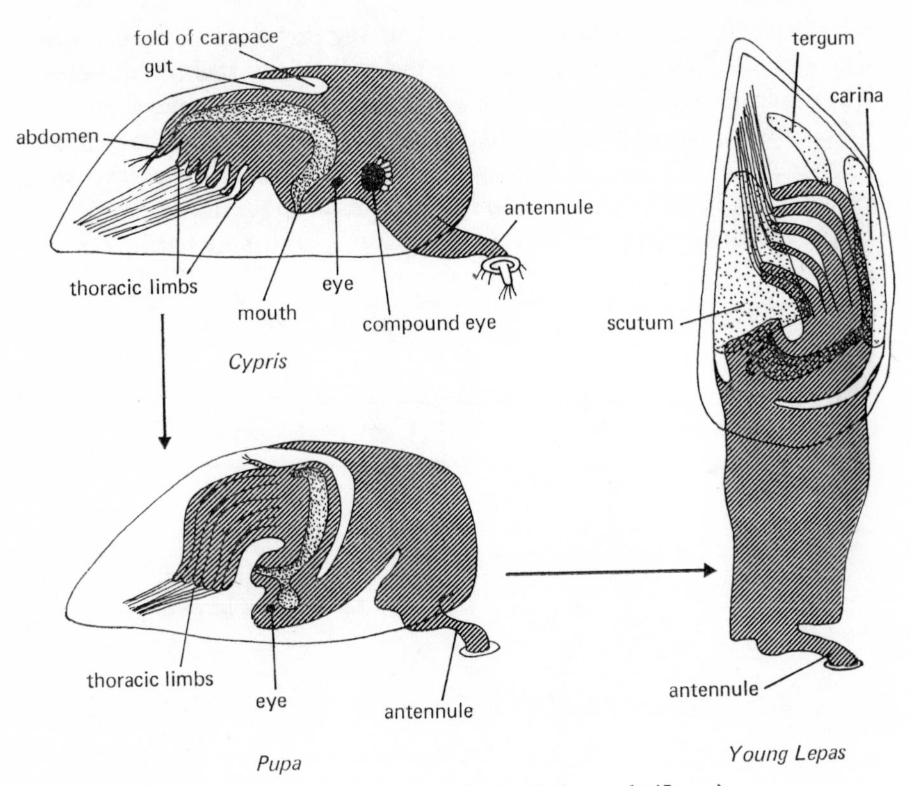

Fig. 14.11 Three stages in the metamorphosis of a barnacle (*Lepas*)

no more appendages than a nauplius (Fig. 14.10). Such gradual development often occurs where adult and larva share the same habitat. A relatively simple life history is illustrated by the development of *Calanus*, a copepod which forms a major part of the diet of herring. Its nauplius may moult five times retaining essentially the same appearance, but on moulting for a sixth time, it changes into a copepodid larva, similar to the adult but lacking full development of the limbs. Limb development follows during five subsequent moults after which the mature adult appears.

Among the Cirripedia, the nauplius moults into a cypris larva (Fig. 14.9c) whose divided carapace extending along both sides of the body gives it a bivalved appearance (it derives its name from its superficial resemblance to the adult *Cypris*, an ostracod). It is this active larva which selects a site for settlement (p. 324) moving over surfaces of rocks until a suitable place is found. Its antennules project outside the valves of the carapace. Cement glands in these appendages are used by the larva for attachment. This initiates changes in the body by which the thoracic appendages become reoriented (Fig. 14.11). Great elongation of the part of the body between antennules and mouth occurs to form the stalk of the stalked barnacles. The whole larva stands upon its head, as it were, when it becomes adult.

It is among the Malacostraca that the greatest range of larval body form is encountered. Development may be direct in some, in Peracarida for example, but in many a series of larval stages make up the life history. Few, except the most primitive genera, have nauplii, and most hatch as zoea

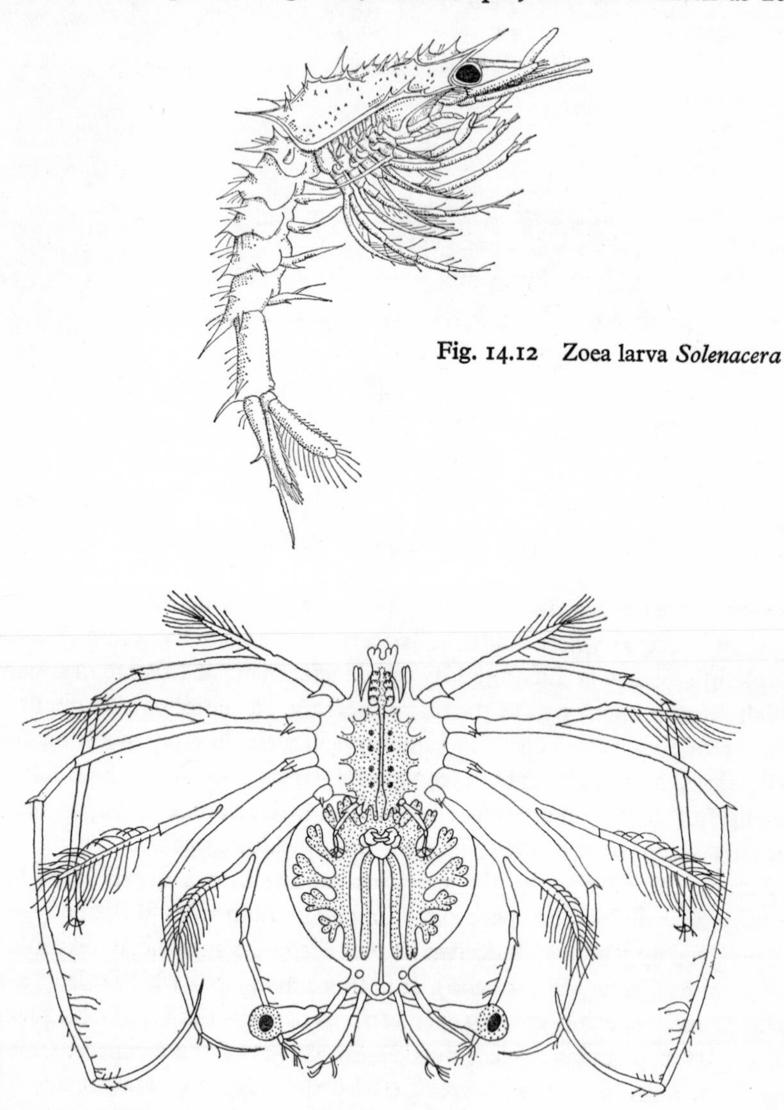

Fig. 14.12 Zoea larva *Solenacera*

Fig. 14.13 Phyllosoma larva of *Palinurus vulgaris*

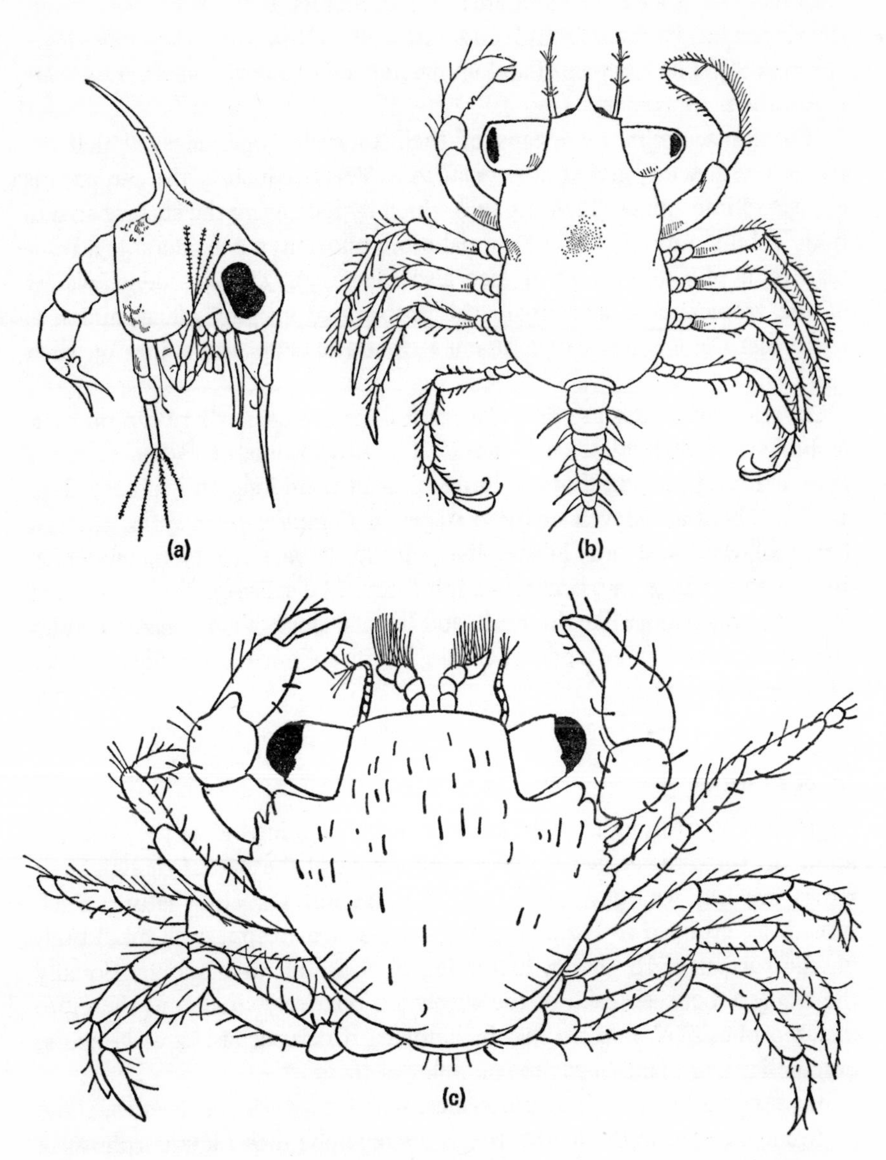

Fig. 14.14 Larval stages of the shore crab, *Carcinus maenas*: (a) zoea, just after hatching; (b) megalopa; (c) young crab

(Fig. 14.12) larvae, even the metanauplius stage being passed in the egg. A zoea has a well-developed abdomen, but at first its thorax is not completely developed at its hinder end and has only a few thoracic appendages (protozoea). The last pair of abdominal appendages have, however, already made their appearance.

The larvae swim by means of their thoracic appendages which are fringed with setae, giving a large surface. Very frequently the carapace is elongated into spines (Plate 14.14a) which by increasing the surface area to body volume ratio makes the animal more buoyant, a considerable advantage to a planktonic animal (see also Plate 19). The less dense waters of the tropical seas accentuate the problem of keeping afloat, and it is there that the greatest development is found of setae and spines on such animals.

A variety of names are given to the following stages where the number of limbs in increased at each moult. The phyllosoma of *Palinurus* has a flattened body and six thoracic limbs, four of them long and spidery (Fig. 14.13). There are eleven of these stages in *Palinurus interruptus*, the last being followed by a more lobster-like peurulus larva, having long antennae. By gradual change this becomes adult form. The schizopod of a common lobster is much more like the adult, and the megalopa of *Carcinus* resembles the parent, even having a pair of chelae, but its abdomen is not flexed under the thorax (Fig. 14.14).

Insect larvae

All insects pass through a series of stages after hatching before they become adult. In some the stages are very reminiscent of the parent though their wings are less well-developed and they are not sexually mature. Such insects are grouped as Exopterygota, for their wings appear as small buds on the surface of the body, increasing in size after each moult. Equally the changes between each of the stages are relatively small and therefore metamorphosis is said to be incomplete. Cockroaches, grasshoppers, dragonflies and plant bugs are examples of these.

But many others, such as butterflies, true flies, bees and beetles, show profound changes in the moult from larva to adult; their metamorphosis is complete. Indeed such wide-ranging changes in body organization and structure come about that a stage, the pupa, is interpolated. This is a resting stage when the insect no longer feeds nor moves far as its tissues become remoulded into the adult form. The wings of these insects develop as pockets thrust into the body and thus these insects are Endopterygota.

Unlike crustaceans, insects are hatched with the complete number of segments (except Apterygota which add some during their life history). Furthermore once they are adult, no further moulting takes place, while adult crabs, for example, continue to cast their skins as they grow.

Though some people prefer to retain the name nymph, for the young stages of exopterygote insects, there is much to be said for calling them larvae. They are immature and though differing only in degree they are structurally different from an adult. Even in this group specialization for larval life occurs (Plate 11). A striking example is the dragonfly larva. This insect spends some months or years in water and is adapted for aquatic respiration. Its mouth-parts are completely different from those of the adult, for its lower lip (labium) is developed into a 'mask' which can be shot out to capture prey (Plate 10).

But the adaptations and variations of larval form are even greater among the Endopterygota. The familiar caterpillars of butterflies, saw-flies and scorpion-flies are typical examples of the polypod larva. They are fully segmented and have six legs, with the addition of pairs of abdominal prolegs all of which take part in movement. They are voracious plant feeders.

More active larvae are found among the oligopod stages in the life history of beetles. They run about actively and feed by using their mandibles to chew plant or animal food. But the most sluggish are the apodous larvae, legless and grub-like. The larvae of honey-bees and of house-flies are of this kind.

All of these forms may have specializations for living in water, or some other means of life. Indeed the specialization of the larval forms to occupy niches different from those of the adult is carried far among the insects.

The changes which occur in the pupa led Berlese to interpret the larval stages of insects with complete metamorphosis as much drawn-out embryonic stages which otherwise would be passed in the egg, and are passed through there by exopterygote insects. On this theory the 'nymphal' stages of the Exopterygota are represented in a highly compressed form by the pupa of the Endopterygota. However, Hinton has argued persuasively that nymph and larva are equivalent and that the pupa is the first of the adult stages which provides a cuticular 'mould' for the proper development of the adult musculature. The adult which emerges is thus the second adult stage.

The deuterostome larva

Though variations occur in the various subphyla of the echinoderms, the larvae of echinoids will serve as a good example of this type of larva and its development. Though the earliest larva (dipleurula) bears little resemblance to the trochophore it possesses an apical tuft and plate. Its ciliation is at first uniform over the whole body, but it later becomes restricted to a clearly defined band which runs between mouth and anus in part of its course over the body. Though originally single, the band becomes highly elaborated and may become broken up into separate circlets of cilia, thus the crinoid doliolaria larva is barrel-shaped with four or five hoops of cilia.

The dipleurula has three pairs of coelomic sacs which bud off the enteron early in development (Fig. 14.15). Parts of these sacs go to form the water-vascular system, unique to the echinoderms, and essential for movement, as it is through that system that the tube-feet are enabled to function. Protonephridia are absent in the larva. The mouth, a new structure, opens into the concave, future ventral side; the anus is formed from the blastopore.

Growth of the ectoderm of the ciliated band is greater than the rest of that layer with the result that it becomes thrown into folds, the larval arms. In this way the typical pluteus larva of Echinoidea and Ophiuroidea comes about (Plate 13). The arms elongate; skeletal support of calcareous rods develops within them. Once again great increase of surface area to body volume is thus brought about. Part of the band forms a pre-oral section, the rest, and largest part, the post-oral (Fig. 14.16).

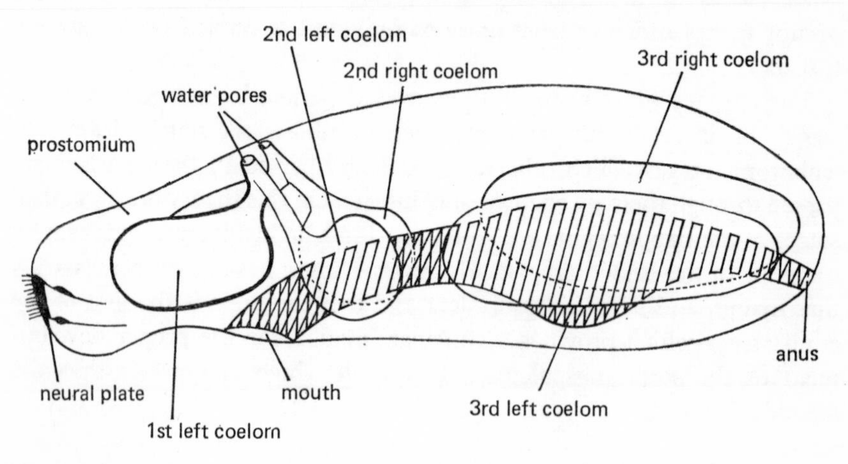

Fig. 14.15 Diagram of an idealised *dipleurula* larva

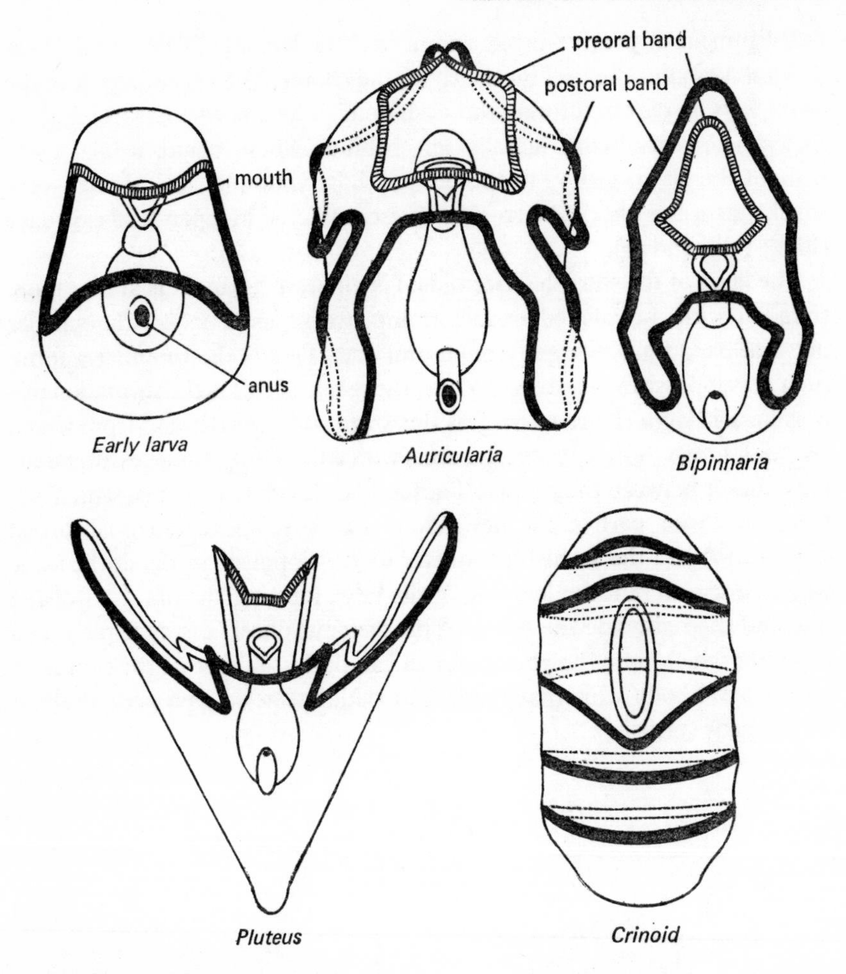

Fig. 14.16 Diagrams of echinoderm larvae

Four principal arms develop in an echinopluteus remaining rather more bunched than those of the ophiopluteus, but other pairs of arms are added, so that in the end the larva has four to six pairs (Plate 13.ii). This is more than in an ophiopluteus in which the pre-oral arms are lacking, but in addition the larva has a flatter appearance because of its wider spread arms.

The change to the adult is a rapid metamorphosis coming at the end of weeks or months in the plankton. The adult is so different from the larva that the changes are necessarily very great yet they may be completed within an hour. This can occur because before the event the basic pattern of the hydrocoel is established by its growth round the gut and the establishment of the five-rayed symmetry. The left side of the larva forms the oral region of the urchin, the right, the aboral. The pre-oral part of the larva plays no part in forming the adult body. The arms are absorbed, their skeletal supports being literally left behind. These changes take place without the larva settling to the bottom. The young urchin, though very small, has a few tube-feet so that it is capable of independent existence (Plate 13.iii and iv).

The larva of the starfish (Asteroidea) is different again. It is at first more rounded with the ciliated band forming two separate loops, the smaller pre-oral band, and a larger circum-oral one. This is the bipinnaria form. As it develops, stubby arms grow out; the result of this is the formation of a brachiolaria larva (Plate 12.ii). It is this stage which settles and from which the adult grows. Three brachiolar arms with adhesive tips appear surrounding a sucker between their bases. The larva settles on to this area, which will form the major part of the new adult. The anterior region is absorbed (Fig. 14.17) and the starfish completes its development in much the same way as a sea-urchin, the left side of the larva forming the oral part of the disc and the right side the aboral. This metamorphosis can take place in a day after which time the young animal can pull itself free of the remnants of the larval body which, as Hardy has said, has acted as a perambulator for the baby starfish.

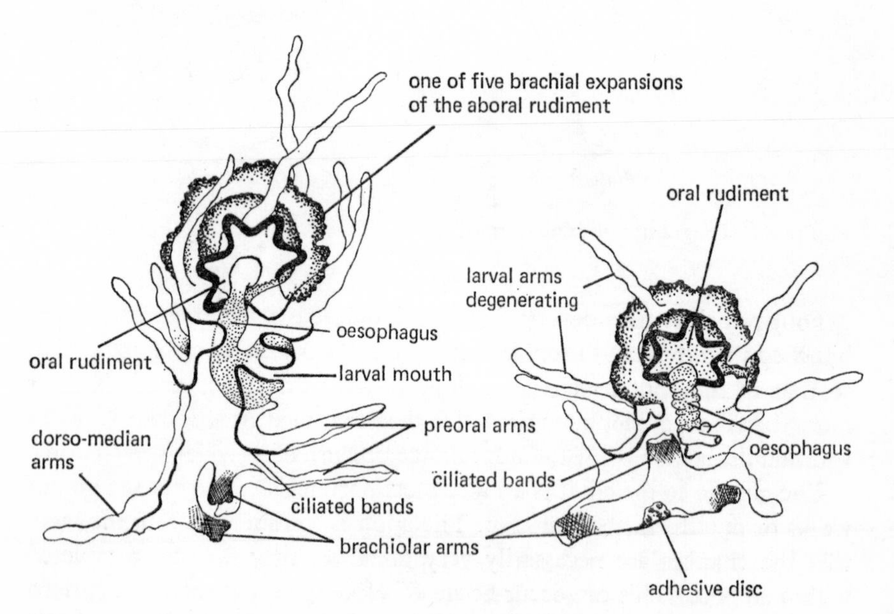

Fig. 14.17 Two stages in the metamorphosis of the brachiolaria of *Asterias*

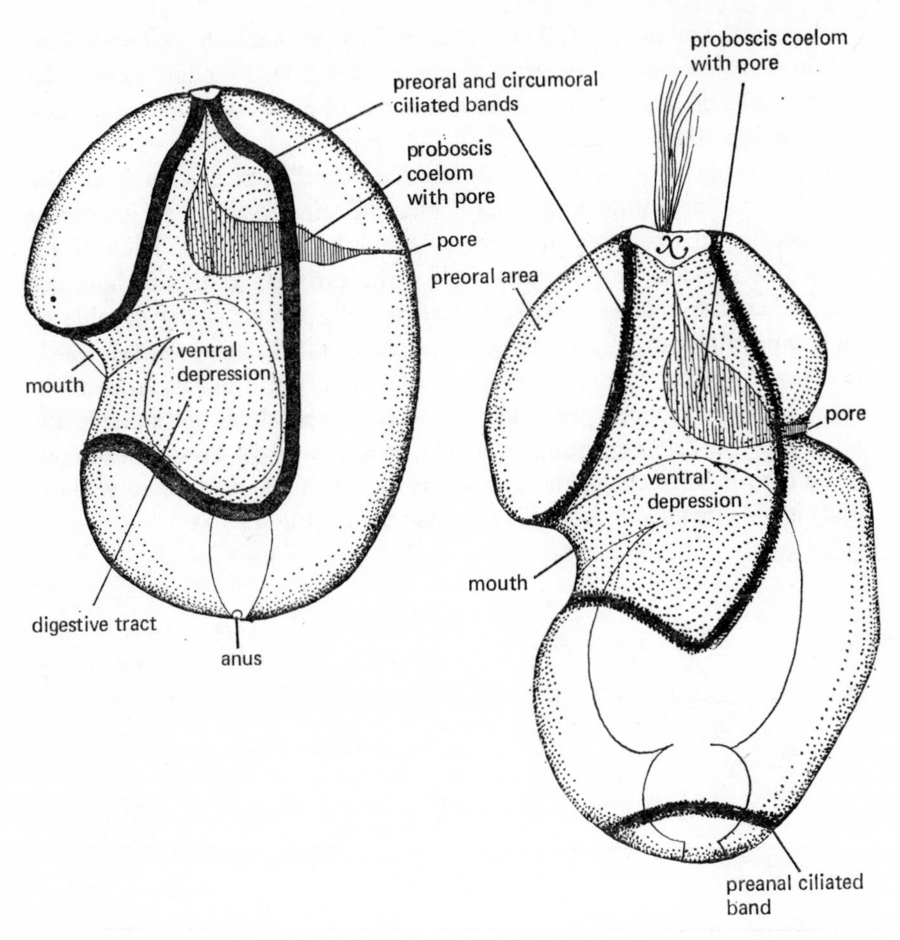

Fig. 14.18 Tornaria larva of a Hemichordate, two early stages in development

The life histories of a number of echinoderms show variations of these main themes often as a result of having more heavily yolked eggs, and less active larval stages than the starfish, brittle stars or sea-urchins. This often leads to a shortening of the larval life, and the resultant morphological changes may well obscure the relationships of the larval forms to others in the same group.

The typical larva of the Hemichordata bears considerable likeness to the pluteus. This tornaria larva (Fig. 14.18) has a ciliated band divided like that of the pluteus into pre- and circum-oral parts, but in addition a posterior telotroch encircles the anus. It contains a coelomic pouch, communicating to the exterior by a pore, which will form the proboscis coelom of the adult worm. There is no sudden metamorphosis as development proceeds but rather gradual change towards the adult form. The likeness of this larva to that of the echinoderms adds evidence to the close relationship of the two groups.

Among the Protochordata (Urochordata and Cephalochordata), the larval forms are quite unlike any other. Their larvae are distinctively chordate in structure, each having a notochord, a post-anal tail and gill slits. The ascidian tadpole is typical of the early stages of urochordates other than the pelagic species. It is extremely mobile being capable of swimming vigorously by means of its tail. At first it swims upwards towards the light but later reverses the sign of its response and moves downwards into cracks and under overhanging rocks, places which are suitable for adult life. On finding a suitable habitat it settles down by the head end. Then considerable change takes place as the tail is resorbed and the whole body is re-orientated to form the sedentary adult (Fig. 14.19).

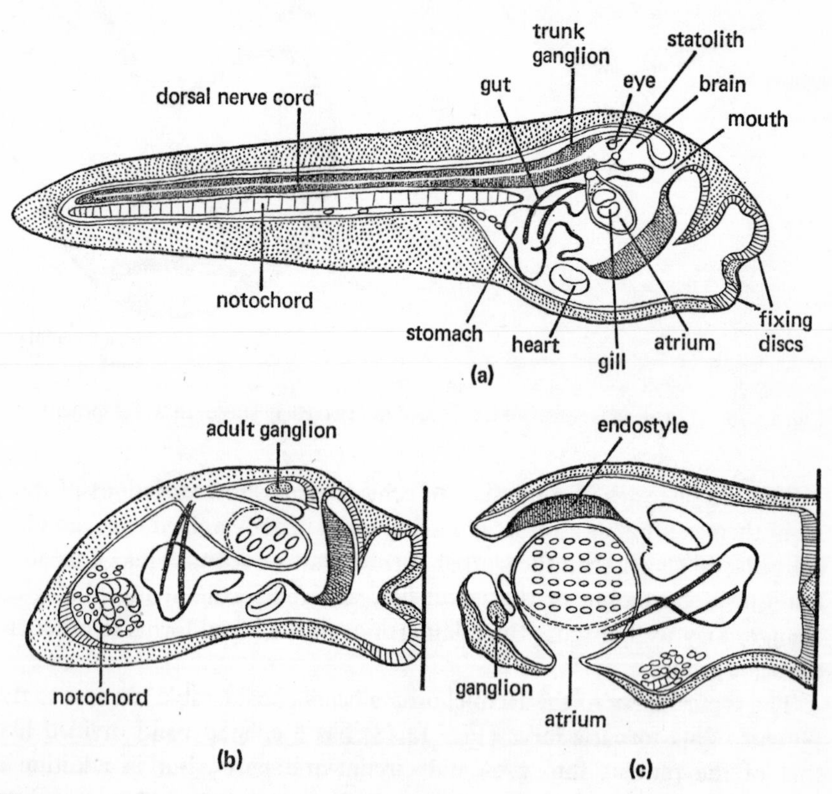

Fig. 14.19 The metamorphosis of an ascidian tadpole larva: (a) at time of fixation; (b) midway in metamorphosis; (c) metamorphosis complete

The tadpole appearance of the ascidian larva gives place to a more fish-like look in the larva of *Branchiostoma* (Cephalochordata). It is a segmented larva foretelling the segmentation of the adult. In its elongated shape and its asymmetry it is plainly highly specialized for planktonic life and feeding. Placed on the left side of the head the mouth is relatively large correlated perhaps with its method of ciliary feeding. The larva is plainly capable of an independent wide-ranging life which it forfeits when it becomes adult, sinks to the bottom and lives in burrows in particular kinds of gravelly or sandy substrata.

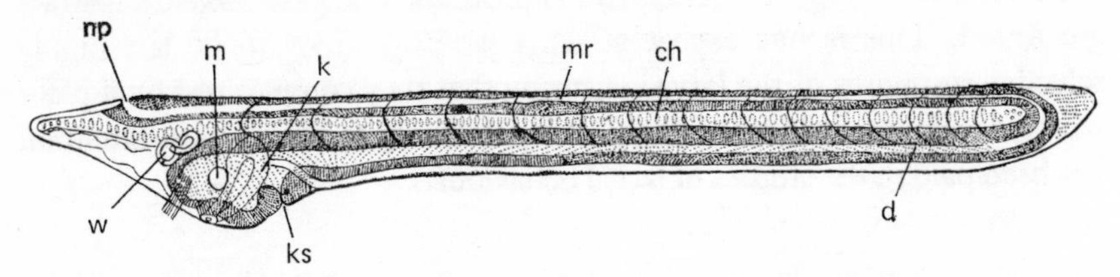

Fig. 14.20 A young larva of *Branchiostoma lanceolatus*. ch, notochord; d, gut; k, club-shaped gland with external aperture; ks, first gill slit; m, mouth; mr, nerve cord; np, neuropore; w, pre-oral pit

Larval behaviour

A number of studies have been made of the behaviour of the larvae of invertebrates; they have all demonstrated how adapted their reactions are to a successful way of life. A simple example is the changes of the light responses of the larvae of house-flies during the final stage of larval life. While they have been mildly photo-negative until then, as pupation time approaches they become very strongly photo-negative, and move away from their food. They seek cracks in the ground which they enter and there pupate. Thus that inactive stage is passed in relative safety. The response to light of many other insect larvae alters even more dramatically, from a positive one to a negative and the change has been shown to be brought about by the endocrine changes which precede the onset of metamorphosis.

Similar changes are known for the larvae of aquatic invertebrates. For example, the larvae of polyzoa, at liberation from the egg, can be seen to gather near the surface of the water on the illuminated side of the container. But this behaviour may change with time. A population of *Celleporella hyalina* will distribute itself mainly in the less illuminated end of a light gradient after 3½ to 4 hours. A possibility could be that, as other animals show a change of light preference after illumination with a certain amount of light, the time of the change would decrease in strong light and be increased in weaker. This proves not to be so, but rather that the effect is brought about by a maturation process in the animal.

But other species remain photopositive. *Flustrellidra hispida* larvae do this throughout the period during which they swim actively, while larvae of *Alcyonidium polyoum* show no clear response. Plainly, however, a response which keeps larvae in the upper layers where they may be carried away by currents is highly adaptive, but before settlement takes place a change to a negative one driving the larvae downwards towards the substrata upon which they can settle will be advantageous.

In a number of species settlement is postponed unless suitable substrata are found. This is one aspect of what we now know to be the highly selective responses of the larvae ensuring that they come to rest on a place where development is more likely. It is selectivity to which most attention has been paid in the studies of larval behaviour.

Selection of substratum

Under experimental conditions the larvae of the polychaete *Ophelia bicornis* tend to choose sand similar to that of their parents' habitat. They will be more likely to metamorphose on particles that are smoothly rounded and of the size of the quartz particles in the sand of their breeding grounds. They do not settle on smaller sharp particles; indeed such a substratum is not a satisfactory one for the burrowing adult, whose distribution is limited to loose, clean sand. An added requirement of the larvae is that the grains should be covered with a film of micro-organisms, sand of the right size which is baked in an oven, thus destroying the algal and bacterial film, loses its attractiveness.

The choice made by the larva may be very species-specific. Thus, three closely related species of *Spirorbis* are each found characteristically on one type of substratum: *S. borealis* on the fronds of the brown seaweed *Fucus serratus*, *S. corallinae* on the coralline alga, *Corallina officinalis* and *S. tridentatus* on rocks and stones. Larvae of the first two species were offered various substrata in the laboratory in order to see which they chose. Thus, larvae of *S. borealis* were put in a dish with *F. serratus* and *Corallina officinalis*; in one experiment 1,297 larvae chose *Fucus* and only 18 *Corallina*. Offered other choices larvae also settled almost as thickly on the green seaweed *Ulva lactuca* as on *Fucus*, and as frequently on a stone which had been allowed to gather a film of algae and so forth. In a direct choice, however, between the green and brown weeds, preference was given to the brown weed. In the places where this worm is found, *Fucus serratus* is much commoner than *Ulva. S. corallinae* on the other hand showed a more restricted choice, choosing the coralline alga in overwhelming numbers.

Theoretically it is possible that the determination of larval choice is by a process of imprinting, the larva being conditioned during its early development so that it selects a substratum similar to that on which it developed. This could explain the restricted choice of the worm larvae, as it has been invoked to explain the very specific choice of food plant by certain monophagous insects. However, such an explanation does not hold, for larvae of *S. borealis* from parents found unusually on *Laminaria hyperboria* still have a preference for *Fucus serratus* over *Laminaria*.

It would be interesting to discover if the selection is genetically determined, but the hybridization experiments which are necessary have so far been unsuccessful. Interestingly enough, these larvae are gregarious until they settle, as are barnacle larvae (see later). As cross-fertilization is more likely to occur between near neighbours, the groups which result will form species-pure interbreeding sets, tending to be genetically isolated from the rest of the population.

Fucus serratus is a favoured site for the settlement of the larvae of a number of polyzoa, for example *Alcyonidium hirsutum*, *A. polyoum* and *Flustrellidra hispida*. Their choices in experiments reflect their commoner distribution in nature. Indeed, not only do the larvae select the plant upon which to settle but also, at least in *A. polyoum*, the part of it as well. For larvae of this polyzoan show a tendency to select the tip of the thallus of the weed, with fewer settlements occurring on the centre and base. Older thalli are less attractive than the younger, possibly because they contain more mucus, some of which exudes over the surface.

Surface texture also seems important for settlement of *Celleporella hyalina* larvae, which occurs most frequently in grooves and depressions. The mechanism by which the very small larva detects the existence of a groove many times larger than itself is quite unknown. Indeed there are still unexplained results in the experiments on choice of substratum which show that other factors yet unknown must be playing an important part.

Light influences the choice of orientation of the surface upon which settlement takes place. Under natural conditions the undersides of overhanging rocks or seaweed fronds will be less well illuminated than the upper surfaces. In laboratory experiments the larvae of *Celleporella hyalina* select the side away from the light whether panels are hung vertically, obliquely or horizontally. Though they choose the underside of horizontal panels, if these are illuminated from below, choice falls on the upper surface. Panels slung horizontally in the sea collect larvae on the underside not only because of their preference but also because the growth of algae and the collection of sediment which occurs on the upper surface makes them unsuitable for settlement.

In many cases, these motile larvae of marine invertebrates explore the surface of a potential site by moving swiftly about over it. Fixation does not come about until the necessary stimuli are encountered. But settlement will not be delayed indefinitely, as is shown by the settlement of larvae on panels of various materials left in the sea. As time goes by the stimuli which can bring about settlements become less and less restricted.

The cyprid larvae of barnacles such as *Balanus balanoides*, *B. crenatus* and *Elminius modestus* search the surface of rocks with their antennules until they make contact with adult barnacles of their own species, or the cemented bases left after barnacles have been torn away (Plate 16). These are recognized by chemical cues, but in addition the larvae show a preference for rough rather than smooth surfaces. In experiments they avoid glass to which other barnacles are attached and settle on stones in the vicinity.

Metamorphosis of the cypris occurs when it attaches by its antennules (p. 312). At this early adult stage it has not formed the plates of its house and therefore can still adjust its position. It does this by reacting to the currents of water passing over the surface, taking up its final position across the current. Orientated in this manner the setae-fringed legs will function optimally if they sweep against the current which, of course, carries food to them. The reaction to current is not shown by the cypris larvae; it only becomes apparent after settlement has taken place.

The behaviour by which the larvae of *Sabellaria alveolata* select their site for settlement is not dissimilar to that of barnacle larvae. The adult worms (honeycomb worms) form large colonies on rocks, particularly near the low-water mark (Plate 14). The larvae swim actively and crawl over any solid surface which they touch (Plate 15). This searching phase may be prolonged for weeks but is brought to an end when they come into contact with the material of the cement used by young and old worms alike in building their tubes of sand grains (Plate 15). The substance responsible has not been isolated but as its activity is removed by cold concentrated hydrochloric acid, unlike that of the barnacle base material, it is not likely to be the same substance. In addition to the chemical cue, the physical stimulus of violent movement of water bearing sand grains is a potent one causing settlement. This is of course typical of the conditions on the part of the shore where the adult colonies are mainly found.

By means of these simple behavioural responses larvae determine the future distribution of their species and maintain a tradition of particular kinds of sites. This in turn tends to ensure genetic isolation of the species and may well be a significant factor in species formation. But their choice equally ensures that the last stage in the life history, which has been accompanied all along by huge losses, will place the young adult where its chances of survival are at least higher than elsewhere.

Chapter 15

Animal associations

EVOLUTION has produced many remarkable adaptations of animals to the physical environment but no less remarkable are those which increase the efficiency of life with another species of animal. These relationships of species which share each other's lives, vary greatly in their degree of intimacy; they are particularly common among invertebrates. For example, almost all parasites are from one or another of the invertebrate phyla, indeed almost every major phylum contains parasitic members. The number of vertebrates which have taken on this mode of life is very small indeed.

It is well-known that certain animals are found associated in communities wherever certain conditions of the environment prevail. In the North Sea, for example, certain bottom communities have derived their names from the animal dominant in them, thus, the *Macoma* community composed principally of the bivalve *Macoma baltica*, or the *Ophiothrix* community dominated by the brittle star and so forth. A striking example is the great variety of fish and invertebrates found in coral reefs. Whatever the reason for these particular collections of animals, they do not necessarily interact directly with each other. The activities of one may, however, alter the environment making it suitable for another. In this way, sabellid worms are often found in mussel beds where the mat of byssus threads securing the molluscs traps silt and sand creating a substratum into which the worms can burrow.

But there is a wide range of associations between individuals of different species in which one or both derive benefit from the other. Sometimes these are permanent relationships, as with some commensals and most parasites, at others they are temporary. Often the association is so close as to be indissoluble but sometimes they are facultative, each partner being able to live without being involved with the other. The grades of association shade into each other, but for convenience they can be divided into commensalism and parasitism. Two commensals may live together to each other's advantage (mutualism), or to the advantage of one only of the partners;

but, by definition, a parasite will always be living to the detriment of its host. One species may only cause discomfort, but equally others will cause debility and even death, though it is clearly better for the parasite if the host survives and nourishes its partner. Certainly parasitism brings in its train morphological as well as physiological adaptations which ensure the efficiency of this way of life; special behaviour patterns may also appear. Commensalism, however, rarely, if ever, evokes the changes from the free-living form that the evolution of a parasite brings about.

The classification of the associations is made more difficult by our lack of real knowledge about so many of these relationships. A vertebrate heavily infested with a parasitic worm may be sickly, sluggish in movement and plainly affected, but what is the give-and-take between a fish and the sea-cucumber in whose gut it lives? We can infer the advantages which accrue to the partners, but the subtleties of the situation are too often lost to us; associations which appear one-sided may have mutual advantages to each animal if we could but discover them.

Commensalism

The apparent dividends of commensal relationships are usually shelter or food, or, very often, both together. How far a *Nereis furcata* hiding in a hermit crab's shell (Fig. 15.1) is protected from predators can only be guessed. It can be seen to take food from between the mouth-parts of the crab, but the crab seems to derive no benefit from the presence of the worm. This verges on parasitism though no positive harm seems to come to the crab by the reduction of its food supply. Perhaps it merely feeds more often to obtain enough for itself and its partner.

It is not surprising that animals which have some well developed means of defence are often partners giving protection to another species. Thus, among the trailing nematocyst-loaded tentacles of the Portuguese man-of-war (*Physalia*) are often found small fishes (*Nomeus gronovii*) related to the horse mackerels who seem not to cause the discharge of the cnidae. They no doubt derive protection from their position, for potential predators on the fish would be deflected by the tentacles, and at the same time they may get scraps of food from the material captured by the siphonophores. Again, the strangely rounded amphipod, *Hyperia galba*, shelters amongst the tentacles of the jelly-fish, *Cyanea*; it is suspected of taking food from the tentacles.

The stinging cnidae of a large coelenterate must be a deterent to many animals, hence the demands for members of this group as partners. A most striking association of this kind is between the fish, *Amphiprion percula*,

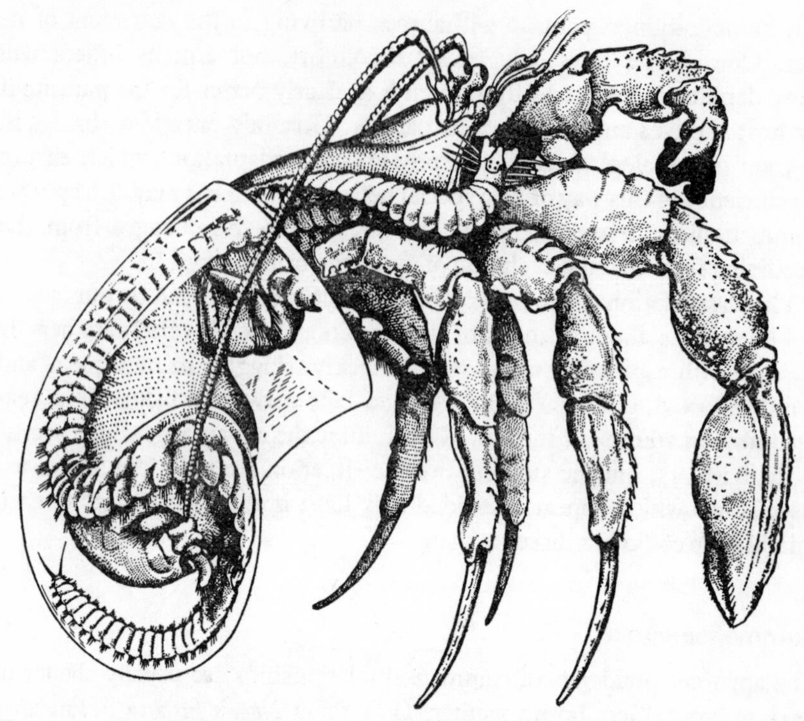

Fig. 15.1 A hermit crab and its commensal *Nereis fucata* inhabiting a glass shell

and a large anemone, such as *Stoichactis*. Each fish remains among the slowly waving tentacles of the anemone. This is its territory and it fights off any other fish which comes near. Once again the relationship has evolved so far that the fish is not affected by the nematocysts of the anemone. Indeed, when the anemone entraps food in its tentacles and draws it down into its mouth, the fish may be engulfed as well, but it emerges unharmed. A factor in the mucus on the skin of the fish reduces the sensitivity of the nematocysts so that they do not respond. When the fish first encounters the anemone it rubs itself more and more vigorously across the disc and among the tentacles. Gradually it is less and less entrapped by nematocysts until finally there is no response from these cells. It is as if the anemone has learned to recognize its particular partner. Offered to another species of anemone the fish are devoured immediately. There is evidence that the shelter offered is real and not merely in our imagination for these fish when kept in an aquarium with predatory fish are soon eaten, but if an anemone is there they go untouched.

Sponges can form ideal habitations for animals of all kinds. Not only are the intruders enclosed and hidden within one but the water currents set up by the sponge carry food to them. In one not over-large specimen of the loggerhead sponge, *Speciospongia vespera*, over 13,500 animals were found. Over 12,000 of them were a small shrimp, *Synalpheus*, while the remaining eighteen species included polychaetes, copepods, amphipods and a fish. Sometimes the animals living in the cavity seem never to leave it, or even cannot do so. The shrimps, aptly named *Spongicola*, which enter the sponge when young are trapped as they grow larger for they cannot pass through the sieve plate covering the osculum of the sponge. They may perhaps browse on the sponge's collar-cell layer, for otherwise they must be dependent upon whatever particles of food remain in the water current after the sponge has taken its food from it.

Other animal bodies supply crevices and orifices which can be inhabited just as rocks on a shore do. Thus, a polychaete, *Flabelligera commensalis*, matches the purple colour of the sea-urchin *Strongylocentrotus purpuratus* amongst whose spines it lives. As adult or half-grown worms only have been found, and always on the urchin, it is still a matter of conjecture where the growth stages are passed. A small hesionid worm, *Podarke pugettensis*, lives on the underside of the starfish, *Patiria miniata*, and crawls into the ambulacral grooves if it is disturbed (see p. 330). The cloacal opening of sea-cucumbers offer yet another site for life, a fish, *Carapus*, lives there having the added advantage of a stream of fresh water passing into and out of the cavity and bathing the respiratory trees within the anterior part of the cloaca. In this case it seems that the fish may live indefinitely away from its host. Species of crab *Pinnixa* are also found in this part of the gut of sea-cucumbers off Japan and California. In fact, as will be seen later (p. 332), this genus is composed of commensal specialists.

Animals which make burrows provide potential dwelling places for others, and they are particularly welcome when the rest of the environment tends to be featureless offering no hospitable cracks and crannies. It is not surprising that burrows made in open mudflats tend to be inhabited by other organisms as well as their makers. The echiuroid *Urechis caupo* of central California is a well-known example of an animal which might be said to keep open house. The U-shaped burrows of this worm-like creature contain scaleworms (*Hesperonöe adventa*), crabs (*Scleroplax granulata*) and gobies (*Clevelandia io*). As the echiuroid feeds by creating a current of water through the burrow the commensals are supplied with oxygen and food along with their host.

Callianassa and *Upogebia* are crustaceans which burrow in habitats similar to those occupied by *Urechis*. Their burrows too serve as shelters for other invertebrates. Some species, such as the crab *Pinnixa franciscana*, may be found in the burrows of any of these species, others are specific in their choice of shelter so that a complex of commensals surround these three hosts (Fig. 15.2).

One of the most easily observed examples of commensalism in European waters is the association of a hermit crab with an anemone. A number of

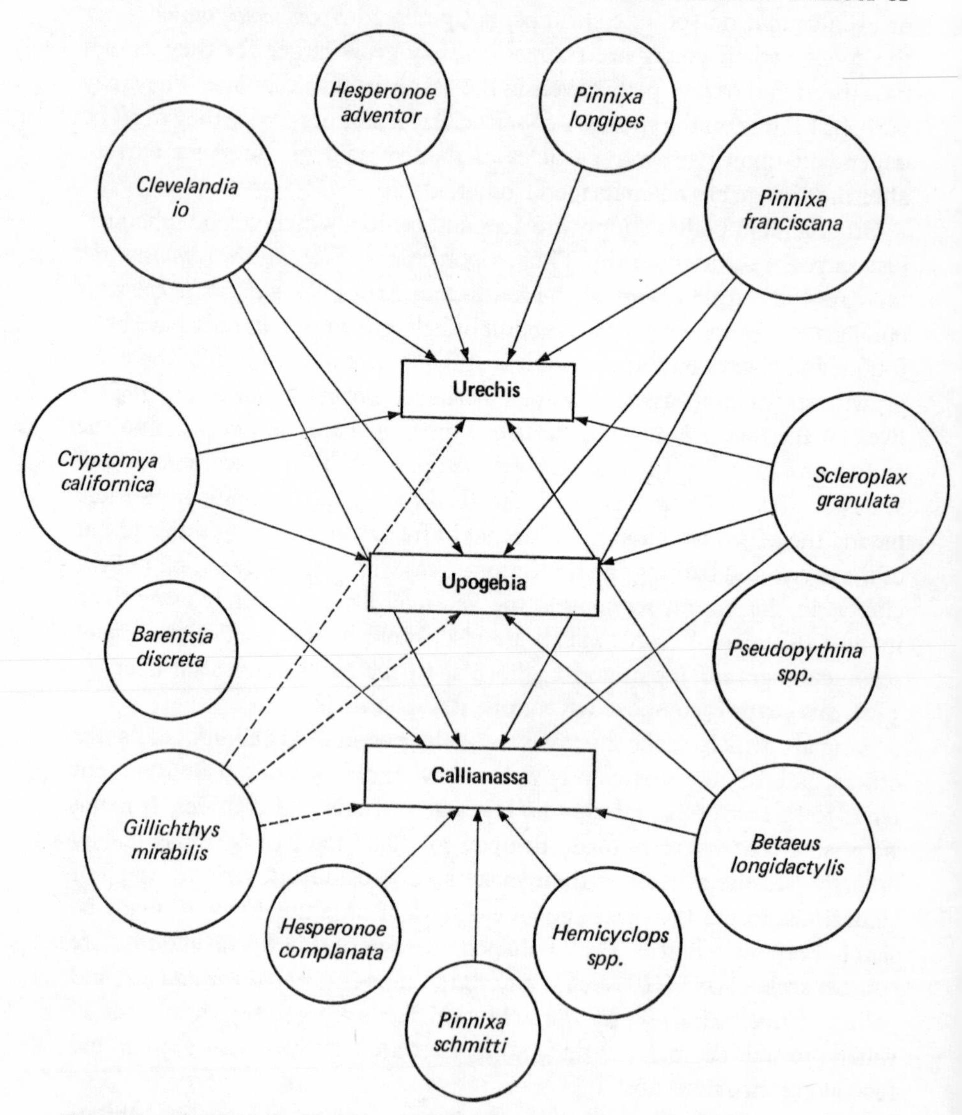

Fig. 15.2 Diagram of the relationships between the animals associated with *Urechis*, *Callianassa* and *Upogebia*

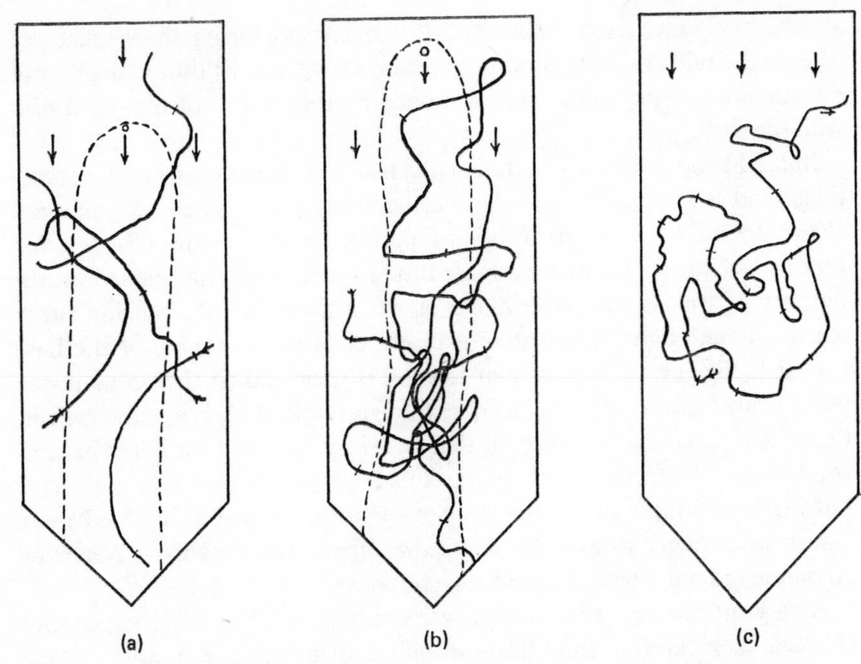

Fig. 15.3 Tracings of tracks of *Pinnixa chaetopterana*. (a) Four tracks made by a single crab in water containing a marker dye only. (b) The same crab's movements when water containing host-factor from *Chaetopterus* was flowing through the chamber. (c) The track of a crab in a trough filled with host-factor but without a current running through. The dotted lines in (a) and (b) indicate the area covered by dye-marked water which contained host-factor

different species of crab utilize anemones, and of these many carry *Calliactis parasitica*. The anemone fixes itself to a whelk shell whether it is occupied by a crab or not, reacting to chemical stimuli from the organic matrix of the shell. It can live without the crab, just as the crab *Eupagurus bernhardus* can and does frequently live without its anemone partner. Probably the tentacles of the anemone are protective for the crab, while in return the anemone has the advantage of receiving scraps when the crab tears its food (Plate 22). On the other hand, the crab *Pagurus prideauxi* has been seen to feed its partner (*Adamsia palliata*) by putting pieces of food into the anemone's mouth.

The crab will detach the *Adamsia*, holding the shell with its large claws and grasping the anemone with its other legs. It then transfers it to another shell. *Pagurus striatus*, a Mediterranean species, stimulates its *Calliactis* partner to detach, after which the anemone searches the new shell with its tentacles, adheres by special nematocysts, releases its base and lifts it over to attach itself. Once fixed to the shell, its tentacles no longer show stickiness (due to the release of nematocysts) in contact with shell; thus in some way the response of the cnidae is conditioned by the contact of the pedal disc with the shell.

Special behaviour such as this marks many of these commensal associations and serves to bring the partners together. The crab *Pinnixa chaetopterana* lives in the tubes of the polychaete worm *Chaetopterus pergementaceus*. If put into a stream of sea water which has passed over a number of worms, the crab moves along a tortuous path making turns which become more frequent as it draws close to the source of the host material (Fig. 15.3). This rate of turning is greater than that in plain sea water (klinokinesis). The behaviour, though undirected, keeps the crab in the area of highest concentration, that is, brings it to the host under natural conditions. The factor which evokes this behaviour is probably a protein, but it is not unique to the one species of worm for the crab would react almost as strongly to material from *Amphitrite*. Material from *Nereis* or *Arenicola* did not, however, produce a response.

A very specific protein seems to be the factor attracting each one species (*Unionicola* species) of mite to its specific host fresh-water mussel. Three species are known, each of which inhabit only one species of host clamber-

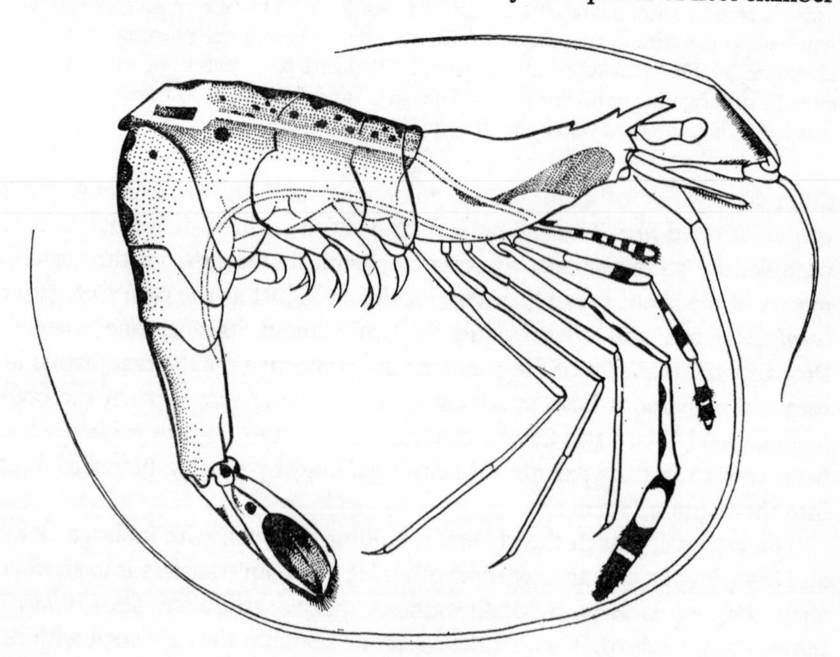

Fig. 15.4 *Periclimenes pedersoni*, a cleaning shrimp

ing about in the mucus-covering of its gills. They are positively phototactic when outside their host (see p. 321) but as soon as the specific factor is added to the water they swim away from the light towards the bottom where their hosts lie buried.

Thus far, most of the examples of commensals have involved invertebrates alone but there are a number of associations of invertebrates with vertebrates. Perhaps one of the most remarkable is that of the shrimp, *Periclimenes pedersoni*, with a number of fish. This shrimp signals when a fish swims near by whipping its conspicuously coloured antennae back and forth while swaying its brightly coloured body (Fig. 15.4). The fish may respond by coming to within an inch of the shrimp, then allowing it to walk over its body eating off scar tissue and removing parasites. The fish even raises its gill covers to allow the shrimp to clean its gill filaments. It has even been reported that fish learn where the cleaner is to be found and queue up for attention!

Parasitism

When the benefits of an association become so one-sided that one partner suffers, we use the term parasitism. Parasites usually must be associated with their host, but not necessarily for the whole of their lives. Some crustaceans, for example, have larvae which are parasitic at certain stages though the adult is free-living; the reverse situation in which the adult is a parasite and the young stages free-living is not at all uncommon (see also p. 309). Since it allows dispersal, this is only to be expected. However, parasites may live within the tissues, or in the cavities of organs, of their hosts. These *endo*parasites stay there throughout that stage, which may be all of their lives. Others such as ticks or leeches visit a host for a meal of blood remaining only long enough on the outside of their hosts before dropping off. These *ecto*parasites are barely separable from such animals as mosquitoes, who, taking a blood meal and flying on, might be described as predators.

The parasite association is very often between vertebrate and invertebrate. The tapeworms (Cestoda), for example, are found as sexually reproducing adults only in vertebrate hosts. Other animals are pressed into service as vectors to disperse the reproductive stages of worms as widely as possible. Although flukes (Trematoda) are typically found in the gut of vertebrates as adults their larval distributive stages often develop in molluscs.

Bird flukes, for instance, often have seven stages in their life-cycle, each adapted to its role in distribution and to its place of living. Thus, the herring gull fluke, *Cryptocotyle lingua*, lives in the bird's intestine where sexual reproduction with the formation of eggs takes place (Fig. 15.5). The eggs are passed out in the bird's faeces and scattered over the moist ground. Ciliated miracidia larvae capable of swimming actively develop from them after they have been swallowed by one of the many common periwinkles, *Littorina littorea*, to be found on the shore. Within the mollusc the miracidium gives rise to a sporocyst within which multiplication takes place resulting in numerous redia larvae, worm-like with a reduced gut and a muscular pharynx. Within each redia, more rediae are formed to burst out and add to the population which feeds on the reproductive and digestive

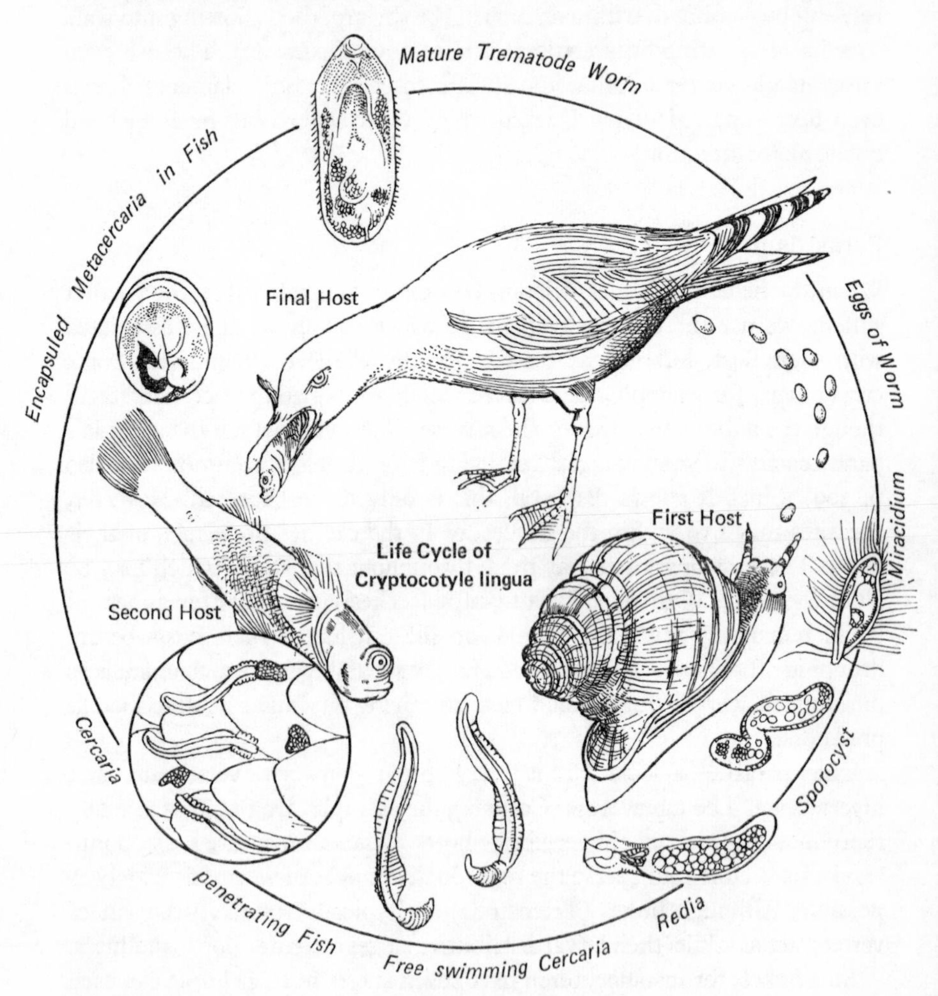

Fig. 15.5 The life-cycle of the herring gull fluke, *Cryptocotyle lingua*

organs of the snail. In this way many of the periwinkles are rendered sterile. After this process of multiplication has been repeated a number of times, in response to some stimulus still unknown, the germ cells give rise to another form of larva, the cercaria. Place a periwinkle in a small tube with enough sea water to cover it, put the tube in a warm place, and in a few days swarms of cercariae will almost certainly be seen swimming about in the water.

These larvae are well suited for active movement in a search for yet another host. For in addition to a gut, a sucker and flame cells, all characteristic of adult trematodes, they have a long tail for swimming, penetration glands and spines for aid in boring through the skin of the next host and eye-spots by which they can respond to shadows cast by a potential host. For the next stage in the life-cycle takes places in a common shore fish such as a goby or a blenny. The cercaria bores into the skin just below the scales where it encysts. The swellings caused by these cercariae are often seen on these fish though it is rare for one fish to have more than a few of them and yet survive. If the fish is eaten by a herring gull, the metacercaria, as the encapsulated form is called, hatches into a sexually mature adult. Thus the cycle is completed.

In this life-cycle, very great numbers of larvae have been produced by the various multiplicative stages. These huge numbers help to ensure that new hosts are infested, though the wastage is very great for many thousands of larvae will fail to reach a new host. Though life-cycles in the diagrams of text-books have an air of certainty, the life of a parasite is a hazardous one in the extreme.

There is no better illustration of this than the reaction of a host to its parasite. Often the cells of the host's tissue divide to form a capsule round the intruder. In insects it is certain of the blood cells which do this. Occasionally the walls of the cyst become calcified, this is one reason for the muscular pain caused by an infection of *Trichinella spiralis* for this nematode's larvae feed upon muscle tissue working their way into it and finally encysting there. The tissue reaction is then to enclose the larva. Later, calcification hardens the walls and may lead to pain. Since the next stage of the life history requires that the muscle be eaten by another host – and as man eats man hardly anywhere in the modern world – the larvae die in their capsules.

But apart from such tissue reactions some hosts show immune reactions to parasites. These are identical with the organism's response to the introduction into it of any foreign material. The active parts of this material are usually large-molecule proteins or carbohydrates called antigens. In response, the body produces antibodies which combat the foreign material often by combining with the antigen part and rendering it harmless. Since the antigens may well be on the cell surfaces of the parasites the worms may be completely and quickly destroyed; in other cases though antibodies are certainly circulating they have little effect.

What is termed 'self-cure' happens in this way. Sheep, for example, may recover from an attack of stomach worm (*Haemonchus contortus*) because the primary infection sets in train a reaction so that, after a peak of egg production by the worm is reached, the infection declines and adult worms are expelled from the body. Another infection is then not possible even if larvae are eaten in quantity. The reactions has been brought about by infecting sheep with X-ray irradiated larvae. These will move through the body but do not mature and produce eggs. Their presence is, however, sufficient to produce the immune response.

Most of the major phyla of invertebrates have parasitic members. The wider the radiation of a phylum into various niches the greater is the possibility of there being parasitic forms. This is admirably illustrated by the Crustacea for in this class there are many parasitic members distributed among different families. The medical and veterinary importance of nematodes, cestodes and trematodes has tended to obscure in the minds of some parasitologists the fascination of crustacean parasites as an experiment in adaptation – and a successful experiment at that.

In the reverse way though the Nematoda and the Platyhelminthes are phyla with numerous species, most of which are parasitic, there are, never the less, some free-living members. It is a remarkable fact that whether they are free-living or parasitic on plant or animal, the body plan of nematodes remains the same even in detail (p. 37). However, specialization for a parasitic way of life, whether inside or outside the host, usually involves the development of an apparatus of hooks or suckers to avoid, for example, being swept down the gut by the waves of peristalsis, or being scratched off by the host when it cleans itself.

The Monogenea are platyhelminths parasitic on the gills and skin of fish. Those which fix themselves among the scales on the surface of their host often have a relatively simple arrangement of hooks and a sucker-like organ for adhering (the opisthapter). But in the environment of a fish's gills a parasite has to be firmly attached for otherwise it may be carried away in the current of water passing over the gills for respiration (Plate 21). Often these species (Fig. 15.6) have six or more suckers and complicated sets of hooks included with them. The elaboration of suckers and hooks on the scolices of Cestodes (Fig. 15.7) bear witness to the importance of anchoring the head ends of these worms in the intestinal epithelium of the host. The crustacean carp louse, *Argulus foliaceus* (Fig. 15.8), has maxillae specialized into suckers by which it sticks to the skin of the fish while it sucks a blood meal through a short piercing proboscis. These fish lice do not remain permanently on a host and frequently they are netted in fresh-water plankton. The suckers of leeches are another example of a specialized attachment organ.

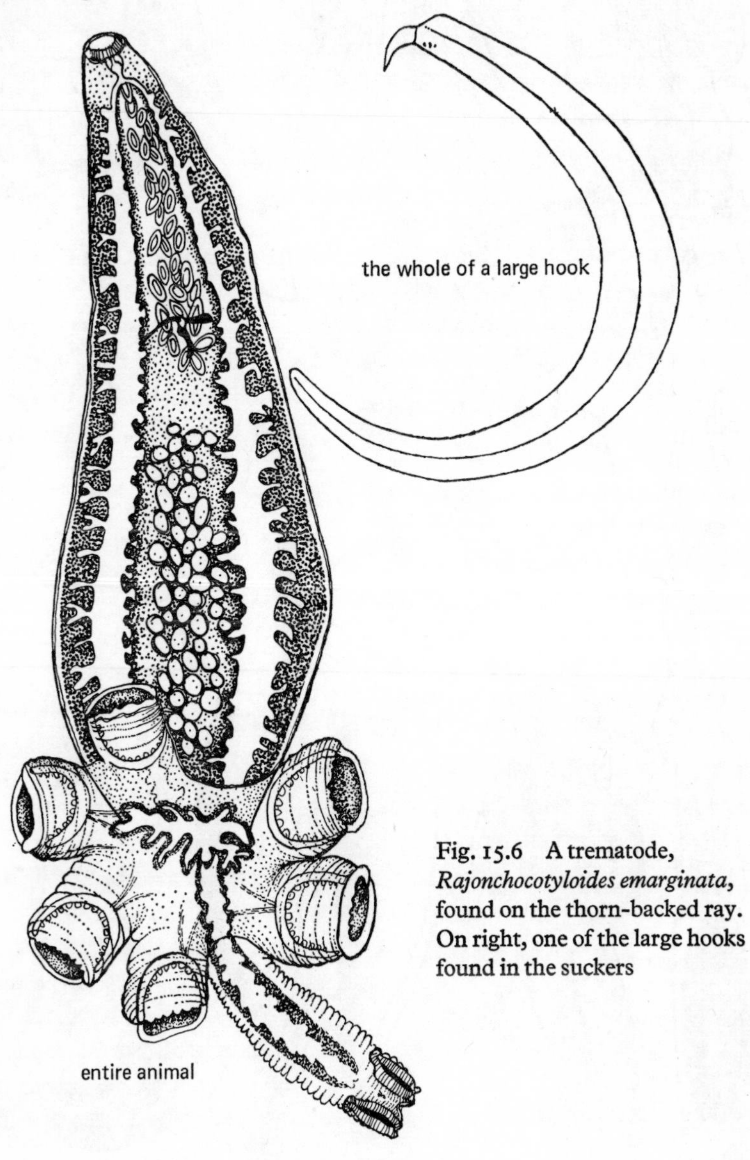

Fig. 15.6 A trematode, *Rajonchocotyloides emarginata*, found on the thorn-backed ray. On right, one of the large hooks found in the suckers

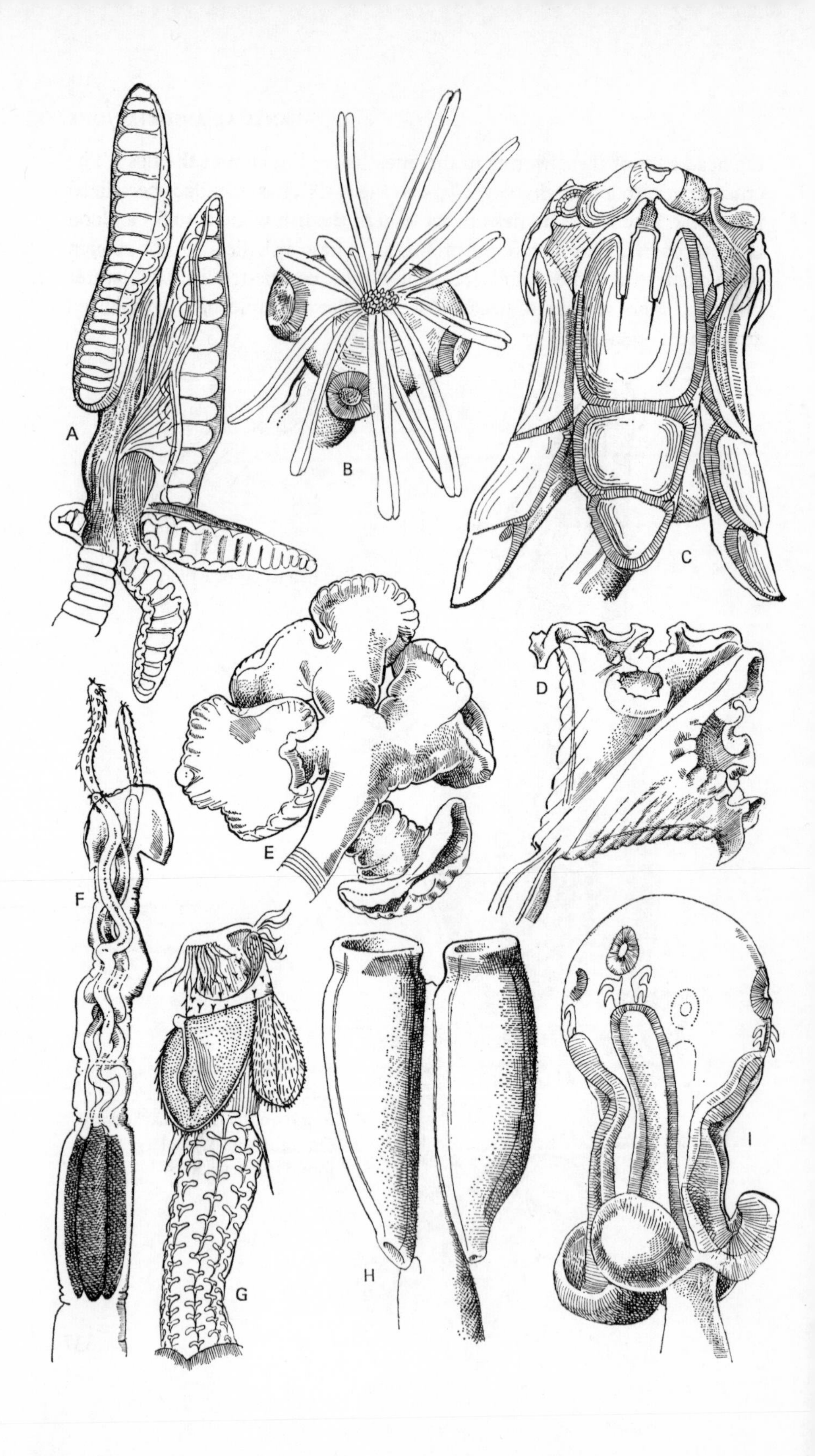
A
B
C
D
E
F
G
H
I

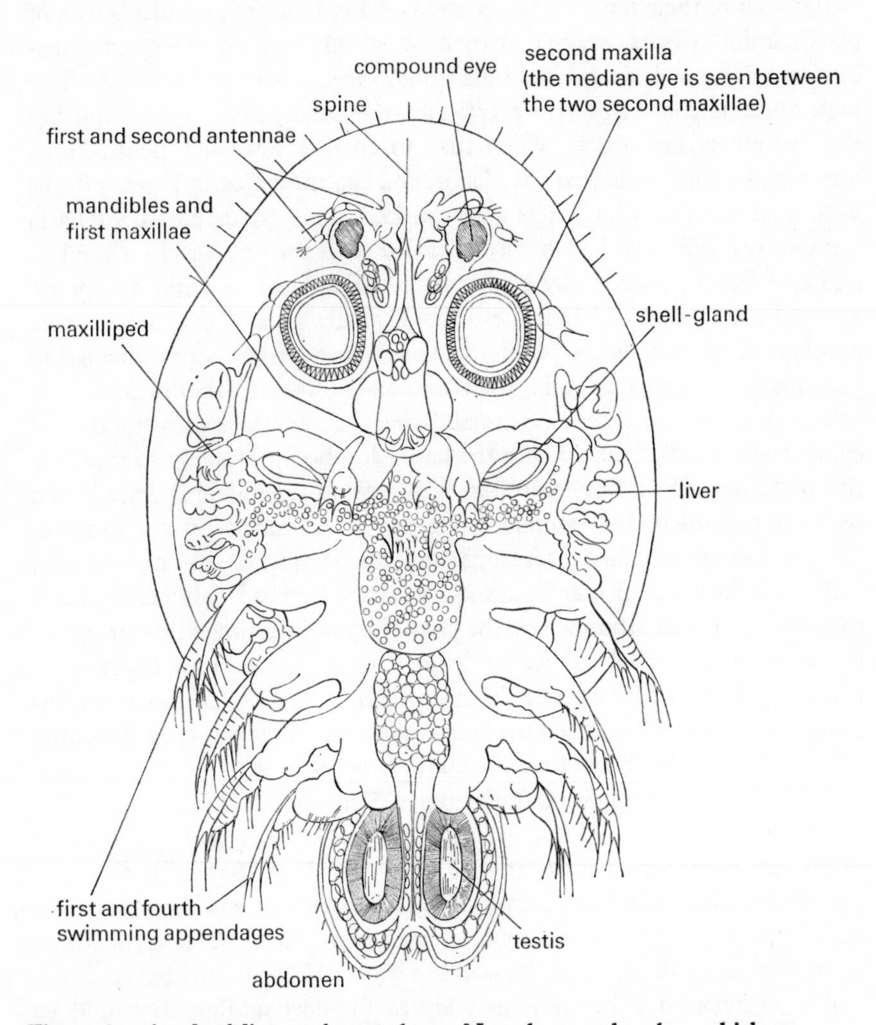

Fig. 15.8 *Argulus foliaceus*, the carp louse. Note the round suckers which are the second maxillae

---

Fig. 15.7 Examples of the variety of form of scolices among tapeworms: A, *Rhinebothrium*; B, *Parataenia*; C, *Acanthobothrium*; D, *Duthiersia*; E, *Phyllobothrium*; F, *Tetrarhynchobothrium*; G, *Echinobothrium*; H, *Bothridium*; I, *Balanobothrium*

No less than the elaboration of attachment devices in parasites is the elaboration of their reproductive systems. This is most particularly true of platyhelminth worms where a tortuous set of tubules from the hermaphrodite set of organs is found in adults of both trematodes and cestodes. The male systems consist of either two larger or many small testes pouring their products into ducts which fuse to enter a muscular penis. It is, however, in the female that the elaboration can most clearly be seen for as well as an ovary in which eggs are produced, there are extensive vitellaria forming the yolk which is finally enclosed with the egg within a shell of tanned protein produced by a special shell gland. The fertilized eggs are stored in a uterus which in cestodes expands greatly to hold its vast numbers of eggs filling the proglottis. One tapeworm has been estimated to produce 36,000 eggs a day and some thousand million in a lifetime.

The parasitic crustacean *Sacculina*, whose larval stages of nauplius and cypris show its relationship with the barnacles, becomes solely a reproductive machines after it has entered its host crab. It loses all appearances of a segmented hard-bodied animal becoming transformed into a formless white mass with root-like extensions ramifying throughout the crab's body.

The life histories of many parasites are such that at the infective stages they will be in a situation where the possibility of infecting a host or vector is greater. The eggs of tapeworm dropped on the ground are likely to be picked up by pigs as they root about for food. The encysted metacercariae of the sheep liver fluke are on the grass which the sheep will crop. Infesting a host which is the food of another host in another stage of the life-cycle is an obvious way of ensuring transference. For example, the metacercariae of the oviduct flukes (*Prosthogonimus*) of birds often enter the rectal respiratory chamber of dragonfly larvae. From there they migrate to the insect's muscles where they encyst. They are carried over to the adult when the nymph undergoes its final moult. These insects in either stage are often eaten by geese, gulls or ducks, as well as a wide variety of other birds.

But behaviour too has become adapted for host finding. It would be expected that ectoparasites of warm-blooded animals would have clear responses to warmth; the human louse, *Pediculus*, shows this behaviour when it moves towards a warm tube. The sheep tick, *Ixodes ricinus*, does likewise but its reaction is even stronger if the tube is covered with sheep's wool (Fig. 15.9); the chemical stimulus from the wool enhances the tube's attractiveness.

This arachnid has a special searching behaviour which it uses when it needs a blood meal. A tick drops off its sheep host after sucking in so much blood that its body becomes greatly swollen. It remains in the humid

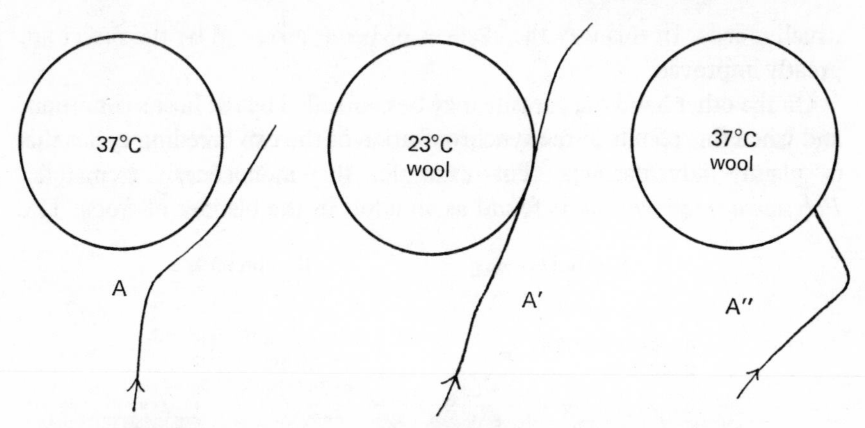

Fig. 15.9 The responses of ticks (*Ixodes ricinus*) to smell and warmth. A, a hungry female approaches an odourless warm tube; A′, the same tick and a tube at room temperature with wool wrapped round it; A″, the same tick and a warm tube with wool

micro-climate of the bases of the grass until the meal is entirely digested, a process which may take weeks. The hungry tick then shows a change in behaviour for it no longer shuns the light but, moving towards the sky, it climbs up a grass stem. At the top it comes to rest. Some stimulus such as vibration or air movements when a sheep comes near causes the tick to move its forelegs in 'questing' movements. Since a tick's main sensory receptors are on these legs they can easily detect their host as it nears them, and letting go of the grass grasp the sheep's wool. In experiments a tick resting on an upright will quest when a warmed tube is brought within $\frac{3}{4}$ to a $\frac{1}{2}$ cm of it. One might almost say that they are too eager for some wave their forelegs so violently that they fall off their support!

Stimuli which arise from the potential host are obvious candidates for a parasite's attention. Simple shadow responses are effective in bringing hungry leeches and many trematode larvae into activity in a pond. They swim upwards from rest in an undirected manner which brings them generally into the vicinity of their hosts – ducks and so forth – which are swimming overhead. Leeches also will respond in this way if they are hungry.

Occasionally the behaviour of the parasite has been synchronized with that of its host. The nematodes infecting the blood of man have larval stages, the micro-filariae, which are found in different parts of the body at different times of the day. Each species has its usual insect vector, and the micro-filariae of a species are always in the peripheral blood system, that is in the capillaries of the skin, just at those times when the biting fly vector usually feeds. In this way the chances of being taken up by the insect are greatly improved.

On the other hand the parasite may be controlled by the host's condition, and when this results in the synchronization of the two breeding cycles this is plainly advantageous. For example, the monogenean trematode, *Polystoma integerrimum*, is found as an adult in the bladder of frogs. The

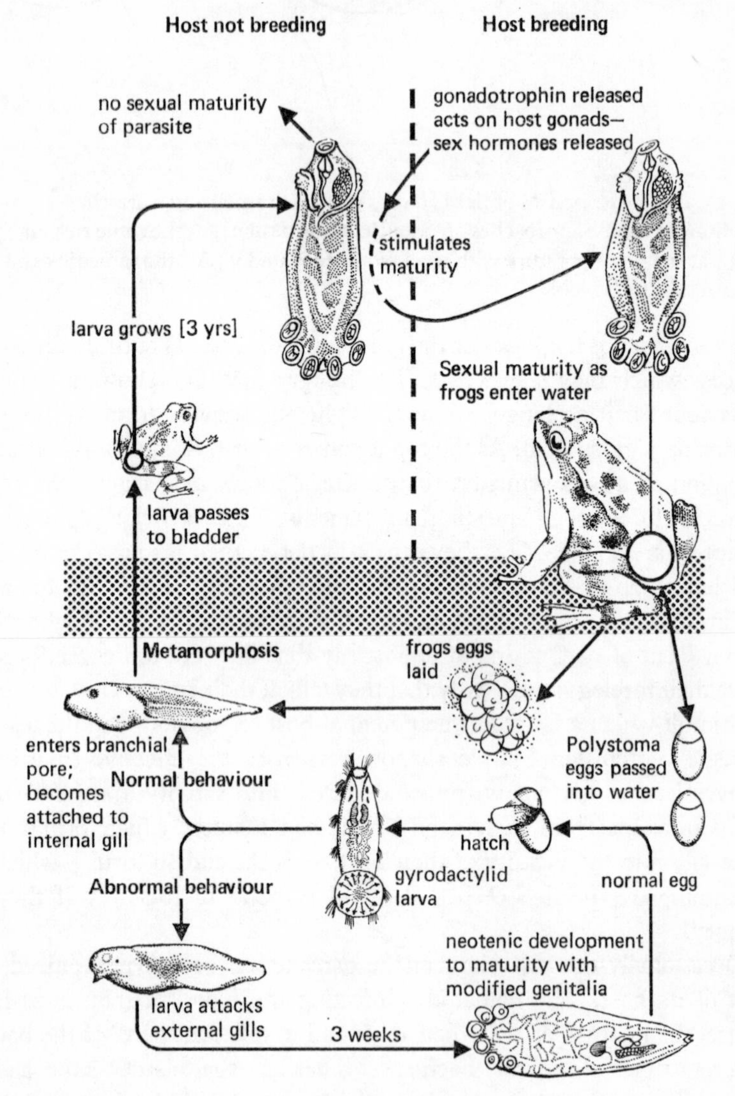

Fig. 15.10 The life of *Polystoma integerrimum* and its relationship to the breeding cycle of the frog host

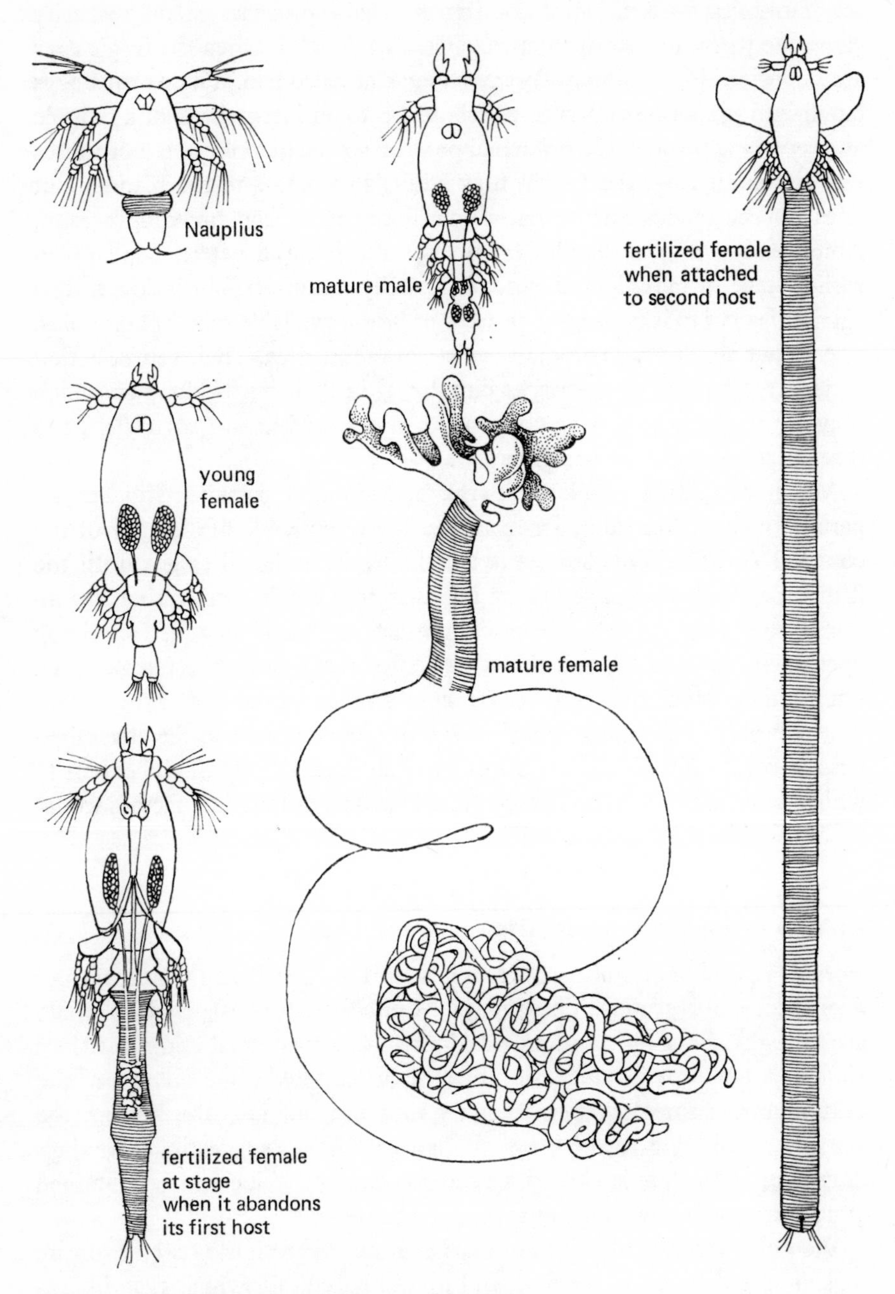

Fig. 15.11 Stages in the life history of *Lernaea*

sex hormones released when the female frog approaches sexual maturity cause the worm to become mature (Fig. 15.10). Thus when the frog's eggs are deposited in the water, *Polystoma* eggs are also laid. The worm's eggs hatch into gyrodactylid larvae which attach to an internal gill of a tadpole after entering through the branchial pore. After the tadpole metamorphoses into the adult frog, the worm, now adult, also passes to the bladder but does not become sexually mature until the next frog breeding season. Abnormally the gyrodactylid larva may attach to an external gill of an earlier stage of tadpole and become sexually mature before its time. The eggs laid serve to increase the number of larvae available to infect tadpoles.

A rather similar state of affairs exists in rabbit fleas. Their reproduction is triggered by the sex hormones circulating in their host's blood on which they are feeding. In this way when the young rabbits are born, fleas can take possession of them straight away.

Many fascinating problems in the behaviour of parasites still remain barely touched. Consider, for example, the remarkable life history of the copepod *Lernaea*. This animal is free-living in its larval stages until the fifth copepodid when it settles on a flatfish in a sessile form rather like an insect pupa (Fig. 15.11). The next and last copepodid stage is free-living once again and moults into the adult. After fertilization the females seek out a cod on which they settle to remain parasitic for the rest of their lives. The complex of changing behaviours with the alterations in the attractiveness of some stimuli must be most intricate and a study of the extent to which behaviour is determined by genetic and environmental factors would be of considerable interest.

Animal groups and social life

For one reason or another some animals tend to collect with others of their own species into groups. These may be quite fortuitous associations which are brought about by common reactions to environmental stimuli. Brittle stars in a tank, for instance, do not remain scattered randomly about the bottom but clump together. If glass rods are put into the bottom the starfishes will cling to these rather than each other suggesting that their clumping behaviour is simply a need for contact which can be obtained equally from their own kind as from inanimate objects.

Woodlice will often be found in large groups beneath the bark of rotting logs. There they are in the dark and humid habitat for which each individual shows a preference in experiments. However, there is a mutual stimulation as well for it appears that they produce a scent which is attractive to other woodlice. So that aggregations are here caused at least in part by stimuli unique to the other members of the group.

Such collections of animals have, however, none of the complex patterns of behavioural interactions which are found in social animals. There is a definite structure of dominant-subordinate relationships in a mammal or a bird group, even out of the breeding season. The groups cohere and their membership remains reasonably stable. But the members of a woodlice group are constantly shifting and changing.

Indeed the most elaborate animal associations are those of insect societies. Probably the best definition of a society is a behavioural one which includes the necessary condition that members of the group shall react to signals from other members. We can include birds and fish for members of flocks and shoals remain together by reacting to the behaviour of their species mates. In the breeding season the mutual behaviour is greatly heightened by courtship with its conflicts between avoidance and attraction behaviour.

Insect societies are unique in that within them work is divided among the members. The castes of hymenopterous and isopterous social insects (the ants, bee, wasps and termites) are morphological reflections of different functions in the colony. The workers clean and care for the brood, carry eggs about, construct the nest and forage for food. A special subdivision of the worker caste may be soldiers responsible for colony defence, whether on the march, as in army ants, or in the nest, as in termites. Their special function may be shown by a greater size than the other workers, but sometimes they have morphological specializations, such as the 'snouts' of nasute termites through which they squirt repellent liquid at their enemies.

The adaptation of parts of the body for their work is best developed in honey-bee workers with their apparatus of pollen combs on the second pair of legs and pollen baskets on the hind legs. Internally their gut has a honey stomach in which nectar for the hive is kept separate from that required by the bee for its own metabolism. There are no comparable specializations in vertebrate societies, though in man, tools have taken the place of modifications of the body.

The reproductives, male and female, form the other two castes found universally among social insects. The queens are long-lived in almost all species and throughout the greater part of their lives continue to produce fertilized eggs utilizing spermatozoa which they store from the single insemination resulting from the mating with a short-lived male. In termites, however, the mate remains with the queen and the royal chamber in a termitery contains the royal pair.

The behaviour of the different castes is as different as their morphology, though this must not be taken to mean that, within the range of their behavioural possibilities, the actions they perform are as rigidly fixed as is their morphological structure. On the whole, honey-bee workers go through a cycle of duties as they grow older, beginning with cleaning the comb for the first few days. After this the worker acts as a nurse feeding the young, a task which requires the secretions of her labial glands. At about the tenth day she becomes a builder making comb with the wax secreted from her abdominal glands. At about the sixteenth day she begins to forage and remains a forager for the remainder of her life. Though these are the tasks predominantly carried out in order in a normal hive, workers can revert to a stage they have passed, or proceed to a stage inappropriate to their age. Thus, an artificially created colony of young bees will nevertheless have foragers, and one of old bees only will have nurses. Much of a 'busy' bee's life is spent in resting and a great deal of it in aimless patrolling, an activity which brings it into contact with stimuli from tasks requiring to be done. In this way efficient use is made of the labour available.

A similar efficiency is obvious in the very important task of foraging. A social insect colony is in essence a reproductive machine, thus a good food-supply steadily delivered is essential for feeding the young insects and those adults which care for them. The dance of a successful honey-bee forager, for example, ensures that the attention of other foragers is concentrated on to food sources which have not dwindled. Again the ability of bees to learn to visit food places at particular times of the day means that their visits are made in the period when flowers are giving their most copious supply of nectar. Or again, the scent trails of ants are only reinforced by workers coming from abundant food supplies so that foraging potential is not channelled along trails which lead to food sources which have been used up.

The wide distribution of social insects, their varied diets and the many different habitats which they occupy show how successful this highly-evolved form of social life has been. One can only speculate why no other invertebrate group has specialized in this way. It is difficult to see any point which distinguishes them from the other invertebrates. Even flight is not essential for many social insects are wingless. Since there seems to be a correlation with social life and the development of the corpora pedunculata of the brain perhaps it is the nervous apparatus for the complex behaviour which is absent in the other groups.

Chapter 16

The past history and evolution of the invertebrates

THE PRESENT day, for all its awesome importance to biologists, is only a random moment in time, and human experience is so relatively short that only the most minute evolutionary change can actually be seen to occur. The fossil record adds that vital third dimension to the invertebrate panorama.

This must not imply that the science of palaeontology can display the entire picture of invertebrate evolution: it can, at most, present a background against which to refer other kinds of evidence obtained from neontology, those which analyse the nature and characteristics of the whole gamut of organisms occupying the biosphere at the present time.

Even so, the fossil record starts far too late in the story to help unscramble the intricate events of early invertebrate evolution. When fossils are first found in any numbers and in a state of preservation which makes them possible to interpret, the main invertebrate phyla are already delimited. Nonetheless, the invertebrate fossil record tells a story of gradual elaboration within phyla, of the dynamic interchange of rise and extinction, of divergence and specialization to new ways of life and exploitation of new habitats. It is a story which is vital to any attempt to view the broad invertebrate story, because it provides the only tangible evidence of the actual course of evolution during earth history. Fickle and fragmentary though it is, and downright conflicting though it often appears to be, the evidence from the fossil record is absolutely vital to the student of invertebrates if he is to view the present deployment of the organic world against the background of its unfolding.

The purpose of this chapter is to relate briefly some aspects of this fascinating story which help to place the invertebrates in historical relationship. We shall consider first the very sparse record of the Pre-Cambrian strata; then, the state of the invertebrate world at the true beginning of the fossil record, that is, during the Cambrian period;

finally, we shall survey the subsequent history of invertebrates in outline, up to the present day; and this will be used in the last chapter as a backcloth against which the facts from modern biological disciplines can be viewed to bring the invertebrate panorama into phyletic perspective.

Pre-Cambrian fossils

The geological record is necessarily more obscure the further back we study it. Naturally, one expects older rocks, more deeply buried and with longer exposure to heat, pressure and mineralizing solutions, to be generally sparser in fossils than more recent ones. But the situation that meets the palaeontologist who searches the Pre-Cambrian rocks is far worse than mere antiquity can explain. The fact is that, except for a few pathetic but gratefully treasured remains in a few parts of the world, the Pre-Cambrian is all but barren of evidence of past life. So marked is the difference between it and the subsequent periods of geological time, from about 570 millions years ago, that the term *cryptozoic* ('hidden life') has been borrowed from its more usual ecological meaning, to describe the period of earth history before the Cambrian, and *phanerozoic* ('visible life') for the subsequent time. It appears that phanerozoic time represents only about one-fifth of the total history of life on the earth.

An additional hazard to the Pre-Cambrian palaeontologist is the difficulty of knowing whether the vague imprints left behind in the rocks are truly organic remains, or whether they are mere inorganic concretions, mud pellets, deformation effects or diagenetically altered sediments. Glaessner has reviewed the physico-chemical effects likely to produce 'pseudofossils' in the Pre-Cambrian and has concluded that some enigmatic moulds in certain ways resembling molluscan or arthropod shell fragments, are indeed most likely inorganic structures.

However, there are now undisputed records of invertebrate remains from the Pre-Cambrian. There are fairly certain records of Porifera, Coelenterata, Arthropoda, Annelida and Echinodermata from regions as far separated as the Arizona Grand Canyon (Nankoweap and Unkar Groups), Southwest Africa (Nama Formation), Siberia (Middle Riphean) and South Australia (Ediacara Sandstone). Of these, the Australian finds are by far the most notable. In addition, there are trace fossils (trails, burrows and other marks) of worm-like or other creeping organisms from the Belt Series of Montana, the Brioverian Series of Brittany and the Lower Vindhyan of India.

The Ediacara fossils deserve special mention, for here we have not merely several different types of organism but the remains of an assemblage of organisms which once lived together in the Pre-Cambrian seas. First discovered by an Australian geologist, R. C. Sprigg, in 1947, they occur in a coarse sandstone in which there are patches of fine mud and silt. The organisms preserved as fossils were apparently stranded and died on the patches of silt to be covered later by sand and transformed in time into a hard sandstone; only the gentle weathering of the arid South Australian climate could expose the delicate fossil imprints still remaining at the sand-silt interface: hammering the rocks will almost certainly not split them across the interfaces, so the chances of collecting many specimens are rather slim. Nonetheless, about fifteen hundred specimens have so far been found, assigned now to some twenty-five species.

The animals (Plate 23) were apparently all soft-bodied, an important fact the implications of which will be considered later. The simplest are probably the jelly-fish-like creatures, assigned at present to six genera, the commonest being *Ediacaria*, *Medusinites* and *Pseudorhizostomites*, but none of these medusa-like animals can be assigned to a living taxonomic category of the coelenterates, principally because no evidence of manubrium or other mouth structures has been found with any certainty, and also the symmetry is not four-fold. Among the Ediacara fossils there are other animals which are also probably coelenterates: *Rangea*, *Pteridinium* and *Arborea* resemble present-day pennatulids (sea-pens, of the coelenterate group Octocorallia). These frond-like creatures were supported by spicules, as are living sea-pens, but the whole structure resembled a simple leaf with grooves running laterally from the central axis, whereas in modern ones the side-branches are separate from one another, like a compound leaf. The Ediacara specimens of *Rangea* and *Pteridinium* somewhat resemble South-West African and Siberian Pre-Cambrian fossils which have also been assigned to these genera.

Of the annelid-like animals, the most important are *Spriggina* and *Dickinsonia*. *Spriggina* has a horseshoe-shaped head and forty to fifty segments; *Dickinsonia* has up to four hundred segments, and in both the number increases with size, in typical annelid fashion. The important points to note about these animals are, first, that they resemble well-known annelid worms of present-day seas: *Spriggina* is not unlike the pelagic polychaete *Tomopteris*, and *Dickinsonia* shows a remarkable resemblance to the ellipse-shaped amphinomid polychaete *Spinther*. Secondly, the position of the parapodia and the apparent sclerotization of the head are features which raise the question whether these animals represent a type of organism somewhere intermediate between the annelids and the trilobite arthropods. Strömer has suggested that *Spriggina* may represent a type of trilobite in which the head, but not the dorsal integument, is sclerotized. True trilobites appear first a little later than the Ediacara fossils, in the Lower Cambrian.

An animal which may be an arthropod is *Parvancorina*, which had a kite-shaped body with an anchor-like ridge lying antero-medially; there are signs of other markings within the outline of the kite, to either side of the mid-line ridge, suggesting that it might have had legs or gills underneath.

A fossil difficult to place in any known phylum is *Tribrachidium*, consisting of a circular platform on which are three equal radiating tentacle-fringed arms. If anything, it resembles some members of the Palaeozoic echinoderm class Edrioasteroidea: three-fold symmetry is not unknown in the echinoderms, and some theories on the origin of the phylum (see next chapter) suggest that a tentacle-bearing lophophorate animal probably gave rise to it. While *Tribrachidium* could be interpreted as such a form, other interpretations can be made of its poorly-known structure.

The skeletons of early invertebrates

As the Cambrian period opens, we see the fairly sudden appearance of a wide range of invertebrate phyla. We are not favoured by 'transition forms', animals linking one phylum with another: rather, the representatives that appear are clearly members of identifiable phyla, however different from present-day members of those phyla. So it seems that delimitation of the phyla appearing in the Cambrian, and probably some others that appear in later periods, took place in the Pre-Cambrian. Why, then, are Pre-Cambrian fossils so scarce? Several explanations have been put forward, most of which are untenable for one reason or another; for instance, some workers invoke a cosmic catastrophe which decimated the entire faunal record either before or after fossilization; others suggest that all Pre-Cambrian organisms lived at the surface of the sea, or in very deep water, or in fresh water, where the chances of fossilization were slim. But the explanation which is most in accord with the known facts is that the ability to build complex hard skeletons was acquired only within the span of so-called phanerozoic time, and that prior organisms could manage to build only 'chitinoid' skeletons, or, at most, to produce only small isolated crystalline spicules. Over the 80 or 90 million years of the Cambrian it seems likely that the ability to produce a calcareous shell of any size was evolved in one phylum after another; as one group 'discovered the secret', this would create selective pressure on competing groups also to produce skeletons or become extinct by competition, as Simpson has pointed out. What factors were responsible for the evolution of the skeleton-building facility is also a matter of debate. Some authorities suggest that the concentration of certain ions may have been different in the Pre-Cambrian compared with later: there may have been a high H^+ ion concentration, or a low Ca^{++}, though this now seems improbable. More likely is the suggestion that a stage was reached as the geological history of the world entered the Cambrian period when invertebrate animals evolved a mechanism for controlling the excretion of calcium, phosphates and carbonates in such a way that these substances could be turned to the animals' advantage as components of a skeleton.

Then, with this new-found structural asset, new habitats and ways of life were opened to a large part of the invertebrate world, and there occurred a period of rapid evolutionary advance as hitherto unexploited niches were filled. So as the Cambrian period progressed the selective interaction between the groups would tend to accelerate the speed of change and intensify the divergent pattern of their evolution, and the earth saw a period of explosive radiation.

The geological background to invertebrate evolution

The basic diversification of the main invertebrate phyla, then, happened during a relatively undocumented period of earth's history, and there is little hope that we shall ever get a true picture of the course of events; as Simpson has put it: 'the paucity of early records is bitterly regretted'. During phanerozoic time, however, the record is more adequate, and as era succeeds era we get a better and better picture of the evolution of invertebrates. In the Palaeozoic era (see Fig. 16.1) we see waves of high evolutionary diversification interspersed with waves of slow advance and even extinction; we see the origin of the chordates from the invertebrates; during its span land plants arose and spread, and we see the first forests appear, opening up enormous new ecological possibilities to the invertebrates; we see waves of inundation of continental land masses, interspersed by waves of marine retreat, and we find captured in the fossil record the effect such orogenic changes had on the invertebrate fauna. Then, as the Palaeozoic era draws to a close, comes a chapter of extinctions and catastrophes, but the forms that managed to survive into the Mesozoic grasped at the ecological opportunities left by the decimated Palaeozoic faunas and produced a faunal pattern strikingly different from anything that had gone before, particularly in the sea.

Time-scale (m.y.)	Eras	Periods	Age of base (m.y.)	Geological conditions	Special fossil bearing depos
	Cenozoic		65	Ice ages	
100		Cretaceous	136	Seas restricted	Chalk
	Mesozoic	Jurassic	190-195	Seas widespread	Corallian
200		Triassic	225		
		Permian	280	Seas restricted	
300		Carboniferous	345	Continents depressed -seas widespread	Coal measures
		Devonian	395	Continents elevated -seas restricted	
400	Palaeozoic	Silurian	430-440		
		Ordovician	500		Arenig beds
500		Cambrian	570	Continental seas widespread	Burgess Shale Poleta formatio
600	Proterozoic	Pre-Cambrian			Ediacara sandstone

lain events in lant evolution	Appearance of invertebrate groups	Important faunal events -invertebrates	Faunal events -vertebrates
cline of forests rise of erbaceous plants			
		Ammonites and belemnites extinct	
lowering plants appear	Irregular echinoids		
	Scleractinian corals		
irst conifers		Decline of many groups	Mammals appear Mammal-like reptiles
	Belemnites Insects		Reptiles appear
and forests appear		Decline of trilobites nautiloids graptolites	Fish radiation Amphibia appear
and plants appear	Ammonites	Reef-building corals abundant Eurypterids prominent	
	Ectoprocts Corals Ostracods Lamellibranchs	Nautiloids and graptolites abundant	Agnathan fish first appear
Rise of marine algae	Protozoa (forams) Archaeocyatha Molluscs Brachiopods Onychophora	Trilobites dominant	
Primitive aquatic plants only	Coelenterates Annelids ?Arthropods ?Echinoderms		

Fig. 16.1

The Mesozoic era saw the earth movements that were to produce the main physiographic features of the modern world; it saw the marine faunas of the world become greater in numbers and more diverse in kind than in the Palaeozoic; it saw the appearance of the flowering plants and their domination of the world's flora; and it saw the phenomenal rise, diversification then eventual extinction of an incredibly successful invertebrate group with important stratigraphical implications, the ammonites.

At the close of the Mesozoic there was a marked retreat of the seas in many parts of the world, to leave the continental patterns very similar to their present form. Many invertebrate groups became extinct, to leave the way clear for the radiation of faunas broadly into the patterns we see today. As the Cenozoic era progressed we can trace the gradual elaboration of present-day faunal provinces by the earth movements which occurred during this time; we see the isolation of what is now the Indo-West Pacific Province from the Atlantic; we see the rise of the Isthmus of Panama to divide the Pacific Tropical American fauna from the Atlantic Tropical American; and finally we see the onset of the Pleistocene Ice Ages, and the appearance of extremes of climate such as the world had probably never known before. Against this backcloth the invertebrate fauna passes into the last scene to be written: the present day, in which the invertebrate zoologist is no longer restricted to a study of only those animals with skeletons, or those animals which happened to die and slump down on a fine silt which preserved the imprint of some of the soft parts.

The Palaeozoic era

The fossils from the Ediacara Sandstone of Australia assume further importance in the story of the early recorded history of the invertebrates, because this deposit is one of the very few in the known outcrops of the world where Pre-Cambrian fossil-bearing strata grade almost without a break into fossil-bearing levels of the Palaeozoic era, about five hundred feet above. The animals that appear in these Lower Cambrian rocks, in addition to the ones that were present in the Pre-Cambrian strata, include Brachiopoda, Mollusca and a sponge-like extinct phylum Archaeocyatha. The important story these fossils tell is one that confirms the claims that biochemical evolution in the sphere of skeletal formation probably occurred over this period. While certain phyla, such as the Archaeocyatha, seem suddenly to appear with a fully-calcified skeleton, others seem gradually to acquire the facility to build one: early trilobites have a shell distinctly less calcareous than later ones; early ostracods and molluscs show a similar state; most of the typically calcareous shell-bearing or reef-building groups, such as the foraminiferans, coelenterates, 'corals' (both hydrozoan and anthozoan), tube-building annelids, ectoproct polyzoans, lamellibranchs and cephalopods, are absent from the Lower Cambrian, and indeed most of the later-named groups straggle into the record either in the Upper Cambrian or even in the early Ordovician period. So the time-span over which invertebrate animals apparently acquired the facility to build a crystalline skeleton was something over 100 million years, and some groups, such as the protozoan Foraminifera, did not acquire the means to secrete a skeleton until a further 100 million years had elapsed, that is, in the Silurian, though they could protect themselves with an accreted skeleton in early Ordovician times.

The early Palaeozoic seas were very widespread. As the era opened, the margins of the stable Pre-Cambrian land masses were becoming inundated by the seas surrounding them, and during the Cambrian vast continental shelves became available for exploitation by the invertebrates. The vast extent of these sub-littoral zones may indeed have contributed to the explosive evolution which occurred then. During Cambrian, Ordovician and Silurian times these extensive shallow seas, though locally disturbed by orogenies and other tectonic movements, were of about the same extent. From the fossil record of these periods the most significant group from the evolutionary viewpoint appears to have been the trilobites (Plate 26), which ranged through the entire Palaeozoic era, though they reached their acme in the early periods and declined in the latter half, becoming extinct in the Permian. Because there are no living representatives, we can but infer their modes of life: many, it seems, wandered over the bottom; some were active swimmers; others burrowed; a few forms could roll up, woodlouse-like, in a protective ball.

In the lowest levels of the Cambrian we find the first representatives of the palaeontologically extensive and important phylum Echinodermata. In the Poleta Formation of Southern California are found together members of three classes of this phylum: Eocrinoidea, Edrioasteroidea and Helicoplacoidea. The phyletic implications of these finds will be dealt with in the next chapter; here, suffice it to say that they all apparently represent animals which fed on the rain of detritus falling to the ocean floor, and it was much later in the history of the phylum that other forms took to a more active mode of feeding, involving a re-orientation of the mouth relative to the sea bed.

In the record of the Middle Cambrian rocks there occurs one of the rare palaeontological windfalls of the geological column, a rock composed of silt so fine that it has preserved not only animals with extensive skeletons, but also carbonized imprints of soft-bodied forms as well. This occurs in the rocky mountains of British Columbia called the Burgess Shale. Though an isolated moment in time, a single-frame 'still' from the cinefilm of earth's history, it provides us, as did the Ediacara sandstone in a slightly less prolific way, with a remarkably full picture of an assemblage of marine organisms at one period in time. The sea bed was a fine silt, with small clumps of archaeocyathids dotted over it (Plate 24); moving over the surface, or possibly burrowing into it, were annelid worms, shrimp-like arthropods, trilobites and an onychophoran called *Aysheaia* (an arthropod retaining some annelid features); in the waters above were swimming trilobites, medusae and a primitive arachnid allied to the later eurypterids. In addition to these forms which we can fairly easily place in conventional taxonomic groups, there are others which resemble no known groups, animals which have left no descendants, and whose relationships we may never know.

In the Ordovician we see the appearance of more new groups. The corals appear here, first mainly tabulates and solitary cups; the foraminiferans, ectoprocts, lamellibranchs and ostracods also enter the record in the Lower Ordovician, though whether this means that only at this time did they evolve the skeleton-building technique, as mentioned above, or whether they really did appear this late is something the record does not tell us. The cephalopod nautiloids, having made an unspectacular entry as small, slightly curved conical shells in the Upper Cambrian, underwent a great radiation. They were a prominent feature of the invertebrate scene throughout the period and during the first part of the succeeding one, the Silurian, but thereafter they declined markedly, except for local bursts at a few horizons; they have teetered on the brink of extinction right up to the present day.

The graptolites, too, became prominent and stratigraphically useful members of the fauna. They are handy tools to the geologist, because many of them were planktonic and had wide distribution patterns, so they help to make lateral stratigraphic correlations easier and more certain. In the graptolites we have one of the most puzzling groups of fossils from a phyletic point of view, for we do not know to which phylum they belong. Are they coelenterates, and therefore rather lowly invertebrates, or are they pterobranch hemichordates, animals with a much higher level of organization? Most graptolites look like mere scratches on the cleaved faces of argillaceous shales (the name refers to this); but some occur in three dimensions, and careful etching out has enabled a surprising amount to be learnt about them, yet still not sufficient to place them in a phylum with certainty. The skeletal tubes of many of them consist of dovetailed half-rings, with a stolon within connecting the individual zooids, in a pterobranch-like way. But on the other hand the branching of the stipe is sympodial, and there were apparently several polymorphic forms of zooid, as in some coelenterates; and no anal opening has ever been shown on specimens in which all other apertures are clearly visible. So this must rank as a problem which cannot yet be solved on the available evidence; most authorities, however, appear to favour a pterobranch (hemichordate) relationship.

During this important Ordovician phase of earth's history, when so much evolutionary radiation was taking place, we see the first signs of the eleutherozoan (free-living) echinoderms. A fantastic fossil record from the Lower Arenig beds of Southern Europe, for instance, enables a remarkably complete picture of the early evolution of the asterozoan (star-shaped) echinoderms to be followed, including the separation of the asteroids and the ophiuroids. At this time, too, we find the first echinoids. The oldest is the Middle Ordovician *Bothriocidaris*, an aberrant, rigidly-plated echinoid; then slightly later in the same period comes the first flexible-tested specimen, an echinocystitoid, thought to lie closer to the stock from which the rest of the great class of the echinoids arose.

In the Silurian there are some changes in the aspect of the fauna. Reef-building corals assume importance; chain-corals, tube-corals, and honeycomb-corals appear and diversify. The articulate brachiopods, ectoprocts and stalked echinoderms radiate extensively. But the trilobites and nautiloids both declined somewhat, possibly due to an increase in the number and efficiency of predators, such as the early agnathan fish, the ostracoderms, which were becoming prominent by this time. Another important predatory group to reach its acme in the Silurian was the eurypterids (Plate 25), primitive arachnid arthropods, some of which grew to a length of ten feet and were probably the largest animals alive at the time. They had a segmented chitinous body with appendages at the head end modified for walking, swimming, defence and/or attack. One further event occurred in the Upper Silurian which was to have a profound effect on the fauna preserved in the fossil record: the ammonites arose from the nautiloids. They differ from the nautiloids mainly in the nature of the septa between the shell compartments: those of ammonites were normally thrown into complex folds and patterns, whereas those of nautiloids were more simple; also, the ammonoid siphuncle, the tiny backward prolongation of the soft parts of the body, was ventral in position whereas that of nautiloids was more central.

One other event of the Upper Silurian is worth mentioning: the first fossils of land plants have been found in rocks of this age in Australia. The conquering of the land by plants must have had a profound effect, albeit badly documented in the fossil record, on the deployment of invertebrates. In subsequent periods of the Palaeozoic, particularly the Carboniferous, we see a distinct fauna of non-marine lamellibranchs from swamp-like regions rich in coal-forming plants.

The Devonian saw a period of comparatively low evolutionary activity. The continental seas were now beginning to become more restricted by orogenies. Though there is understandably no proof, it is possible that the restriction of the shallow-water environment helped to cause the set-back of the invertebrates. The decline of the trilobites and nautiloids continued, and the graptolites too became sparse. On the other hand, some bottom-living animals such as sponges, corals, ectoprocts, brachiopods, lamellibranchs, gastropods and echinoderms continued to radiate into many different new forms, and pelagic organisms like the foraminiferans and ammonites enjoyed an evolutionary burst. The period, of course, saw the radiation of highly efficient vertebrate predators, the fish, creating a selection pressure which was to show its efficiency in the evolution of new and better adapted types within the invertebrate groups that had already been delimited.

In Carboniferous seas the graptolites finally became extinct, and so, in the latter part of the period, did the trilobites. Some groups, such as the stalked echinoderms (Plate 27) continued to thrive, and the period saw the origin of one group that was to become a prominent (and stratigraphically useful) representative of the fossil record: the belemnites arose from the nautiloids. Belemnites were straight-shelled dibranchiate cephalopods which formed the basal stock from which the modern squids, cuttlefish and octopods arose. They had a chambered shell with a thick, heavily calcified region, the phragmacone, posteriorly, probably acting as a counterweight to the main mass of living tissue of the body at the front end. The period also saw an event of profound significance to the later terrestrial organisms, for here we find the first evidence of the insects. When they first occur in the record, they are quite highly evolved, indicating that their origin was probably a good way back in time.

The last period of the Palaeozoic era, the Permian, was probably the most disastrous and ill-fated for the animal kingdom. The reduction in the shallow seas was still continuing, but this factor alone could probably not account for the vast extinctions of marine organisms which occurred as it drew to a close. Whatever the reason, whole classes of invertebrates were decimated, leaving, in many cases, such as the trilobites, nothing to carry on into the Mesozoic, or in others, such as the echinoids, perhaps a single genus only. Cosmic catastrophe has again been invoked as the cause; some writers have suggested that a particularly fierce burst of sun-spot activity or other cause of severe radiation may have occurred. But this seems unlikely, since it is the marine forms that appear to suffer, whereas the land faunas seem to be virtually untouched.

The Mesozoic and Cenozoic eras

So the earth entered the Mesozoic era with a very much depleted fauna, for some reason. The foraminiferans, sponges, coelenterates, ectoprocts, brachiopods and gastropods are very rare in Lower Triassic strata, and very few groups seem to have survived the transition with any success. Such a disaster had the effect of vacating many ecological niches, and we find a striking change in faunal composition as the invertebrate world deployed again.

When the Triassic period opened, the world's continents had risen from the sea in basically the same form as we know them today. As the Mesozoic era advanced, however, there was a progressive marine inundation until towards the end of the Cretaceous period, when the seas were at their maximum. Then the seas again retreated, to reveal once more the rough continental outline of today.

The Triassic chapter in invertebrate history shows a rapid re-establishment of structural complexity within those phyla and classes that had struggled across the Palaeozoic-Mesozoic boundary. Only very few genera of ammonites entered the era, but by its end there were about four hundred genera. Again, among the echinoids only one genus is thought to have survived the crisis, yet by the beginning of the next period, the Jurassic, there were some nine new orders, which, on the rather slim palaeontological evidence we have, all appear to have evolved from the one genus that managed to straddle the Palaeozoic-Mesozoic boundary.

The story of Mesozoic marine life, so well documented as it is, is too complex to be told in detail here. The reader is referred to papers listed in the references for a more detailed treatment. Here, the broadest outline must suffice. The dominant phylum is probably the Mollusca: the ammonites enjoyed a phenomenal radiation, the belemnites became widespread (Plate 28), and both the lamellibranchs and gastropods evolved a large and diverse fauna. Ammonites and belemnites suffered severe set-backs at the end of the Triassic, but rallied again in Jurassic and Cretaceous seas, before becoming finally extinct at the end of the Cretaceous.

The corals, recovering from their Permian set-back, became prominent again in the Middle Triassic with the appearance of the true modern reef-building forms, the Scleractinia; which Palaeozoic group gave rise to them we do not know. The corals have probably always been temperature restricted, as they are today, yet in the Jurassic period they occur as far from the equator as 55° North, and fairly extensive reefs are found in British Jurassic rocks, 20° of latitude further north than their present distribution.

The Jurassic period also saw an important event in the great echinoderm class Echinoidea: on at least two separate evolutionary lines the sea-urchins lost their radial symmetry and started a trend which was to lead to the highly successful burrowing groups of irregular echinoids. The first group to appear in the record with this new bilateral symmetry was an atelostomatous form, that is, one in which the characteristic feeding apparatus of the regular echinoids, called Aristotle's lantern, is absent; then, very soon afterwards, a gnathostomatous group (retaining the lantern) became bilateral. The atelostome stock gave rise through subsequent time to the spatangoids, highly specialized and successful burrowers; and the gnathostomes gave rise to the clypeasteroids (sand-dollars), also burrowers but mainly restricted to warmer waters.

The Cretaceous period was the most spectacular of the Mesozoic. The marine transgression of the continents, begun as the era opened, continued until many regions of land which had been inundated previously only far back in time were under water. Invertebrate groups radiated as never before. There were vast periods in some parts of the world when calm seas and low-lying adjacent land masses allowed the deposition of animal skeletal remains almost completely unpolluted by terrigenous material. These conditions continued for many millions of years, and produced the very pure limestones we call Chalk. Within the Chalk lived spatangoid echinoids in vast numbers, and so complete is the fossil collection which can be made from successive levels in these deposits that an almost unbroken sequence can be obtained of a single evolutionary line continuing for many millions of years. Such examples (and the Chalk spatangoids present one of the best, enable the student of evolution to trace the mode of change in a single line, the directional evolution of individual characters (micro-evolution), on which the greater changes in higher taxa (macro-evolution) depends.

At the end of the Cretaceous there was a drastic reduction again in the seas, leaving the land pattern very similar to today's. The effect on the fauna, as might be expected from the happenings of previous phases of marine regression, was a series of set-backs to marine life, but the crisis was nothing like so serious as that at the close of the Palaeozoic. One group which suffered fatally was the ammonites which had been such prominent members of the shallow-water marine fauna for so long.

Marine invertebrate faunas became almost completely 'modern' in constitution in the Cenozoic. As mentioned previously, during this era the tectonic movements that helped to form the zoogeographical provinces of recent seas can be traced, and the establishment of modern faunal types can be seen in the record. Of course, in common with almost all other levels of the geological column, animals with extensive skeletons, and particularly those living in certain parts of the sea, assume perhaps an unjustified prominence. One must not forget that soft-bodied animals were as important in the general economy of any habitat as were the ones more prone to fossilization, and evolutionary radiation was progressing in those places where the chances of fossilization were slim. In this account we have of necessity over-emphasized the skeletally-protected invertebrate animals from marine sedimentary rocks; an equally important and intricate story will have gone by unrecorded in other environments.

Chapter 17

The phylogeny of the invertebrates

THE SPAN of life on earth during which the radiation of living things occurred is believed to be at least two thousand million years, but we have seen in the last chapter what a relatively small part of this is recorded in the geological column. We have seen also that when fossils start to appear the main invertebrate phyla have already been evolved, if one can generalize from those having members capable of leaving a record. So the phyletic picture must be reconstructed mainly from living representatives of the groups, and the basic assumption must be accepted that among present-day animals are ones retaining primitive form. Hanson has given a poignant simile: the phyletic biologist has to reconstruct a tree in every detail of its branching when he is given only the outermost leaves. Admittedly, the leaves are not all identical and contain clues as to their position; and again, in the upper part of the tree a few bits of the branches themselves may be available. But the task of relating the animal phyla in a natural scheme is beset with enormous problems, and it is small wonder that so many interpretations, tolerably consistent with the scanty facts, are possible.

One point must be clearly made: no arrangement of the animal phyla in evolutionary sequence can yet be anything but conceptual: this is a realm of theories and debate, and all interrelationships of phyla are the result of loading the facts for and against a particular arrangement on the scales of reason and hoping that one side will be heavier than the other. If we are scientifically honest, as Kerkut has pointed out, we will conclude that we have no direct proof of the relationships of any phylum of the animal kingdom. Nonetheless, the subject is very necessary for our complete understanding of the process and course of evolution. Further, it demands as complete a knowledge of animals as it is possible to get to date; all the available facts on all groups are barely sufficient, so anything less than a complete knowledge is that much worse.

The main problem inherent in phyletic zoology is that of deciding between the effects of parallel evolution and of relationship. It remains true that the more complex a structure the less likely it is to have evolved twice in the same way; the final form of the structure will be reached by different embryological pathways from different starting points, and these differences will be reflected in the finer details of construction. In the classic examples, such as the convergence between the vertebrate and cephalopod eyes, such differences are fairly easy to see, but in the great majority of cases no such easy clues are afforded, as is pointed out later (p. 365).

The other great difficulty is of deciding what sort of evidence should carry most weight. Is biochemical similarity more likely to reflect true relationship than morphological? Should embryological evidence weigh more heavily on our scales than evidence from the adult form? The answer to the first question must be that at present there is no evidence suggesting that biochemical similarity is any more likely to reveal relationship; there has been a sad tendency, even recently, for biochemical evidence to assume undue importance in phylogenetic speculation, in some cases leading to unjustified rejection of long-held theories which were backed up by fossil evidence. The second question, on the relative importance of embryological evidence, is harder to answer. If we accept von Baer's principle that the early stages in the life history of animals are more alike than later stages, then one is tempted to give more significance to observed similarities in the young stages of groups whose relationship is suspected. But at the same time we must be aware that comparatively minor differences in the nature of the egg, such as the amount of yolk it carries, can exert a profound effect on the subsequent nature of the embryo. Further, embryos and larvae themselves will be subject to the pressures of natural selection which might result in modification away from the ancestral pattern. So here again considerable caution must be shown.

One might think that students of invertebrate zoology, when faced with the same evidence, would come up with roughly the same conclusions as to its application to the evolutionary picture – that there would be a broad similarity in the phylogenetic trees produced from the available facts. But this is far from the case. The views of individual writers may cut right across those of their predecessors; their own lines of evidence will assume peculiar importance, and cogent arguments will be brought up in their defence. The reader is forewarned: the present writer is no exception. Later in this account a new interpretation is presented on the significance of the lophophore in the evolution of the higher invertebrates, and here, as in previous essays on invertebrate relationships, the writer's own view is defended vehemently, because such a view does not appear to have been put forward in quite the same way before.

In summary, the phyletic zoologist must be aware of the possibilities of convergence. Phylogenetic statements must, in consequence, be based on a balance of probabilities, and must be liable to disproof and modification as new evidence becomes available.

The origin of the invertebrates

The usual view is that the autotrophic bacteria probably gave rise to those single-celled (or, better, acellular) organisms feeding in a plant-like way, the Protophyta, and that these gave rise not only to the higher plants but also to the Protozoa (Fig. 17.1). Though no bacterium resembles closely a possible ancestor, some bacteria have many of the protophytan characters, such as vacuoles, mitochondria, nucleus and chromosomes, flagella, reproductive spores and a form of sexual process; some also appear to have a cell membrane similar to that of the Protozoa. There are points for the other side of the scales too, such as the dissimilarity between the bacterian flagellum, with its single or double axial filament, and that of the Protozoa and all subsequent animals, with the typical 'nine-plus-two' axial structure; but it would appear highly possible that the autotrophic bacteria, or something very like them, were the group from which the invertebrates arose.

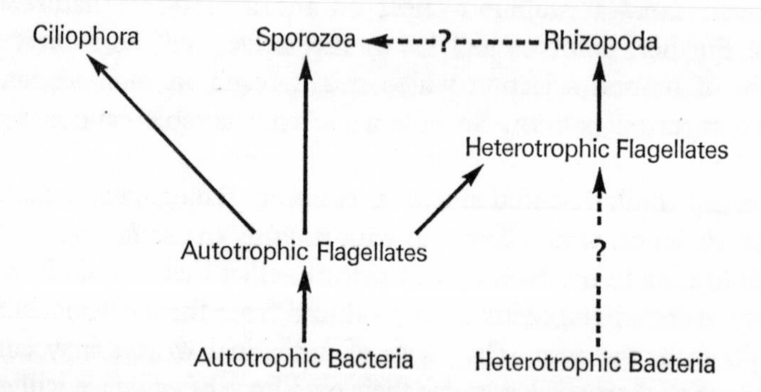

Fig. 17.1 A possible scheme for the origin of the more important 'protozoan' groups, with alternative derivations of some groups shown as dotted lines. This figure and Figs. 17.2 to 17.6 are highly tentative and give the more usually accepted relationships only; the possibility of the polyphyletic origin of groups is not indicated in the diagrams

This being so, it is highly probable that the most primitive of the 'protozoan' classes is the Flagellata, since so many of its members are autotrophic. Further, a link with the lower plants is suggested by the fact that some filamentous brown algae produce 'swarmers', single-celled flagellated dispersal agents, almost indistinguishable from chrysomonads such as *Chromulina*. This view appears to be most generally in vogue at the moment, yet the classical view, around the turn of the century, was that the Rhizopoda are primitive, since it was thought that the earliest protoplasm did not contain chlorophyll. In the 'classical' scheme, the Mycetozoa, a group containing the slime-moulds, were the progenitors of the Protozoa. What is still not clear is whether the acellular invertebrates have had a single origin from the pre-protozoan organisms, or whether they are polyphyletic. This raises an important general point which will arise again and again in our discussion of the status of invertebrate phyla: whether a particular phylum is really a natural entity with a single origin, or whether it is a 'grade of organization' reached independently from more than one group of lower organisms. Knowing the efficacy of selective pressures as we now do, we are justified in suggesting that when one group of organisms acquires a new and advanced character that puts it into a higher grade, the chances are that a similar event will occur in other groups, similar to the first, to raise them to the new grade too. In fact, where there is good fossil evidence at the time of a group's origin, as happens, say, in the therapsid reptiles and the mammals, then it does appear that polyphyletism is the rule rather than the exception, and there is no reason to think that the same principle does not apply in the lower animals. This point assumes importance in such evolutionary attainments as a metazoan, coelomate and lophophorate condition, and the acquisition of such important invertebrate structures as nematocysts, striated muscle, respiratory tracheae and compound eyes, all of which, and many others, may have originated many times.

To return to the origin of the Protozoa, it is not clear whether the flagellates arose several times, though such a possibility seems likely. That there are close links between the Flagellata and the Rhizopoda is, however, strongly suggested, again probably along several lines: the rhizopods sometimes have flagellated larvae, and there are rhizoflagellates which alternate between a pseudopodial and a flagellar mode of locomotion, according to the ecological situation in which they find themselves. Among the Sporozoa, some have flagellated sporozoites, to suggest that they too arose from a flagellate ancestor, but others have amoeboid sporozoites, suggesting affinities with the rhizopods. The Ciliophora are difficult to place: though there are holotrichous ciliates with uniform ciliation and a terminal pharynx from which the higher ciliates could have arisen, yet their unique nuclear dimorphism, sexual process of conjugation and transverse binary fission cannot be linked satisfactorily with any non-ciliate protozoan group.

The metazoan condition

This is probably the most hotly debated stage in invertebrate evolution. A full assessment of the enormous literature is clearly impossible, and a brief summary of the main ideas must suffice. Broadly, there are three theories, two of which see the Protozoa as ancestors and one the Metaphyta, or higher plants. The first theory, proposed by Haeckel, Lankester and others, and developed by Metschnikoff, Jägersten, Marcus and others, derives the Metazoa by integration and specialization of protozoan colonies. Many flagellates show a tendency to colony formation, sexual reproduction and regional differences in individuals of the colony, such as special cells set aside for reproduction and others forming a distinct front end to the colony. According to the Haeckelian view, the hollow colonial phytomonadines, such as *Volvox*, are the most likely candidates, but according to Lankester it was the solid, morula-like colonies such as *Synura* or *Pandorina*, and according to Metschnikoff a hollow colony became solid by an inwandering of ectodermal cells to form an endoderm.

So Haeckel saw the original metazoan animal as a hollow sphere of cells, his so-called 'blastaea', which became a two-layered 'gastraea' by inpushing of the originally single surface layer to form an archenteron. Lankester and Metschnikoff saw it as a solid animal called the 'parenchymula' and 'planula' respectively; this solid animal fed by phagocytosis and only later evolved a gut by splitting the endoderm. The phylum of metazoan animals which is considered by many to be morphologically the simplest is the Coelenterata, and the two-layered adults do indeed resemble Haeckel's gastraea; but in their development they strongly support Metschnikoff's idea (Fig. 17.2), because they often pass through a planula stage and form the gut by an inwandering of cells which become the endoderm, later developing a hollow lumen and a single opening to the exterior.

The second theory of protozoan origin sees the Metazoa as originating by cellularization of a syncytial ciliate, that is, by the formation of cell walls in a multinucleate, acellular organism. Though this idea appears to have been suggested first in the middle of the nineteenth century, and to have been reiterated in none too forceful a way in succeeding years, its

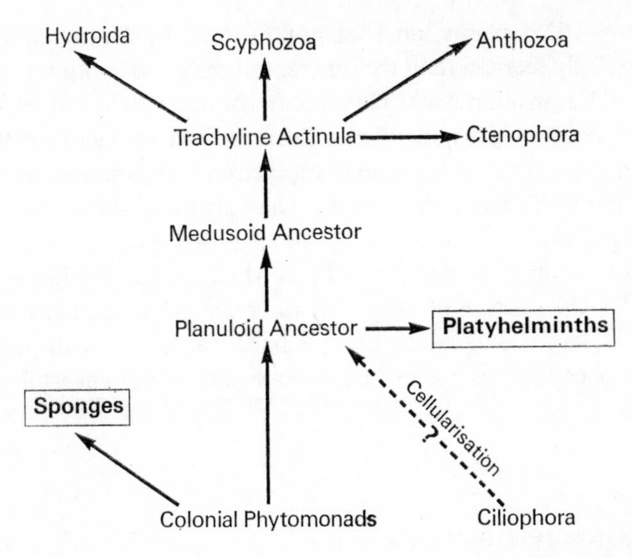

Fig. 17.2 One possible scheme for the origin of the coelenterate groups from protozoan ancestors

main champion was Hadži, who saw the most primitive metazoan resulting from this process as an animal like the present-day Acoela, a class of the turbellarian flatworms. He considers these animals likely candidates because in the adult form they lack a true gut, are syncytial and the external surface is ciliated. Hadži's version of the theory then goes on to derive the coelenterates from the flatworms, and he sees the Anthozoa as the most primitive class of the cnidarian coelenterates, mainly on the grounds that there is incipient bilateral symmetry in this class alone, and the range of structure of the nematocysts is simpler than in the other classes. But there are strong objections not only to the primitive position of the Acoela, with their highly specialized mode of nutrition, but also to the derivation of the coelenterates from the flatworms. Steinbock, while accepting the primitive position of the acoelans, sees the coelenterates as a separate derivation from the Protozoa, and this modification of the theory makes it slightly easier to accept. Some authorities suggest that the occurrence in so many coelenterates of the planula larva, sometimes with an incipient gut, may indicate a derivation of the platyhelminths from the coelenterates, or alternatively a common origin from a planula-like organism. But however the platyhelminths arose, it seems most likely that it is not the Acoela which are primitive; this point is mentioned again below (p. 369).

The third main theory for the origin of the Metazoa states that they arose from many-celled plants, an idea put forward by Hardy, not so much as a more likely method than the others, but rather as another possible way in which the transition could have occurred. After they had evolved from single-celled autotrophic organisms, the metaphytes, as a result of shortage of phosphates and nitrates, began to capture and feed upon small organisms in much the same way as do insectivorous plants. A difficulty inherent in the other theories which this idea by-passes is how a metazoan animal could have evolved from a protozoan colony in which each individual was already adapted for the capture of prey. But the main point against it is that the cellulose cell wall of the metaphytes is hardly adapted for the extrusion of pseudopods capable of dealing with the digestion and absorption of small organisms.

Coelenterate relationships

There are such clear similarities between the Ctenophora and the Cnidaria, such as construction with two principal layers of cells and tetraradiate symmetry, the occurrence of a planula larva and the presence of nematocysts, that most authorities group them together in the phylum Coelenterata, a name which emphasizes the presence in both of a gastro-vascular cavity with no anus. In this phylum, as in others, it helps in suggesting a possible origin if we first deduce which group is primitive; in the coelenterates there occur two temporal polymorphs, the hydranth and the medusa, and the first problem is to deduce, if possible, which form evolved first. The view of Hyman is widely accepted, that the medusoid form evolved before the hydroid (Fig. 17.2), principally because the coelenterates normally have separate sexes, and such a situation is more applicable to a free-swimming form like the medusa than to a sessile one. Further, a suitable model for the primitive form is still seen in some of the trachyline cnidarians, in which the actinula larva gives rise to a sexually-reproducing medusa whose eggs develop into the actinula again. Then the next stage is seen in those trachylines which interpose a hydroid phase between the actinula and the medusa. So in this type of life history we see a possible starting point for the radiation of groups emphasizing the different polymorphs in this cycle. There is also a very convenient 'passage-form' among the trachylines called *Hydroctena*, with a pair of tentacles arising in sunken pits, which provides a possible link with the Ctenophora.

The origin of the Porifera

Most authorities say they stem from the flagellate protozoans (Fig. 17.2), though an origin from a planula was suggested by Lankester, an origin from the radially-symmetrical gastraea by Haeckel and an origin from a bilaterally-symmetrical gastraea-type by Jägersten. If the ancestor were a flagellate, it may have been a colonial choanoflagellate, with similar type of cell, or it could have been a volvocine colony with similar type of development.

Bilateral symmetry – the Platyhelminthes

We have already mentioned that an organism similar to the present-day planula larva of coelenterates provides a convenient and widely-accepted ancestor to the Platyhelminthes; it provides a link between a phylum showing almost perfect radial symmetry and one showing bilateral symmetry; it is a level of organization which shows how the archenteron could have formed from a mass of endoderm cells. There is no dispute that the Turbellaria among the platyhelminth classes is the primitive one, having as it does mostly free-living members, but it is far from clear whether the almost gutless Acoela or some other order is primitive. The acoelan epidermis is often syncytial, and the epidermal cells sometimes have basal muscle fibres, both coelenterate-like features; but on the other hand, the mode of nutrition would appear to be highly specialized, relying on a symbiotic relationship between the animal and acellular algae which live and multiply in the interior of the body. Possibly the rhabdocoels which also have a number of supposedly primitive features, such as a reproductive system sometimes lacking discrete ducts, are better candidates, and the acoelans specialized derivatives from them. It does seem likely, however, that the rhabdocoels were ancestral to the other classes of the platyhelminths, which are all parasitic, since endocommensal rhabdocoels occur from whose structure that of the fully parasitic flukes and tape worms could have been derived. The suggestion is that flukes and tape worms arose from the rhabdocoels independently, possibly along several lines each.

The account so far has necessarily emphasized the concept of grade of organization – of the probable acquisition of features appropriate to a higher evolutionary level sometimes along several different lines. We have seen that the Protozoa have reached an advance on the pre-protozoan level probably along several independent lines; that the early metazoans represent the next higher grade, again probably on several different lines; and that the attainment of bilateral symmetry in at least one of these lines can be considered the next.

Now we come to a further grade, the Nemertina, where we see for the first time the provision of an anal opening to the gut, a hydraulically-operated proboscis, and, for the first time, a blood vascular system. As it happens, not only is this a good example of a group of animals representing

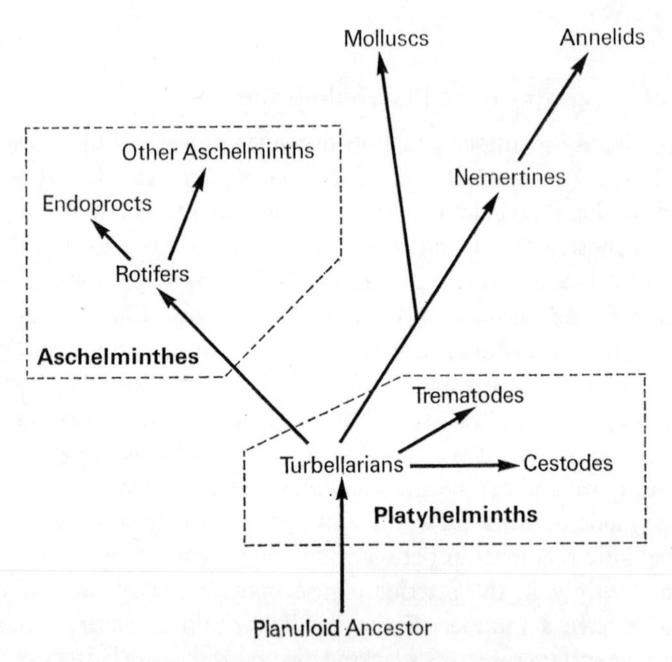

Fig. 17.3 One possible scheme deriving the higher worm-like groups (aschelminths and annelids) from the tubellarian-nemertine complex

a higher grade, but there is also a convenient group of turbellarian flatworms, the kalyptorhynchoids, from which at least some members of the phylum could well have evolved (Fig. 17.3). This turbellarian group, too, has a rapidly-eversible proboscis at the anterior end, but its members do not show any evidence of the other features which raise the nemertines to a higher grade. At the nemertine grade we reach the culmination of invertebrate evolution of animals with a solid body: at this point the important advance in grade was the acquisition of some sort of body cavity.

The pseudocoelomate grade – the Aschelminthes

In the acquisition of a secondary body cavity came a great advance which was to have enormous implications not only in its immediate effects on the individuals possessing it but in subsequent evolutionary events. It would appear most likely that the pseudocoel of the phylum Aschelminthes and the true coelom of subsequent groups are very different in nature, so unlike is their embryological derivation and structure: the pseudocoel is a persistent blastocoel and is not normally bounded by an epithelium, while the true coelom forms *de novo* in the mesenchyme and has a limiting coelomic epithelium.

It seems to be generally agreed that the Rotifera are the most primitive members of this phylum, and so we must seek an acceptable origin for this group. Of the various theories put forward the most acceptable seems to be that the creeping benthic forms are primitive among the rotifers, and may well have evolved from early creeping metazoans, such as the turbellarian flatworms (Fig. 17.3). In support, the protonephridial systems of both groups are similar, there is a tendency in both for the formation of cuticularized structures, and the female reproductive systems are said to be similar. The relationships of the other groups within the phylum are obscure, to say the least. The aschelminths are almost certainly polyphyletic, the blastocoel having persisted into the adult stage as a body cavity on more than one occasion, but the possible ways in which the groups within this phylum are related need not concern us.

One further point deserves mention in the pseudocoelomate groups: though some authorities claim that the ectoproct and endoproct 'bryozoans' are related, the evidence does seem to be against such a conclusion, mainly because the body cavities and the tentacle systems of the two groups are so different. It seems far more likely that their superficial external resemblance is the result of convergence and that the Endoprocta properly belong within the aschelminths, possibly fairly close to the rotifers.

The coelomate grade – the Annelids

We now pass on to the origin of a completely different type of body cavity, the true coelom, and to the origin of probably the most important group of invertebrate animals from a phyletic viewpoint, the Annelida (Fig. 17.4). They are regarded as of prime importance because from them several great groups of invertebrates have arisen; some of the derived phyla have undergone specialization within themselves to produce highly advanced and specialized types, and others, though in themselves rather

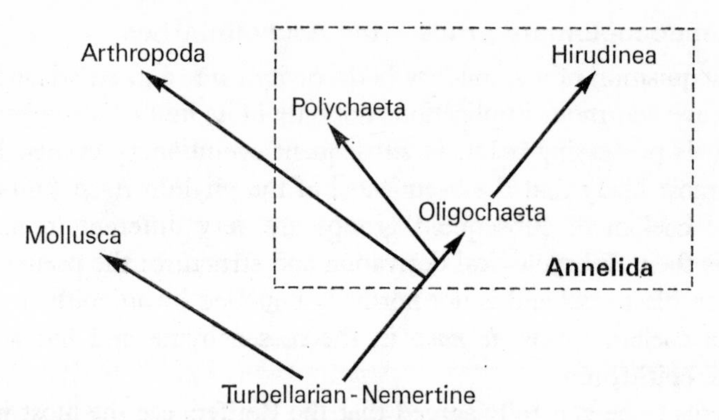

Fig. 17.4 Possible relationships of the annelid groups

insignificant, have provided a spring-board to the organization of the chordates. The annelids, then, are here considered paramount in the phylogeny of the invertebrates.

In animals alive today there are two main ways in which the coelom develops in ontogeny, one by a splitting of the mesoderm (schizocoely) and the other by enlargement of diverticula of the gut (enterocoely). Both these developmental processes have figured in theories on the phyletic origin of the coelom. But strangely enough neither appears as attractive an idea to explain the origin of the coelom as one based on a process which is not well supported by embryological evidence, namely, the gonocoel theory, the origin of the coelom from enlarged gonadial sacs. The evidence for and against the various theories has been reviewed by Clark, and though all methods may possibly have occurred during evolution, only the gonocoel theory will be mentioned here, because it is one that can be closely identified with a plausible origin for the annelids.

Briefly, the theory envisages the pseudometamerically-arranged gonadial sacs of a lower invertebrate, such as a turbellarian flatworm or a nemertean, becoming enlarged and remaining fluid-filled, so that muscular distensions produced in one part of the body could be transferred via the fluid to other parts. The theory helps to explain the close association of the gonads with the coelomic wall and the opening of the gonoducts from the coelom. It assumes an association of the coelom and of metameric segmentation from the start, an association seen as a serious disadvantage to the theory by many authorities, but seen as a distinct advantage to its acceptance by the present writer, chiefly because the idea put forward here on the origin of the non-segmented coelomates sees them as derivatives of the segmented forms, not as precursors.

The origin of the molluscs

The difficulty in deciding whether the serial repetition seen in several of the supposedly primitive molluscan groups is the remnant of true metamerism, or is nothing more than unconnected repetition of certain organs, lies behind the difficulty in assessing the position of this group in the invertebrate scale. On the one hand, the almost identical early embryology of the amphineuran and gastropod molluscs and that of the polychaete annelids suggests that molluscs and annelids should be fairly closely related; on the other, the points of resemblance between the molluscs and the turbellarian flatworms, such as the general body plan, locomotory mode, presence in both of intracellular digestion and similarity in the development of the nervous system, suggest affinity here too.

The ancestral mollusc is generally seen as a creeping animal with ciliated foot but using muscular ripples as its main propellant; with specialized head end containing a rasping organ for feeding; and with a single low-conical shell laid down by a mantle of tissue shrouding the dorsally-held visceral mass. This archimollusc can best be regarded as resembling a primitive prosobranch gastropod before torsion, though such an animal does not exist.

What, then, is the true nature of the serial repetition of organs, particularly in the monoplacophoran molluscs? Does the presence in *Neopilina* of five pairs of gills, six pairs of renal organs, two pairs of gonads and auricles and a variable number of shell muscles represent the legacy of a metameric ancestor? It appears not. It is, first, difficult to 'balance the segmental arithmetic', in Morton's words; secondly, there are other examples of organ repetition elsewhere in the molluscs; and, thirdly, the single shell of *Neopilina* is surmounted by a tiny spiral protoconch, suggesting a derivation from an unsegmented gastropod-type. The fairly extensive coelom of this intriguing animal is an awkward complication, and, in this view, must be regarded as a specialization from the more modest coelom of other members of the phylum. The molluscan coelom, in fact, strongly supports a gonocoel origin, because the only cavity that can be regarded as coelomic, the pericardium, is generally interpreted as a dilation of the gonadial ducts, with which it is confluent. But if this is so, it is a gonocoel probably unconnected with metamerism.

A reasonable conclusion on the relationships of the molluscs, then, would seem to be that they stemmed from a non-coelomate grade independently of the annelids, possibly, like them, from within the turbellarian-nemertean complex; they may have shared with the annelids an ancestor showing their common mode of early development and a trochophore larva, and the two phyla may have attained a coelomate grade independently, the one developing a single cavity to permit pulsation of the heart and the other developing a paired series to improve locomotor and burrowing efficiency.

The arthropod grade

There is little doubt that the Arthropoda arose from the early annelid grade, but the pattern of their evolution and the annelid stem-group or groups which gave rise to the phylum is far from clear. It is claimed, with a wealth of comparative detail in support, that the advance to an arthropod grade occurred along more than one line. If this is so, then it means that those features which add up to an arthropod-type of organization, such as the exoskeleton, tracheal respiration, malpighian excretory tubules, compound eyes, etc., have probably originated at least twice, a degree of convergence which some find difficult to credit. But there are no grounds for escaping this conclusion: the trilobite-chelicerate-crustacean complex on the one hand is clearly a separate division of the arthropods from the onychophoran-myriapod-insect complex on the other. There may also have been evolutionary lines from the early onychophorans or myriapods to the small phyla (or sub-phyla) Tardigrada and Pentastomida (Fig. 17.5).

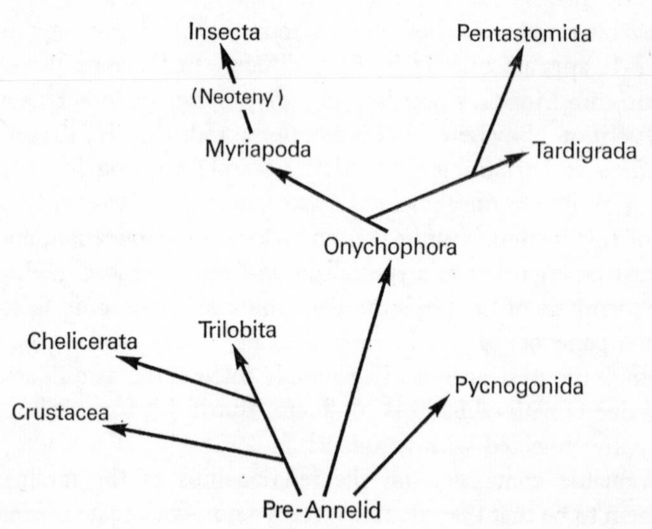

Fig. 17.5 Possible derivation of the arthropod and related groups, based on Tiegs and Manton

The arthropods represent *par excellence* a grade of organization. While the ancestral grade, the annelids, exploited the potentialities of the very flexible body form, with a hydrodynamic 'skeleton', the arthropods have advanced into a grade using a true skeleton, including endophragmal components, and skeletal muscles. Either as a consequence of this, or independently, the coelom has become reduced and the segmentation modified. But we do not have a clear picture of the steps along which this evolutionary advance was made; it seems that an unknown early annelid with parapodia probably gave rise to a protonychophoran, something like the Middle Cambrian *Aysheaia* (Chapter 16, p. 356) on one line, and a different early annelid, probably a detritus or particle feeder, gave rise to the trilobites and chelicerates along two different lines; possibly the crustaceans came off the common trilobite-chelicerate stem early on, but there is little evidence to say more than this.

The lophophorate condition – the Minor Coelomates

The annelid trochophore larva consists, in general terms, of an almost spherical body with an equatorial ring of cilia and sometimes an extra ring of cilia near the lower pole. Cilia from the equatorial ring continue a little way into the mouth, which was formed at the site of the blastopore (protostomatous), and food is wafted by the adoral cilia into a recurved gut, opening at an anus at the lower pole. The coelom, formed by schizocoely within the mesoderm, has a pair of nephridia within it, opening to the exterior. To form the segmented adult annelid, segments are cut off at the larva's lower pole, and an elongated vermiform animal results which is adapted either to moving across a surface or to burrowing within a substratum.

Its general structure and phylogenetic position suggests that the trochophore offers a most suitable starting-point for the evolution of a branch of invertebrates consisting of a number of somewhat insignificant phyla called for convenience the Minor Coelomates. This collection consists of the Sipunculoida, the protostomatous lophophorates (Phoronida, Ectoprocta and Brachiopoda), Pogonophora and Hemichordata, and within the same complex of phyla we can include the rather more successful phylum Echinodermata. Somewhere in this grouping too, perhaps, belongs the phylum of pelagic arrow-worms, the Chaetognatha. The basic mode of life of all these phyla, with the exception of the chaetognaths, is sedentary and filter feeding, relying heavily on the rain of organic detritus falling to the bed of the sea or pond.

The relationships of these phyla have hitherto taxed the ingenuity of phylogenists, because they represent a group of non-segmented phyla – or, to some, a group with three segments only (see p. 379) – with a typical coelom of mixed schizocoelous and enterocoelous formation. Further, they show a mixed protostomatous and deuterostomatous fate of the blastopore, and some retain the trochophorea-type larva while others have another, the auricularia-type which is said to be very different. The difficulty in placing them in an evolutionary scheme arose because they had either to have evolved before the annelids (if so, from what?), or from the annelids, in which case it was hard to see why no trace of segmentation remained in any of them to betray their ancestry. In the scheme presented here, they are seen as arising from the annelids, but from the unsegmented annelid trochophore larva.

The transformation from this larva involves an acceptance of the evolutionary efficacy of *paedomorphosis*, linked with *neoteny*. Simply, this process envisages the postponement of metamorphosis to the adult form, so that the larva can become precociously mature sexually – that is, can reproduce while still in the form of the larva to give rise to individuals which inherit this trait. In this way, as Garstang, Hardy and many others have emphasized, an evolutionary line leading to a specialized adult can form a branch and initiate a new line (the word neoteny means this) with the larval morphology of the previous line as a foundation on which to built it. Garstang, showing his flair for verbal economy, sums up the relation between the Haeckelian view, the principles of von Baer and the concept of neoteny by writing: 'Ontogeny does not recaptulate phylogeny: it creates it'.

In the origin of the Minor Coelomates we see a situation which can best be explained by neoteny, one to add to the many examples already postulated. If the trochophore larva of an annelid were to stretch in one direction, taking the loop of the recurved gut with it, and were to become sexually mature, then an animal with the basic form of a Minor Coelomate would result. As such an animal increased in size, the sparse provision of feeding cilia in the oral region would hardly suffice, so it is not difficult to see such an animal developing folds of the body wall orally simply to increase the ciliary surface area. Such an animal is unlikely to have retained a pelagic existence, because of the difficulties of movement, and once it has sunk to the ocean floor, and possibly also burrowed into it so that only its mouth, anus and nephridiopores protruded, its ciliary folds would become target for predation. Any form of retraction requires subsequent extension, and in the case of the ancestral Minor Coelomate this would have been solved

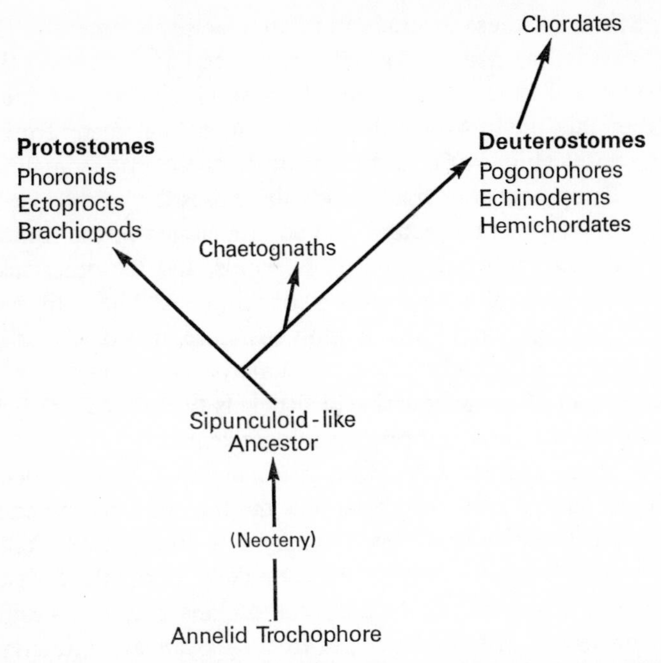

Fig. 17.6 Suggested origin of the Minor Coelomate phyla. A sipunculoid-like animal, a 'protolophophorate', is seen as arising by neoteny from the annelid group, then dividing into two main groups, the protostomes and the deuterostomes, both retaining the lophophore in the adult but differing mainly in the mode of development

by increasing the hydraulic pressure of fluid in that part of the coelom within the ciliary folds; fluid pressure is the only way a soft structure can be extended when no skeleton is present. But if this protraction were brought about using the perivisceral coelom then all other regions of the body would react to the change in coelomic pressure during eversion. Not only would this make the animal unstable in its burrow, but the activities of, for instance, its alimentary tract would be impaired. So there would appear to be good reasons why selection should favour subdividing the coelom in such a way that a special anterior compartment of it is set aside solely for protraction of the feeding organ, or lophophore. An important fillip to this theory is that there exists today an animal group which provides a convenient model for this stage in the evolution of the Minor Coelomates: the Sipunculoida.

The early cleavage process in the sipunculoids is indistinguishable from that of the annelids, and they share protostomy, schizocoely and trochophory with the annelids – indeed, they were almost always included within the annelid phylum on these grounds until comparatively recently. But now they are regarded as Minor Coelomates, possessing as they do the non-annelid features of recurved gut, lophophore and lack of segmentation.

It is from this phylum that the other Minor Coelomate groups have most likely arisen (Fig. 17.6). In the case of the three protostomatous phyla Phoronida, Ectoprocta and Brachiopoda this raises few problems, because the general developmental pattern and adult anatomy of the sipunculoids is almost identical with those of the phoronids, and the other two phyla can be derived from the phoronids with fair ease. The ectoprocts and brachiopods are highly modified in adult structure, it is true, and some of the brachiopods apparently adopt an enterocoelous mode of coelom formation instead of the ancestral schizocoelous type; but nonetheless the close relationship of these four phyla is seldom denied.

But it is quite another story with the remaining Minor Coelomates. Unless the remaining group of phyla, plus the echinoderms and chordates, represent an entirely separate evolutionary line from an acoelomate (or pseudocoelomate) ancestor (which seems unlikely, to say the least), then a change from a schizocoelous to an enterocoelous coelomic origin must occur somewhere in the phylogenetic tree of coelomate animals; so too must a change in the fate of the blastopore from protostomy to deuterostomy, a change in larval type from a trochophore to an auricularia and a change from regulation to mosaic determination in the zygote. To be fair, this is a formidable list of difficulties, and to many such a drastic series of modifications is unacceptable. But it is submitted that such a strong case can be made out for the evolution of the coelomic system of lophophore protraction that the conclusion is inescapable that the minor coelomate line divided early on into two, called, for easy reference, the protostome line (to the phoronids, ectoprocts and brachiopods) and the deuterostome line (to the echinoderms, hemichordates and chordates), and that these two lines had a common origin in a sipunculoid-like group.

Furthermore, not one of the so-called diagnostic characters for each line can stand close scrutiny on its own: first, within two widely-separated phyla of the Minor Coelomates there is a change from one type of coelom formation to the other, showing that this feature is far from immutable; secondly, the fate of the blastopore has been shown to have less phylogenetic significance than previously thought; thirdly, larvae are notoriously misleading in assessing relationship; and, lastly, the egg type does not always maintain integrity within a group. So perhaps it is not altogether unreasonable to lay more emphasis on general mode of life and adult morphology.

To pursue the adult characters briefly, the lophophore protraction mechanism of the sipunculoids consists of a circum-oral coelomic compartment with branches to the lophophoral tentacles. Such a system is found also in the pogonophores, echinoderms and pterobranch hemichordates; in the pogonophores the lophophore is apparently the entire feeding system of the body and arises from a coelomic compartment at the front end. In echinoderms the lophophore coelomic system has become the water-vascular system, since here the 'tentacles' (now called brachioles or arms) are permanently extended and the protractive function is relegated to a lower level in the system, the tube-feet. In the pterobranchs the lophophore is double and is operated by a pair of compartments in the middle of the body.

This idea circumvents the difficulty of trying to explain the origin of the echinoderms' single water-vascular ring. Previous suggestions have invoked a process of torsion to explain the origin of this ring, or have merely conjured up, but without a convincing pedigree, a convenient ancestor with such a system already. The idea also supersedes the much-criticized suggestion that the echinoderms and hemichordates arose from a creeping bilateral animal, the dipleurula, with three pairs of coelomic compartments, the middle one of which on one side only formed the ubiquitous lophophore coelomic system.

We are now left with the Chaetognatha. This is the phylum which is probably the most difficult to place in any phylogenetic scheme, its relationships being masked by specializations to pelagic life and its mode of development giving nothing away. That it is a coelomate group seems fairly certain; yet after its larval coelom has been formed by 'a modified form of enterocoely' as a pair of sacs, the entire coelomic space is secondarily filled with mesenchyme cells and then a fresh cavity is formed by coalescence of vacuoles in these cells, so that the final cavity comes to resemble a pseudocoel, with no coelomic epithelium. The regionation of the larval coelom is said to indicate affinities with the Minor Coelomates, and the 'modified form of enterocoely' may place the phylum somewhere along the deuterostome line. But the evidence is slim.

The question of regionation of the body of some Minor Coelomates deserves some mention. Much is made in some schemes of the three basic compartments, protocoel, mesocoel and metacoel, which are said to occur in most of the Minor Coelomate phyla. We are told that the mesocoel forms the lophophoral coelom and the metacoel the perivisceral coelom, but there is hesitation when it comes to the protocoel. It does seem that in some forms, such as the pterobranchs, ectoprocts and phoronids, the protocoel is the cavity within the epistome or its equivalent and functions either as an operculum over the mouth or as a device for holding the animal in its tube. In the pogonophores, unfortunately, the coelomic compartment that grips the tube is not the most anterior one, but lies between the lophophore coelom and the perivisceral coelom. This has given rise to the suggestion that in pogonophores the lophophore coelom arises from the protocoel, and is therefore not homologous with lophophore systems of other Minor Coelomates, which arise from the mesocoel. This view does not seem attractive; it is better to regard the pogonophore protocoel as missing, and the cavity operating the tentacles as a mesocoel homologous with other lophophores, in which case the metacoel has divided secondarily to form two functionally separate components.

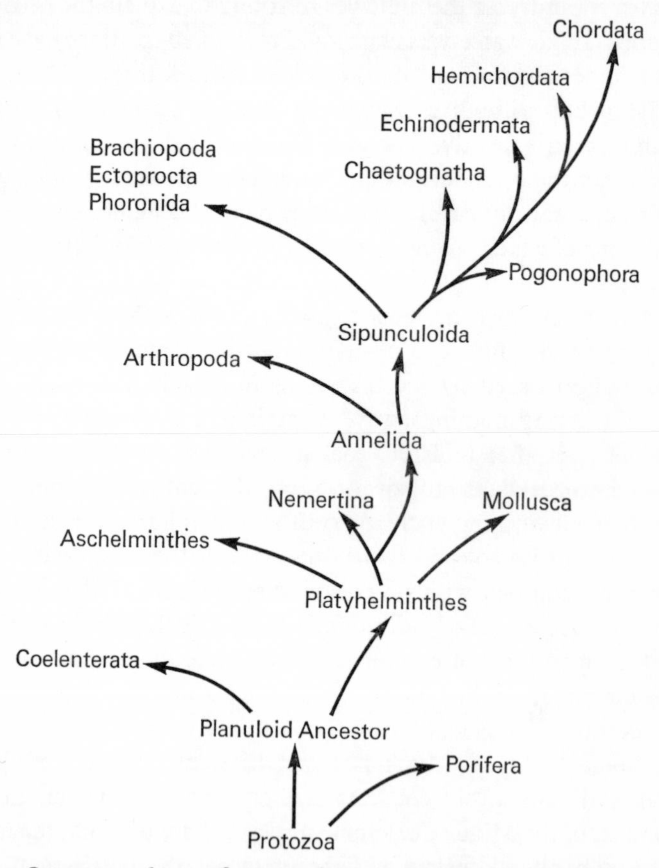

Fig. 17.7 Summary scheme of one possible way in which the main invertebrate phyla can be related. The relative lengths and attitudes of the lines have no significance

At the echinoderm-hemichordate level of the enterocoelous line we have reached a stage at which a neotenic transformation probably occurred which allowed the evolution of chordate organization. This marks one of several peaks of evolutionary attainment in the invertebrates; other outstanding levels of advancement are the cephalopod molluscs and the higher arthropods, both these groups showing an impressive level of nervous co-ordination and concomitant behavioural complexity. The peak of the enterocoelous line is in itself disappointing from the viewpoint of 'success', on whatever criterion this is defined, except for the evolutionary potential engendered by the transformation to a fish-like form.

Some authorities view all the major phyla of the invertebrates as separate derivations from the protistan level. Such a view cannot be discredited on present knowledge, yet it is basically unsatisfactory. In this account the phylogeny of the invertebrates is seen as a series of radiations (Fig. 17.7), one or more branches of one level attaining features which initiate a new level of attainment. In some cases, such as the transformation from a platyhelminth to a nemertean grade (p. 370), the method of advancement appears to have been a 'normal' evolutionary process of gradual elaboration of an existing adult animal. In other cases, such as the origin of the Minor Coelomates (p. 376) and later the origin of the chordates, the method may well have been one using the phenomenon of paedomorphosis, the gradual attainment of sexual maturity in a larval form.

When a group undergoes advancement resulting in a new grade, subsequent evolution will cause the new grade to radiate into new habitats hitherto unattainable. Sometimes, the very features which appear to have provided the foundation for the new grade are secondarily modified and even lost; often, intense specialization results in too narrow a range of available habitats, leading to extinction. The resulting picture is depressingly hard to interpret in phyletic terms; this has been one attempt which, like all others, must be viewed as highly tentative.

Bibliography

Chapter 2: A history of the study of invertebrates

Nordenshiöld, E. (1946), trs. L.B.Eyre, *The history of biology: a survey*. Tudor Publishing Co.: New York.

Raven, C.E.R. (1942) *John Ray, naturalist: his life and works*. Cambridge University Press.

Singer, C. (1931) *A short history of biology*. Oxford University Press.

Chapters 3 and 4: The personae of the invertebrates: The major phyla: The minor phyla

Barradaile, L.A., Eastham, L.E.S., Potts, F.A., and Saunders, J.T. *The Invertebrata* (G.A. Kerkut, ed.). Cambridge University Press.

Hyman, L.H. (1940) *The invertebrates: Protozoa through Ctenophora*. McGraw-Hill: New York.

(1951a) *The invertebrates: Platyhelminthes and Rhynchoela*. McGraw-Hill: New York.

(1951b) *The invertebrates: Acanthocephala, Aschelminthes and Entoprocta*. McGraw-Hill: New York.

(1955) *The invertebrates: Echinodermata*. McGraw-Hill: New York.

(1959) *The invertebrates: smaller coelomate groups*. McGraw-Hill: New York.

Morton, J.E. (1967) *Molluscs*, 4th ed. Hutchinson University Library.

Sedgwick, A. (1909) *A student's textbook of zoology*, vol. 3, *Arthropoda*. Swan Sonneschein: London.

Chapter 9: Movement and locomotion of invertebrates

Allen, R.D. and N.Kamiya (Eds.), (1964) *Primitive motile systems in cell biology*. Academic Press, New York.

Chapman, G. (1950) *J. exp. Biol.*, **27**, 29–39. Of the movement of worms.

Chapman, G. (1958) *Biol. Rev.*, **33**, 338–371. The hydrostatic skeleton in the invertebrates.

Chapman, G. (1967) *The body fluids and their functions*. Arnold: London.

Clark, R.B. (1964) *Dynamics in metazoan evolution*. Clarendon Press: Oxford.

Gray, Sir J. (1968) *Animal locomotion* (World Naturalist Series) Weidenfeld and Nicolson; London.

Harris, J.E. and H.D. Crofton (1957) *J. exp. Biol.*, **34**: 116–130. Structure and function in the nematodes: internal pressure and cuticular structure in *Ascaris.*

Holwill, M.E.J. (1966) *Physiol. Rev.*, **46**. 695–785. Physical aspects of flagellar movement.

Manton, S.M. (1958) *J. Linn. Soc. (Zool.)*, **44**, 58–72. Habits of life and evolution of body design in Arthropoda.

Robson, E.A. (1966) *Symp. zool. Soc. Lond.*, **16**, 331–360. Swimming in Actiniaria.

Ross, D.M. and L. Sutton (1961) *Proc. Roy. soc. Lond. (B)*. **155**, 266–281. The response of the sea anemone *Calliactis parasitica* to shells of the hermit crab *Pagurus bernhardus.*

Smith, J. E. (1950) *Symp. Soc. exp. Biol.* **4**, 196–220. Some observations on the nervous mechanisms underlying the behaviour of starfishes.

Trueman, E.R. (1968) *Symp. zool. Soc. Lond.* **16**. 167–186. The burrowing activities of bivalves.

Trueman, E.R. and Packard, A. (1968) *J. exp. Biol.* **49**, 495–507. Motor performances of some cephalopods.

Wallace, H.R. (1958) *Ann. app. Biol.* **46**, 74–85. Movement of eelworms. I. The influence of pore size and moisture content of the soil on the migration of larvae of the beet eelworm, *Heterodera schachtii* Schmidt.

Wigglesworth, Sir V. 1964. *The life of insects.* (World Naturalist Series) Weidenfeld and Nicolson: London.

Chapter 10 *: Sensory equipment and perception*

de Bruin, G.H.P. and Crisp, D.J. (1957) *J. exp. Biol.* **24**: 447–63 (effect of pigment movement on acuity in crustacea).

Bullock, T.H. and Horridge, G.A. (1965) *Structure and function in the nervous system of invertebrates.* Freeman: San Francisco.

Carthy, J.D. (1958) *An introduction to the behaviour of invertebrates.* Allen and Unwin: London.

Carthy, J.D. and Newell, G.E. (Eds.) (1968) *The structure of invertebrate receptors.* Symposia of the Zoological Society of London No. 23. Academic Press: London.

Cohen, M.J. and Dijkgraaf, S. (1961) Mechanoreception in *Physiology of Crustacea.* Vol. II (T.H. Waterman, Ed.). Academic Press: New York.

Dethier, V.G. (1963) *Physiology of insect senses.* Methuen: London.

Haskell, P.T. (1961) *Insects sounds.* Witherby: London.

Hoffmann, C. (1967) *Z. vergl. Physiol.* **54**: 290–352 (sensory hairs on scorpions).

Hughes, D.A. (1966) *J. Zool.* **150**: 129–43 (sound in *Ocypode*).

Milne, L.J. and Milne, M. (1959) Photosensitivity in invertebrates in Handbook of Physiology, Section 1. Neurophysiology, Vol. I. American Physiological Society: Washington.

Rutherford, J.D. and Horridge, G.A. (1965) *Q. Jl microsc. Sci.* **106**: 119–330 (rhabdom of lobster).

Walcott, C. and Kloot, W.G. van der. (1959) *J. exp. Zool.* **141**: 191–244 (vibration reception in spider).

Young, J.Z. (1960a) *Nature, Lond.* **186**: 836–9 (retinal organization in the octopus).

— (1960b) *Proc. R. Soc.* **(B) 152**: 3–29 (statocysts of octopus)

Chapter 11 *: Nervous systems and co-ordination*

Barnes, G E (1955) *J exp. Biol.* **32**: 158–74 (ganglionic function in *Anodonta*).
Bullock T.H. and Horridge, G.A. and others (1965) *Structure and function in the nervous system of invertebrates.* Freeman: San Francisco.
Horridge, G.A. (1963) *Proc. R. Soc.* (**B**) **157**: 199–222 (organization of parapodial innervation of *Harmothöe*).
Jones, W.C. (1962) *Biol. Rev.* **37**: 1–50 (existence of nervous system in sponges).
Nicol, J.A.C. (1948) *Q. Rev. Biol.* **23**: 291–323 (giant axons of annelids).
Pantin, C.F.A. (1935) *J. exp. Biol.* **12**: 119–38, 139–55, 156–64 and 389–96 (nerve net of anemones).
Passano, L.M. (1963) *Proc. natn Acad. Sci.* **50**: 306–13 (the origin of nervous systems).
Roeder, K.D. (1963) *Nerve cells and insect behaviour.* Harvard University Press: Cambridge, Mass.
Romanes, G.J. (1885) *Jellyfish, starfish and sea urchins, being a research on the primitive nervous systems.* Paul Trench Trubner: London.
Smith, J.E. (1937) *Phil. Trans. R. Soc.* (**B**) **227**: 111–73 (nervous system of starfish).
— (1950) *Phil. Trans. R. Soc.* (**B**) **234**: 521–58 (motor nervous system of starfish).
— (1957) *Phil. Trans. R. Soc.* (**B**) **240**: 135–96 (nervous system of polychaete body segments).
Wells, M.J. (1962) *Brain and behaviour in cephalopods.* Heinemann: London.
Wilson, D.M. (1961) *J. exp. Biol.* **38**: 471–90 (origin of locust wing beat rhythm).
Young, J.Z. (1964) *A model of the brain.* Oxford University Press: Oxford.

Chapter 12 *: Aspects of behaviour in invertebrates*

Bainbridge, R. (1961) Migrations in *The Physiology of Crustacea.* Vol. II (T.H. Waterman, Ed.). Academic Press: New York.
Bartels, M. and Baltzer, F. (1928) *Revue suisse Zool.* **35**: 247–58 (light orientation of spiders).
Carthy, J.D. (1958) *Introduction to the behaviour of invertebrates.* Allen and Unwin: London.
— (1965) *The behaviour of arthropods.* Oliver and Boyd: Edinburgh.
— J.D. (1966) *Insect communication in insect behaviour.* Symposia of Royal Entomological Society, No. 3. *Insect Behaviour.*
Clark, R.B. (1965) *Anim. Behav.* Suppl. 1: 89–100 (learning in polychaetes – this Supplement is entirely on learning in invertebrates).
Cold Spring Harbor Symposia on Quantitative Biology, Vol. XXV (1960) Biological clocks (various papers on sun compass and time senses of animals).
Crane, J. (1943, 1949) *Zoologica* **28**: 217–23; **42**: 69–82 (display in *Uca* spp.).
— (1949) *Zoologica* **34**: 31–52 and 159–214 (courtship of *Corythalia* etc.).
Dress, O. (1952) *Z. Tierpsychol.* **9**: 169–207 (feeding behaviour of *Salticus*).
Evans, F.G.C. (1951) *J. Anim. Ecol.* **20**: 1–10 (behaviour of *Lepidochitona*).
Fraenkel, G.S. and Gunn, D.L. (1961) *The orientation of animals.* Dover: New York.

Gidholm, L. (1965) *Zool. Bidr. Upps.* **37**: 1–44 (courtship of *Autolytus*).
Hardy, A. (1956) *The open sea : the world of plankton.* Collins: London.
Harker, J.E. (1958) *Biol. Rev.* **33**: 1 (diurnal rhythms in animals) also papers in Cold Spring Harbor Symposium.
— (1964) *The physiology of diurnal rhythms.* University Press: Cambridge.
Harris, J.E. and Mason, P. (1957) *Proc. R. Soc. Lond.* (**B**) **145**: 280–90 (vertical migration of *Daphnia*).
Kennedy, J.S., Booth, C.O. and Kershaw, J.S. (1961) *Ann. appl. Biol.* **49**: 1–21 (host finding by aphids).
Levandowsky, M. and Hodgson, E.S. (1965) *Comp. Biochem. Physiol.* **16**: 159–61 (amino acid and amine receptors of lobsters).
Millott, N. (1954) *Phil. Trans. R. Soc. Lond.* (**B**) **238**: 187–220 (light responses of *Diadema*).
— (1956) *J. exp. Biol.* **33**: 508–23 (covering reaction in *Lytechinus*).
Pardi, L. and Papi, F. (1952) *Naturwissenschaften* **39**: 262–3. Papi, E. and Pardi, L. (1953) *Z. vergl. Physiol.* **34**: 490–518 (sun compass and sense of time in *Talitrus*).
Wells, M.J. (1962) *Brain and behaviour in cephalopods.* Heinemann: London (useful review of the work on *Octopus* and *Sepia*).
Wieser, W. (1956) *Limnol. Oceanogr.* **1**: 274–85 (substratum selection by *Cumella*).
Young, J.Z. (1964) *A model of the brain.* Clarendon: Oxford (octopus learning and brain structure).

Chapter 14: Larvae and larval behaviour

Barrington, E.J.W. (1967) *Invertebrate structure and function.* Nelson: London.
Crisp, D.J. and Stubbings, H.G. (1957) *J. anim. Ecol.* **26**: 179–96 (the orientation of barnacles to water currents).
Dawydoff, C. (1928) *Traité d'embryologie comparée des Invertebrates.* Masson: Paris.
Green, J. (1961) *A biology of crustacea.* Witherby: London.
Gurney, R. (1942) *Larvae of decapod crustacea.* Ray Society: London.
Knight Jones, E.W. (1954) *J. exp. Biol.* **30**: 584–98 (laboratory experiments on gregariousness during settling in *Balanus balanoides* and other barnacles).
Ryland, J.S. (1959) *J. exp. Biol.* **36**: 613–31 (experiments on the selection of algal substrates by polyzoan larvae.
— (1960) *J. exp. Biol.* **37**: 783–800 (experiments on the influence of light on the behaviour of polyzoan larvae).
de Silva, P.H.D.H. (1962) *J. exp. Biol.* **39**: 483–90 (experiments on choice of substrata by *Spirorbis* larvae (Serpulidae)).
Thorson, G. (1946) *Meddr. Kommn Danm. Fisk og Havunders. Serie: Plankton,* **4**: 1–523 (reproduction and larval development of Danish marine bottom invertebrates).
(1950) *Biol. Rev.* **25**: 1–45 (reproduction and larval ecology of marine bottom invertebrates).
Wilson, D.P. (1955) *J. mar. biol. Ass. U.K.* **34**: 531–43 (the role of micro-organisms in the settlement of *Ophelia bicornis*).
— (1968) *J. mar. biol. Ass. U.K.* **48**: 387–445 (the settlement behaviour of the larvae of *Sabellaria alveolata*).

Chapter 15 : Animal associations

Baer, J.G. (1951) *Ecology of Animal Parasites*. University of Illinois Press: Urbana.

Caullery, M. (1952) *Parasitism and symbiosis*. Sidgwick and Jackson: London.

Dogiel, V.A. (1964) *General parasitology*. Oliver and Boyd: Edinburgh.

Davenport, D., Camougis, C. and Hickok, J.F. (1960) *Anim. Behav.* **8**: 209–18 (behaviour of commensals in response to host factor).

Davenport, D. and Norris, K.S. (1958) *Bio. Bull.* **115**: 397–410 (commensalism of *Stoichactis* and *Amphiprion*).

Henry, S.M. (Ed.) (1966) *Symbiosis*. Academic Press: London (expecially the contributions by R.Phillips Dales and D.Davenport).

Hesse, R., Allee, W.C. and Schmidt, K.P. (1951) *Ecological Animal Geography*. Wiley: New York.

Lees, A.D. (1948) *J. exp. Biol.* **25**: 145–207 (behaviour of the sheep tick).

Ross, D.M. (1960) *Proc. zool. Soc. Lond.* **134**: 43–58 (relationships of hermit crabs and sea-anemones).

Rothschild, M. and Clay, T. (1952) *Fleas, flukes and cuckoos*. Collins New Naturalist: London.

Chapter 16 : The past history and evolution of the invertebrates

Bulman, O.M.B. (1955) Graptolithina. In *Treatise on invertebrate paleontology* (R.C.Moore, ed.), Part V. Kansas University Press.

Donovan, D.T. (1964) Cephalopod phylogeny and classification. *Biol. Rev.* **39**: 259–87.

Durham, J.W. (1964) The Helicoplacoidea and some possible implications. *Yale scient. Mag.* **39**: 24–8.

Durham, J.W. and Melville, R.V. (1957) A classification of echinoids. *J. Paleont.* **31**: 242–72.

Fell, H.B. (1963) The phylogeny of sea-stars. *Phil. Trans. R. Soc.* (**B**) **246**: 381–435.

Glaessner, M.F. (1961) Pre-Cambrian fossils. *Scient. Am.* **204**: 72–8.

— (1962) Precambrian fossils. *Biol. Rev.* **37**: 467–94.

Glaessner, M.F. and Wade, M. (1966) The late precambrian fossils from Ediacara, S. Australia. *Palaeontology* **9**: 597–628.

Harrington, H.J. (1959) General description of Trilobita. In *Treatise on invertebrate paleontology* (R.C.Moore, ed.), Part O, vol. 1. Kansas University Press.

Hill, D. (1964) The phylum Archaeocyatha. *Biol. Rev.* **39**: 232–58.

Howell, B.F. and Dunn, P.H. (1942) Early Cambrian 'Foraminifera'. *J. Paleont.* **16**: 638–9.

House, M.R. (1967) Fluctuations in the evolution of Palaeozoic invertebrates. In *The Fossil Record* (W.B. Harland *et al.*, eds), London (Geological Society), 41–54.

Kay, M. and Colbert, E.H. (1965) *Stratigraphy and life history*. Wiley: New York.

Kummel, B. (1961) *History of the earth; an introduction to historical geology*. Freeman: New York.

Moore, R.C. (1955) Invertebrates and geologic time-scale. *Geol. Soc. Am.* special paper, **62**: 547–74.

Nichols, D. (1959) Changes in the Chalk heart-urchin *Micraster* interpreted in relation to living forms. *Phil Trans. R. Soc.* (**B**) **242**: 347–437.

Nichols, D. (1969) *Echinoderms*, 4th ed. Hutchinson University Library: London.
Oakley, K.P. and Muir-Wood, H.M. (1949) *The succession of life through geological time*, 2nd ed. British Museum (Natural History): London.
Richter, R. (1955) Die ältesten Fossilien Sud-Afrikas. *Senckenberg leth.* **36**: 243–89.
Romer, A.S. (1966) *Vertebrate paleontology*, 3rd ed. Chicago University Press.
Rubey, W.W. (1951) Geologic history of sea-water – an attempt to state the problem. *Bull. geol. Soc. Am.* **62**: 1111–48.
Simpson, G.G. (1950) *The meaning of evolution*. New American Library: New York.
— (1960) The history of life. In *Evolution after Darwin* (Sol Tax, ed.) **1**: 117–80. Chicago University Press.
Sprigg, R.C. (1947) Early Cambrian 'jellyfishes' of Ediacara, South Australia, and Mt. John, Kimberley District, Western Australia. *Trans. R. Soc. S. Aust.* **71**: 212–4.
Stirton, R.A. (1959) *Time, life and man – the fossil record*. Wiley: New York.
Twenhofel, W.H. and Schrock, R.R. (1953) *Principles of invertebrate paleontology*, 2nd ed. McGraw-Hill: New York.
Walcott, C.D. (1911) Cambrian geology and paleontology. II. Middle Cambrian annedlids. *Smithson misc. Collns* **57**: No. 5, 109–44.
— (1931) Addenda to description of Burgess Shale fossils. *Smithson. misc. Collns* **85**: No.3.

Chapter 17 *: The phylogeny of the invertebrates*

Baer, K. E. von. (1828) *Über Entwicklungsgeschichte der Thiere*. Königsberg.
Barrington, E.J.W. (1965) *The biology of Hemichordata and Protochordata*. Oliver and Boyd: London.
Bather, F. A. (1900) Echinodermata. In *A treatise on zoology* (E.R. Lankester, Ed.), Vol. 3. Black: London.
Bury, H. (1895) The metamorphosis of echinoderms. *Q. Jl microsc. Sci.* **38**: 45–135.
Carter, G.S. (1965) Phylogenetic relations of the major groups of animals. In *Ideas in modern biology* (J.A. Moore, Ed.) **15**: 427–45. Natural History Press: New York.
Clark, R.B. (1964) *Dynamics in metazoan evolution. The origin of the coelom and segments*. Clarendon Press: Oxford.
Fell, H.B. (1940) The origin of the vertebrate coelom. *Nature, Lond.* **145**: 906–7.
Fretter, V. and Graham, A. (1962) *British prosobranch molluscs*. Ray Society: London.
Garstang, W. (1922) The theory of recapitulation: a critical restatement of the biogenetic law. *J. Linn. Soc.* **35**: 81–101.
Gröbben, K. (1923) Theoretische Erorterungen begriffend die phylogenetische Ableitung der Echinodermen. *Schr. Balkankommn Akad. Wiss. Wien* **132**: 263–90.
Hadži, J. (1953) An attempt to reconstruct the system of animal classification *Syst. Zool.* **2**: 145–54.
— (1963) *The evolution of the Metazoa*. Pergamon: London.
Haeckel, E. (1874) The gastraea theory, the phylogenetic classification of the animal kingdom and the homology of the germ lamellae. *Q. Jl microsc. Sci* (n.s.) **14**, 142–65.
Hanson, E.D. (1961) *Animal diversity*. Prentice-Hall: New Jersey.

Hardy, A. C. (1953) On the origin of the Metazoa. *Q. Jl. microsc. Sci.* **94**: 441–3.
— (1954) Escape from specialisation. In *Evolution as a process* (Huxley, Hardy and Ford, Eds.), 122–42. Allen and Unwin: London.
Hyman, L.H. (1940) *The invertebrates. Protozoa through Ctenophora.* McGraw-Hill: New York.
— (1951a) *The invertebrates. Platyhelminthes and Rhynchocoela.* McGraw-Hill: New York.
— (1951b) *The invertebrates. Acanthocephala, Aschelminthes and Entoprocta.* McGraw-Hill: New York.
— (1955) *The invertebrates. Echinodermata.* McGraw-Hill: New York.
— (1959) *The invertebrates. Smaller coelomate groups.* McGraw-Hill: New York.
Ivanov, A. V. (1963) *Pogonophora* (trans. and ed. Carlisle). Oliver and Boyd: London.
Jägersten, G. (1955) On the early phylogeny of the Metazoa. The bilaterogastraea theory. *Zool. Bidr. Upps.* **30**: 321–54.
Kerkut, J. A. (1960) *Implications of evolution.* Pergamon: London.
Lankester, E. R. (1877) Notes on the embryology and classification of the animal kingdom; comprising a revision of speculations relative to the origin and significance of the germ layers. *Q. Jl. microsc. Sci.* **17**: 343–6.
Llewellyn, J. (1965) The evolution of parasitic platyhelminths. *Symp. Brit. Soc. Parasitol.* **3**, 47–78.
Marcus, E. (1958) On the evolution of the animal phyla. *Q. Rev. Biol.* **33**: 24–58.
Metschnikoff, E. (1883) Researches on the intracellular digestion of invertebrates. *Q. Jl microsc. Sci.* **24**: 89–111.
Morton, J.E. (1963) The molluscan pattern: evolutionary trends in a modern classification. *Proc. Linn. Soc. Lond.* **174**: 53–72.
Nichols, D. (1969) *Echinoderms*, 4th ed. Hutchinson University Library: London.
— (1967) The origin of echinoderms. *Symp. zool. Soc. Lond.* **20**: 209–29.
Novikoff, M. M. (1953) Regularity of form in animals. *Syst. Zool.* **2**, 57–62.
Nursall, J. R. (1962) On the origin of the major groups of animals. *Evolution, Lancaster, Pa.* **16**: 118–23.
Osche, G. (1963) Die systematische Stellung und Phylogenie der Pentastomida. Z. Morph. Ökol. Tiere **52**: 487–596.
Riggin, G. T. (1962) Tardigrada of Southwest Virginia. *Tech. Bull. Va agric. Exp. Stn* **152**: 1–145.
Schmidt, O. (1849) Einige neue Beobachtung über die Infusorien. *Natizblat. Hebeite Natur u. Heilkunde* **9**: 6–7.
Semon, R. (1888) Die Entwicklung der *Synapta digitata* und die Stammesgeschichte der Echinodermen. *Jena. Z. Naturw.* **22**: 175–309.
Steinbock, O. (1932) Eine Theorie über den plasmodialen Ursprung der Vierzellen (Metazoa). *Arch. exp. Zellforsch.* **19**: 343 only. (Published in title only – cited by Hyman, 1959 [21].)
Tiegs, O.W. and Manton, S.M. (1958) The evolution of the Arthropoda. *Biol. Rev.* 33: 255–337.
Whitaker, R.H. (1959) On the broad classification of organisms. *Q. Rev. Biol.* **34**: 10-226.
Williams, J.B. (1960) Mouth and blastopore. *Nature, Lond.* **187**: 1132.

Index